THE NEW SOLAR ELECTRIC HOME

the photovoltaics how-to handbook

Joel Davidson

aaTec

To the patient, hardworking people of the PV industry.

Ninth printing 1995

Copyright © 1987 by **aatec publications**
PO Box 7119, Ann Arbor, Michigan 48107
313/995 1470 (phone & fax)

Library of Congress Cataloging-in-Publication Data
Davidson, Joel.
    The new solar electric home.

    Includes index.
    1. Solar houses.   2. Photovoltaic power generation.
I. Title.
TH7414.D36   1987   621.31'244   87–1410
ISBN 0-937948-09-8

Manufactured in the United States of America

Illustrations by F. B. Orner
Cover Design by Carl Benkert

Preface

Photovoltaics (PV) has gone through many changes since the manuscript for the first *Solar Electric Home* was begun in 1978. Some of these changes are the result of new equipment not then available. More importantly, the real world in which PV exists has changed.

The progress of civilization is closely linked to, and strongly influenced by, the energy resource needs of the developing nations. Ultimately, our future will be decided by rapid population growth and the stress that growth puts on society. We are being forced to live on dwindling resources that must be shared with more and more people. Conservation has become a necessary economic consideration for everyone.

But conservation is not enough. As we shift from a consumer society to a conserver society, we merely prolong the inevitable depletion of natural resources. We need to become producers. That is possible with the unlimited input of energy from the sun. This book will help you learn to conserve and produce energy. It offers a shining solution to the bleak problem of global resource depletion.

Oil production in the United States peaked in 1971. Global oil production will peak sometime in the next few decades. Yet population continues to increase. Add the fact that developing nations must be brought beyond bare survival to decent standards of living, and the gluts of rich nations are obviously temporary. Global peace and freedom can only be achieved when everyone has the opportunity to rise above bleak existence.

Does PV offer a better future? I think it does.

PV users perform three important tasks. First, they provide a market encouraging manufacturers to further develop and use the technology and thereby reduce costs. Second, PV users develop practical applications that are cost-effective. Third, they are working toward our energy independence.

Practical and cost-effective mean energy conservation. PV users are champions of energy conservation. They not only use significantly less energy than most people in industrialized nations, they are also examples of how comfortably we can all live on a frugal energy budget.

Practical use of PV is universally beneficial. As PV users learn to use less energy while maintaining a high standard of living, we help slow down resource depletion. By showing the people in energy-rich nations how not to be energy hogs, PV users show the people of developing nations that living better does not have to mean wasting and squandering our planet.

In my work I have met thousands of people, just like you, who have heard about solar electricity and want to know more. The information presented in this book is essentially what I provide to my clients so that they can better understand PV, power requirements, options, the equipment available, what others have done, and how they can do it for themselves.

One of the pleasures of working in photovoltaics is seeing how people change once they begin using PV. After years of being at the mercy of the utility company, it is amazing to know that you can do it yourself. What's more, you can do it quietly, without pollution, with an electric generator that will last virtually forever.

The Solar Electric Home was popular because of its how-to approach to PV. Thousands of do-it-yourselfers were inspired and guided by it. An excellent primer, its scope was limited to small PV systems in remote locations. With *The New Solar Electric Home*, I hope to take PV out of the backwoods and into mainstream society.

About the Author

Joel Davidson started using photovoltaics in 1978 on his farm in the mountains of Arkansas. He has traveled and lectured widely on solar energy, energy conservation, and appropriate technology, and has worked in electronics, engineering, the building trades, as a teacher, and as a farmer. Joel has been involved in international PV manufacture and distribution for more than ten years, has operated his own PV business, and has extensive experience with PV systems. He is an acknowledged industry pioneer. Joel and his wife, Fran Orner (this book's illustrator), now live in Los Angeles, where he is the Director of Marketing at Solec International, Inc. With the goal "to get as many PV systems into use as quickly as possible," Joel has helped thousands of people around the world get started in photovoltaics.

For further information regarding specific systems in this book, send a self-addressed stamped envelope to Joel Davidson, care of **aatec publications,** PO Box 7119, Ann Arbor, Michigan 48107.

Acknowledgments

There are many people who deserve thanks: Edd Jeffords who had an open mind about PV; Steve Baer who sold me my first good solar modules; Steve Willey, Val Bertoia, Robert Sardinsky, Paul Wilkins and all of the many hands-on people who shared clear and simple answers to technical questions; Greg Johanson who is my best PV man; Fran Orner for special reasons. And a very special thanks to Bill and Marge Lamb.

Contents

A Word of Caution

The purpose of *The New Solar Electric Home* is to encourage the wise use of photovoltaics. To this end, information regarding equipment and methods is presented in a clear, straightforward way. The reader is expected to use safe practices and good judgment in the selection of equipment, methods, and workers.

The application of any technology, especially photovoltaics, continually changes. This book will be updated as new and better equipment and methods evolve. However, it is up to you to verify the safety and validity of any photovoltaics work to be done. Be sure to consult with a photovoltaics professional and the *National Electrical Code*.

The author and **aatec publications** disclaim responsibility for any injury, damage, or other loss suffered related to the information presented in this book.

THINK SAFETY!

Chapter 1

Introduction to Photovoltaics

The Age of Photovoltaics is here. Almost everyone has used, or at least seen, solar-powered calculators and watches. Almost everyone has seen photos of the satellites powered by solar panels. Now people are discovering that PV can be used "earthside." From tiny modules that replace throwaway batteries to acres of modules that replace obsolete electric power plants, PV is a fact of modern life.

To put the magnitude of the sun's power into perspective, at noon the solar energy striking an area 70 miles long by 70 miles wide, if converted into photovoltaic electricity, would equal the peak capacity of all earth's existing power plants. A solar cell power plant covering only 1% of the Sahara Desert would produce all the electricity consumed on this planet.

To put this global data into down-home language, if half the surface of the average home roof were covered with solar cells, they could power an all-electric home loaded with every possible modern appliance. Far fewer solar cells are needed to power an energy-conservative home. The do-it-yourselfer or handy homeowner will find that solar electricity is practical right now.

FIGURE 1.1: Solar advocate Nobby Wakumoto enjoys solar electricity at his home in Lakewood, Colorado, a suburb of Denver. Ten 45-watt solar modules charge eight 6-volt batteries for 24 volts at 900 ampere hours to power a 2500-watt inverter. Mr. Wakumoto can switch from utility to solar energy to power lights, appliances, and tools. This system also provides emergency power. Installer: Dr. Nobby Wakumoto. (Courtesy Rocky Mountain Solar Electric)

A BRIEF HISTORY

Ever since 1839 when French physicist Edmond Becquerel discovered that copper oxide electrodes in a liquid could produce an electric current when exposed to light, the direct conversion of sunlight into electricity has been an exciting goal. Charles Fritts, who made the first selenium photovoltaic cells in the 1880s, predicted that roofs covered with solar cell arrays would generate electricity. Interest in photovoltaics waned, however, when it became impossible to produce an amorphous selenium solar cell with even 1% efficiency. Except for some work done on cuprous oxide cells in the 1930s, this interest was revived only when the solid-state researchers at Bell Labs stumbled upon the single-crystal silicon cell in the 1950s.

Although the 1954 silicon solar cell was only 6% efficient in converting sunlight into electricity, it turned out to be the perfect electric power generator for the fledgling space program. Simple to use and extremely dependable, the single-crystal silicon cell could operate indefinitely on the continuously available sunlight in space.

WHAT SOLAR CELLS ARE AND HOW THEY WORK

Radiated light energy—either direct from the sun, diffused through the atmosphere, or from an artificial source (such as a light bulb)—consists of a stream of energy units called photons. These photons strike the solar cell and create an electron flow (electrical energy) in the cell.

Most commercial solar cells are made of crystalline silicon, one of earth's most abundant elements. It is a constituent of sand, but a sand very different from what we see at the beach. Solar cell silicon is a semiconductor with few free electrons. Unlike copper wire with many free electrons, pure silicon is a good insulator. By doping the silicon crystals with elements that have different numbers of electrons (such as boron and phosphorus), the material is made more conductive of electricity. When phosphorus is added, each phosphorus atom contributes an additional free electron, making what is called an n-type semiconductor. If boron atoms are added to the crystal, each one has one less electron than do the silicon atoms, and each can create a "hole" (a place where an electron should be but isn't), making a p-type semiconductor.

To make silicon solar cells, a large crystal of pure silicon containing a tiny bit of boron is grown under exacting laboratory conditions of heat and vacuum. The resultant crystal is sliced into extremely thin wafers, which are then treated in a diffusion furnace, adding phosphorus impurity atoms in a thin layer on the top of each wafer. The zone between the p-type material and the bulk of the wafer and the n-type surface layer creates the barrier called a p-n junction.

When light energy (photons) is absorbed in the doped wafer, negative and positive charges are created in the cell. Since the doping has created dissimilar properties in the wafer, this charge can flow as an electrical current in only one direction. The output voltage produced across the cell under full-sun conditions is determined by the height of the internal barrier. By connecting wires to each side of the wafer, the current produced can be used in an electrical circuit.

For those who would like more detailed information, including discussion of other types of solar cells, I refer you to *Practical Photovoltaics* by Dr. Richard Komp (**aatec publications**).

PRODUCTION AND COSTS

At one time, only solar cells that did not meet the strict requirements of the space program were available for terrestrial use. That is no longer true.

The production of solar cells is complex and requires much capital and sophisticated equipment. Special conditions must be maintained, the wafer-cutting process is extremely wasteful, much hand labor is necessary (even in an automated factory), and there is a growing shortage of high-grade silicon. This all adds to the costs. However, mass production of solar cells and modules by automated equipment has put PV within the budget of many.

I'm often asked, "Shouldn't I wait until solar cells are cheaper? I've just read about a major breakthrough...." Solar cells are expensive and it would be disappointing to buy a PV system only to find that you could have saved a lot of money by just waiting a couple months. But that's not going to happen.

Figure 1.2 shows U.S. Department of Energy projections for solar cell prices in dollars per watt purchased in large quantities. Actual prices for the end user are also shown. The price of solar cells fell annually until 1980 when inflation outdistanced technical development. For a while the price picture was artificial and complicated by legal and market factors. Prices stabilized at around $8 to $10 per watt. Since then there has been a slowdown in price reductions.

Real price breakthroughs will require radical departures in the production process of silicon solar cells. Researchers have developed new ways to produce silicon wafers from cubes or ingots of lesser-grade material and by the ribbon-growth method that produces rectangular cells which require fewer processing steps.

Amorphous silicon cells offer the most likely avenue for significant price reductions. The Japanese have been producing megawatts of amorphous silicon cells for calculators and watches. However, these cells, designed for room light (200 lux or 750 times less intense than sunlight) degrade in sunlight. Only recently have

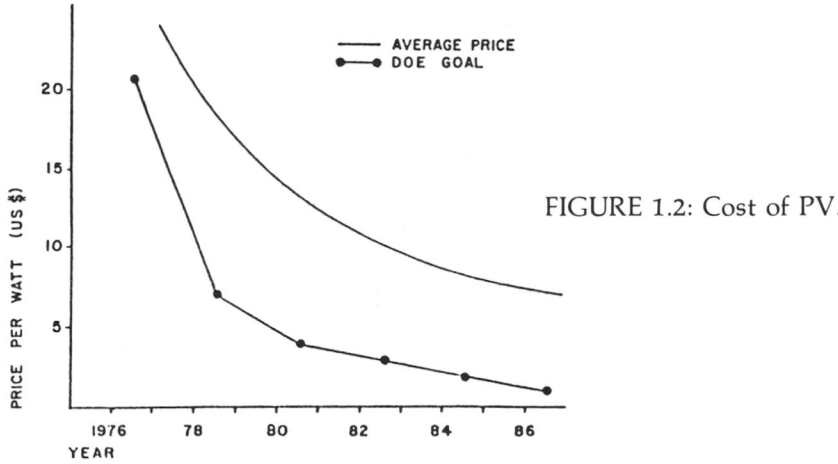

FIGURE 1.2: Cost of PV.

amorphous modules suitable for long-term outdoor use begun to appear. No doubt there will be further cost reductions and improvements in amorphous technology.

Utility companies are very interested in "the solid-state power plant," an electric generating facility that produces no pollution, has no moving parts (except for trackers), requires almost no maintenance, and has an indefinite lifetime. What stops utilities from installing megawatts of PV is cost and the debt they have tied up in obsolete technologies.

The short-term goal of the utility companies' Electric Power Research Institute (EPRI) is to use concentrator technology for lower-cost PV plants. Ultimately, they see amorphous technology with its lower materials costs as the best option to help meet the world's growing demand for electricity (see Figure 1.3).

How research in utility-sized PV power plants will bring prices down is hard to say. Utilities use a different set of economics for determining the technology they use. They have economy of scale on their side, not to mention tax breaks, lower interest rates, and longer time frames in which to amortize costs.

1970 prices for PV were around $100 per watt. Just 15 years later, the Sacramento Municipal Utility District paid $5 per watt for one megawatt of PV. However, most utilities will not begin to use PV until it can rival the cost of conventional power generation. EPRI projects the economical crossover to occur in the 1990s at $1 per

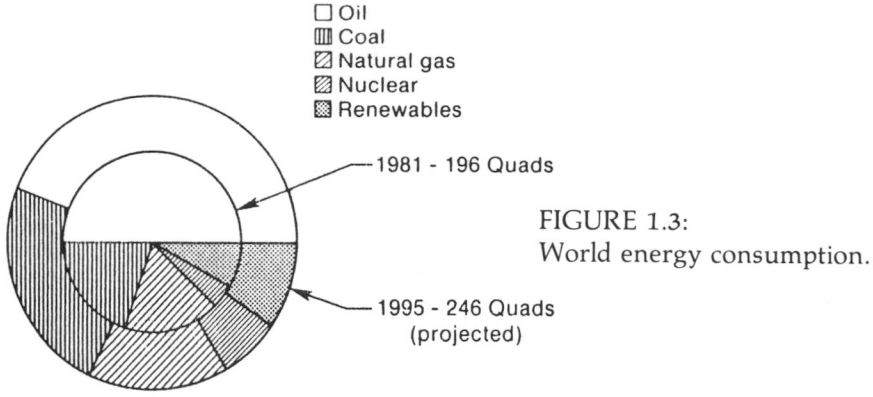

□ Oil
▥ Coal
▨ Natural gas
▨ Nuclear
▨ Renewables

1981 - 196 Quads

FIGURE 1.3:
World energy consumption.

1995 - 246 Quads
(projected)

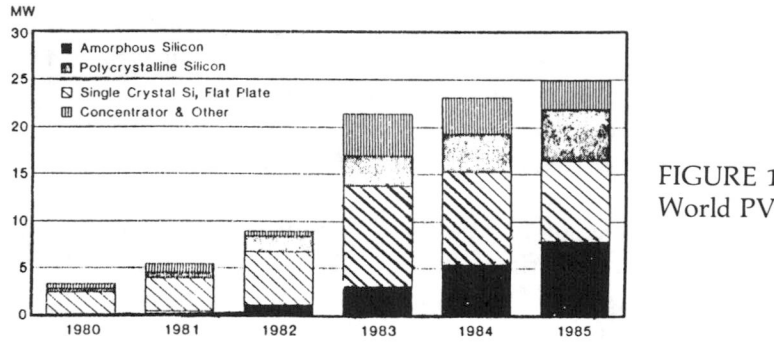

FIGURE 1.4:
World PV shipments.

watt. This can happen only if the technology continues to improve, multi-megawatt prices drop, and 30-year cost levelization is used.

Prices for solar cells *will* come down. How soon and by how much is anybody's guess. (The prediction of 70 cents per watt by 1986 was off by a factor of ten.) More important than relying on guesses about future prices is the fact that PV systems are practical and dependable and affordable now. Why wait?

ELECTRICAL POWER SOURCES—THE OPTIONS

When we consider all the electrical devices we use every day, it's easy to understand how electricity has become a major building block for modern society. Unfortunately, electricity requires special

Table 1.1: Current PV Technology Summary and 1985 Production Figures

Source: P. Maycock and E. Stirewalt, *Photovoltaic News*, 5:2.

Cell Type	Efficiency Lab. Record	Efficiency Prod. Range	Advantages	Disadvantages	1985 Production by Company (MWe)			
Single Crystal Silicon	19.1%	10-13%	• Well established and tested technology • Stable • Relatively efficient	• Uses a lot of expensive material • Lots of waste in slicing wafers • Costly to manufacture • Round cells can't be spaced in modules efficiently	Arco Solar (US) Sharp (Japan) CEL (India) Hoxan (Japan) Solec Int'l (US) BP Solar (UK) Pragma (Italy) Ansaldo (Italy) Nippon Elec (Jap) Solarex (US)	4.7 .9 .6 .5 .4 .4 .3 .3 .3 .2	Helios (Italy) Hitachi (Jap) Mitsubishi (Jap) Kyocera (Jap) Heliodynamica (Braz) Bharat (India) Silonex (Canada) Siemens (Germany) Isophoton (Spain) Komatsu (Japan) Other Total	.2 .2 .2 .2 .2 .2 .1 .1 .1 .1 .4 10.85
Polycrystal Silicon	18%	10-12%	• Well established and tested technology • Stable • Relatively efficient • Less expensive than single crystal Si • Square cells for more efficient spacing	• Uses a lot of expensive material • Lots of waste in slicing wafers • Fairly costly to manufacture • Slightly less efficient than single crystal	Solarex (US) Photowatt (Fr) AEG (Germany) Solavolt (US) Kyocera (Jap) Pragma (Italy) Other Total	1.9 1.0 .8 .5 .4 .2 .1 4.9		
Ribbon Silicon	15%	10-12½%	• Does not require slicing • Less material waste than single and polycrystal • Potential for high speed manufacturing • Relatively efficient	• Has not been scaled up to large-volume production • Complex manufacturing process	Mobile Solar (US) Westinghouse (US) Total	.1 .05 .15		
Amorphous Silicon	11½%	4-8%	• Very low material use • Potential for highly automated and very rapid production • Potential for very low cost	• Pronounced degradation in power output (Staebler-Wronksy effect) • Low efficiency	Sanyo (Jap) Fuji (Jap) ECD/Sharp (Jap) Arco Solar (US) Chronar (US)	3.9 2.8 .5 .2 .2	Kaneka (Jap) Taiyo Yuden (Jap) Solarex (US) Sovonics (US) Other Total	.2 .15 .1 .1 .3 8.45

Table 1.2: Outlook for PV Technology Classes
with Respect to Eight Major Technical (T) and Management (M) Characteristics

	Photovoltaic Technology				
	High Concentration	Tandem Amorphous Silicon	Crystalline Silicon Sheet	Poly-crystalline Thin Films	Novel Concepts
Probability of meeting cost targets (T)	Medium	High	Medium	High	?
Probability of meeting efficiency targets (T)	High	Medium	High	Low	?
Probability of meeting reliability targets (T)	Medium/high	Medium	High	Medium	?
Margin for meeting targets (T, M)	Medium	High	Low	Low	?
Time to resolve technical uncertainties (T, M)	< 10 years	> 10 years	< 10 years	> 20 years	> 20 years
Level of complementary development efforts (M)	Medium	High	High	Low	Low
Degree of private R & D investment (M)	Low	High	Medium	Low	Low
Availability of near-term markets (M)	Low	High	High	Medium	?

Source: *Science* 20 April 84, p. 249.

means of production and that usually means the consumption of non-renewable resources such as coal, oil, and gas. These raw materials are being depleted at a rapid rate.

Nuclear energy is used for the production of electricity, though recent controversy has clouded faith in its practicality and safety. Indeed, the disposal of spent radioactive materials alone poses such a serious threat to life that it makes other forms of pollution seem relatively harmless.

Research into fusion energy (thermonuclear power) has not yet produced a safe, practical alternative to the dirty technology of current-day nuclear power production.

Wind, geothermal, and water power are safe and sane energy solutions. Unfortunately, these natural resources are not everywhere and at all times available. While it has been estimated that hydroelectric technology could produce 25% of U.S. electrical needs, few other nations are sufficiently endowed to provide adequate sites for hydroelectric plants.

Though some natural geothermal sites are being tapped for electrical production, current locations and methods are too limited to compete with fossil fuels. Ocean and tidal power generation remains on the drawing board. Wind power, once a major contributor to the energy picture before rural electrification, is again depositing power into the national grid.

The subject of energy production and consumption is a complex issue with many political, economic, and social implications for both present and future generations. As fossil fuels are depleted, the scramble for the few remaining pockets of conventional fuels will continue to destablize the world political and economic situation, leading to ever greater conflict and confrontation. Meanwhile a virtually limitless and globally available source of energy—solar electricity—offers a solution.

THE POLITICAL SITUATION

No longer do we hear from government and industry that solar energy on a large scale can be implemented soon. In fact, there is a vigorous publicity campaign funded indirectly with tax dollars to convince the public that solar will not be practical until the year 2000. Misleading propaganda presented "in the public interest" is used to prop up the troubled nuclear power plant construction industry. While it is possible for the heavy construction industry to convert to large-scale PV power plant production, such a change-over would cut costs and profits.

We no longer hear government and industry predicting 50 cent per watt PV by 1986. As research dollars dried up and schemes for cheap PV died somewhere between the drawing board and the production line, prognostications for low-cost PV have been re-placed with so-called economic realities. As we shall see, PV is cost-effective immediately in some cases. In others, it has a reasonable pay-back period.

The perception that residential photovoltaics is not practical now is only part of the problem. The practitioners of residential photovoltaics (and most small-scale or appropriate technologies) are viewed skeptically, at best, by the mainstream scientific and business communities and, at worst, as modern-day Luddites who threaten existing systems and institutions. Until the people working with small-scale technologies are recognized, and until they have credibility with the mainstream and technological communities, residential photovoltaics will remain on the fringes of our society.

The fear that PV would decentralize electrical power produc-tion should no longer inhibit its adoption. Large-scale power production, PV or otherwise, is a fact of modern life. Not everyone will want a solar array on the roof.

Seventy percent of the population of the United States lives on two percent of the land. We are an urban society. Central power generation has one major advantage: by producing power for millions at one location, utilities have economy of scale. They can buy technology in large quantities at a much lower price than the individual. All things being equal, the utility price per kilowatt hour for PV will be lower than the individual power producer's kilowatt hour price.

This does not mean that we must wait passively for the utility company to get rid of its old power plants and adopt PV. Utility companies have a lot of money invested in obsolete technology, and it will be years before we see widespread use of large-scale PV. In the meantime, we can install our own personal PV power plants.

If history teaches us anything, it has shown that the individual gives direction and leads society. Those who begin using PV now will gain a greater degree of independence, offset their present electric bills, and have the option of unplugging when utility rates exceed the cost of their individual PV systems.

The belief that solar cells are too expensive is misleading. Solar cells are a practical way to produce electricity for many uses, but present attempts to power energy-wasteful all-electric homes with PV are defeatist. On the other hand, through the use of low-wattage, solid-state equipment, energy-efficient appliances and good design, homeowners can significantly reduce their power consumption, and make PV a very practical alternative.

There are several reasons for what seems to be foot-dragging on the part of industry, government, utilities, and the public.

The monopolistic nature of the solar cell industry is partly to blame. Vast amounts of capital are required to begin cell production: only large companies, such as oil corporations, seem to have the interest or the money. Some oil companies have backed off from PV, giving the impression that big oil is unwilling to support an alternative that may compete with its main product.

Of course, industry has accused government of failing to help solar cell development. During the Carter administration this was not so, but in 1980 solar was forced to take a back seat and fewer dollars were mandated for its research. What little government-funded research that has survived emphasizes space rather than domestic or terrestrial applications.

There is, however, a real need for continued research. It is unfortunate that laboratories must use press releases about imminent breakthroughs to get money to continue research. Perhaps if we had a ten-year national effort, we would see lower-cost PV. That's not so far-fetched. It only took ten years to go to the moon. In the meantime, PV research and development labs go begging and it seems we must either wait for a miracle breakthrough or another energy crisis to spur us to a serious search.

Japan, on the other hand, has adopted a national program for PV research, development, and marketing that may make it the world leader in PV sales. With this threat, some American manufacturers are beginning to wake up. At present, American and Japanese technology appear equal, but lower production costs are beginning to give the Japanese the edge, especially in PV micro-generators for calculators and watches.

The utilities and coal, oil, and gas companies are concerned with how solar cells will be used. The power and profits of these companies come from their control of centralized energy production plants. Solar cells installed at the point of electrical consumption could replace many of the large energy production units that now provide our electricity. But reducing or eliminating the need for centralized power plants and their associated power lines, equipment and personnel, poses a real threat to those in control of energy production.

Utilities burdened with debts from nuclear plant closing or construction cost overruns look at PV as a cheap source of power. However, it seems they prefer to wait until private firms build PV plants, then buy the power produced at wholesale rates. Thus, the utilities win in at least two ways: they will not have to pay the price of the equipment or building the plant, and they buy PV power cheap and sell it at a profit. Investors, consequently, are unwilling to foot the bill for equipment that will immediately profit others.

But "the big boys" are not entirely to blame. The average American knows only two things about electricity—the power companies make it and it costs too much. Most Americans never think twice about the electricity they use. If they did, they would begin to see that the equipment and appliances they use are energy-inefficient. Not only are the electrical devices wasteful, but our very patterns of use are wasteful. Almost everyone everyday leaves lights and equipment on when not needed. Such behavior, and the bargain basement attitude which keeps us from buying energy-conserving equipment, are prime reasons why solar cells are not in widespread use. As long as we fail to assume the responsibility of learning the basics of power usage and production, we must accept high utility bills.

Look at refrigerators, for example. The recent appearance of energy-saving appliances has helped reduce electrical consumption in the average home. However, the wise use of any appliance is what makes the most impact on power consumption. It does not matter if you buy a well-insulated refrigerator/freezer if you leave the door open.

On the other hand, if you add four inches of foam board insulation around the cabinet and door of the average refrigerator, you can cut the duty cycle in half. What does this mean in dollars and cents?

FIGURE 1.5: The Sacramento Municipal utility one-megawatt central station PV array located adjacent to the Rancho Seco nuclear power plant. This is the first stage of a 100-megawatt PV power plant. (Courtesy California Energy Commission)

On average, home refrigerators are responsible for up to 20% of the total electric bill (excluding electric heating and air conditioning). The immediate savings gained by retrofitting more insulation onto the refrigerator can be as much as 10% of the total monthly bill. (More about this in chapter 3.)

Why all the fuss over a few dollars to run a refrigerator? As I said, refrigerators use more electricity than any other home appliance. If energy-efficient refrigerators were in general use today, there would be no need for any nuclear power plants. The savings would be in the billions of dollars. More importantly, reducing the power consumption of major appliances in your home eases the transition from consumer to producer. If the cost of a PV installation is the major hurdle, then it makes sense to (1) reduce your load, (2) cut your electric bill, (3) save money, and (4) install PV

FIGURE 1.6: The PV Breeder plant dedicated in 1982 at Frederick, Maryland. Power from the solar array is used to make solar modules. Its 200-kW array produces electricity to supplement that supplied by the local utility. (Courtesy Solarex Corp.)

with the savings, thus (5) saving more money to (6) add more PV and (7) eventually produce all your own power.

THE PV USER/INDUSTRY CONNECTION

Since there wasn't much information about residential PV when I started using the technology back in the 1970s, I improved my system through trial-and-error and the exchange of ideas with others who were doing the same thing. This exchange grew into my founding the PV Network and the *PV Network Newsletter.*

Through the newsletter, bulk purchases of PV equipment at discount prices were organized. Buyers got reduced equipment prices and the shared knowledge of what worked and what did not. With over 1000 PV users field-testing new equipment and methods, the Network, and the industry, had a reliable source of information on real-life applications.

At first, words like photovoltaics (still a mouthful), array, balance-of-systems, and barrier strips scared people off. It seemed

FIGURE 1.7: The current editor of the *PV Network Newsletter*, Paul Wilkins, beside his mobile office and workshop. In these tiny, efficient, PV-powered quarters, Paul can travel, build his Charge-a-Stat controls, and edit the newsletter. (Photo: A.D. Paul Wilkins, Solar Works!)

too technical. I guess the same would hold true for automobiles if we only read engineering reports.

Through the PV Network we were able to go around the jargon and get to the hardware. In fact, a growing number of technical people were trying to demystify the technology and make it understandable to everyone. The more people know about something, the more inclined they are to use it.

Those early PV systems consisted of locally available batteries and a few solar modules. Wiring and mounts were homemade designs. Often there was no regulator in the system. For trickle charging a battery, this was fine. A few lights and a 12-volt DC car stereo were all that was powered by most early cabin systems.

In the meantime, government and industry were pumping millions of dollars into PV. Laboratory research explored ribbon, semi-crystalline, and amorphous PV. Commercialization of PV was being investigated. All-electric PV-powered homes were field-tested. Agricultural water pumping with PV was being implemented throughout the world.

At one extreme, back-to-the-landers with technical skills were finding low-wattage applications for PV. At the other, government and industry were trying to expand the use of PV into all areas of electrical production and consumption. The government was betting that industry could come up with cheap, clean power. Industry labs were releasing new schemes for reducing PV costs on a weekly basis.

As it turned out, many of those proposed schemes did not work out. Industry cut back and focused on those applications that were immediately cost-effective. Government backed off. The investment community backed off. For remote power applications PV was ideal, but investors felt that neither utility-sized PV power plants producing megawatts of power nor suburban PV was going to sell very soon.

Proponents pointed out that we could not have low-cost PV until production was automated. The businessmen countered with the argument that the market was not yet viable. We were at a stalemate. Prices wouldn't drop until people bought more PV, and people wouldn't buy PV until the prices dropped. It was the old dog-chasing-its-tail.

A way out of this deadlock was to combine the excellent and expensive research done by government and industry with the bootstrap know-how developed by actual users. Some PV users, seeing the practicality of small tracking arrays, began selling trackers. These simple, passive devices were cost-effective, paying for themselves by increasing electrical production 30% while simplifying mounting.

Another PV user developed a high-efficiency inverter, now successfully marketed worldwide. Others have developed better regulators and monitoring equipment. A weeding-out process uncovered which standard house wiring breakers and control boxes would work with PV.

The result of all this activity was a gleaning of the best from each area of research.

GRID-CONNECT VS THE AUTONOMOUS HOME

The 1978 Public Utility Regulatory Policy Act (PURPA) requires utility companies to buy power from independent producers. Excess power is sold to the utility company during the day and bought back at night or on cloudy days. With the added advantage that batteries are not used, eliminating some costs and maintenance, this scheme looks pretty attractive. It led many people to believe they could set up a PV system and sell energy to the power company.

Much publicity has been given to the few homes with large PV arrays that are tied to the grid. Most of these homes have been funded or sponsored by the government or utility companies eager to show what they have done with your money. Another reason for all the publicity is that almost everyone would like to get even with the utilities: the thought of sending an electric bill to the power company seems like sweet revenge.

If it were that simple, lots of people would set up the equipment necessary to become a small power producer or cogenerator. Alas, utility companies are in business to make a profit. They typically buy homegrown electricity at wholesale rates and sell it back at retail. Thus, it may be wiser to simply keep the electricity you produce, at least until you can get a better price for it.

There are progressive utilities that will give you parity for your power. Some will even hook you up with one meter that runs backward as well as forward; this was often the case in the early days of wind and PV utility interconnect homes. Now almost all utilities install, at your expense, two meters. Since law requires that they buy your power at the avoided cost rate—that is, the cost to produce power during the peak periods of grid consumption which you are supposedly helping to offset—that has become the norm. Basically, they charge you to build the power plant they did not build.

Not only do the utilities define some of the costs of your interconnect, they also set the rules. The equipment you install to make the connection must meet with their approval. Now, there are good reasons for this. Your equipment must produce quality power within the engineering limits set by the utility company. In addition, you must have safety equipment and systems. And finally, your system must be approved by their engineers. All this can be costly.

Another factor just beginning to appear is oversupply. This is the short-sighted viewpoint that the growing number of cogenerators will produce more power than can be put into the grid economically. The argument is that too much electricity, especially that produced by large manufacturing firms mostly for on-site use, will force rate increases for everyone else. A similar excuse for rate hikes was used in the 1970s when people began to seriously conserve energy. The utility companies told homeowners in the

FIGURE 1.8: This all-solar grid-interconnect residence was the first of its kind. Located in Carlisle, Massachusetts, this home paved the way for a growing number of systems which sell excess electricity to the local utility. The home has a 7.5-kW array. (Photo: Solar Design Associates)

northwest that they should use more electricity or their rates would go up. No doubt juggling numbers can justify these arguments. However, it would take a real stretch of the imagination to believe that the utilities could be acting in the public interest with such policies. That being the case, we may see an end to PURPA.

But let's say that you do have the money and the desire to be the first on your block to be a cogenerator. In the event of a utility company power failure, your system must shut down to protect workers repairing the power lines. That means that even if it is bright and sunny and your equipment is working perfectly, you might experience a power outage because of problems at the power company. For real energy independence you still need a battery bank and a separate stand-alone inverter, just in case.

My recommendation to those considering a utility interconnect is to move gradually in that direction. First, install a stand-alone or autonomous PV system with batteries and a stand-alone inverter. Then decide—after you have had a taste of independence—if you still want to be a cogenerator. If so, you can always use your storage system and equipment in the event of a power company brownout or blackout.

Grid-connect systems are further discussed in chapter 9.

TAX CREDITS

On January 1, 1986, the 40% federal tax credit, worth up to $4000 of the cost of a residential solar electric power system, became history. For a while it looked like this direct energy subsidy would be extended, but all efforts to revive it failed. Few in Congress wanted to touch the solar tax credits because of the growing deficit and the bad name some crooked solar water heater salespeople have given solar. We had a good law on the books and now it is gone.

Taxpayers have been subsidizing non-renewable energy businesses for years. Worse, we are forced to pay billions of dollars each year to prop up a failing nuclear power industry. Needless to say, to compete with other forms of energy production in the marketplace PV needs the same tax breaks allowed non-renewables. It's not a bad idea to write your representatives in Congress and let them know how you feel. I also want to encourage you to support and use the various state and federal energy information services. For information regarding PV, solar, and energy conservation, contact the National Appropriate Technology Assistance Service (NATAS), Department of Energy, PO Box 2525, Butte MT 59702-2525 (800 428 2525) or the Conservation and Renewable Energy Inquiry and Referral Service (CAREIRS), Renewable Energy Information, Box 8900, Silver Spring MD 20907 (800 523 2929). And while you're at it, find out what your state is doing to promote PV. Many state programs are excellent.

FUTURE TRENDS IN PV

With the end of the federal tax credits, relatively lower oil prices, and no major breakthroughs forthcoming, the photovoltaics industry has stabilized and matured. Since almost half the existing cells are being used in stand-alone applications such as communications relay stations, water pumping and village power in developing countries, and telemetry devices, the future of PV does not require tax subsidies or another oil crisis. The growing use of PV in specialty applications and consumer products assures continued growth of the industry.

Oil companies continue to dominate manufacturing due to the intense capital investment required and long lead time before profit may be realized. Approximately 15 American companies and over 45 foreign companies manufacture PV cells or modules. Some independents have a strong footing. On the other hand, major players such as Exxon's Solar Power Corporation have pulled out completely. It will be a few years before multi-megawatt utility-sized PV systems begin to be installed again. At that time we will probably see major international corporations return to PV. Until then, the market will grow steadily.

While more than half the PV now produced is amorphous, single-crystal technology will dominate stand-alone PV. Should amorphous efficiency reach 10%, single-crystal will only be used in special applications. Semi-crystalline, ribbon, and concentrator PV will remain special market and special application technology. Tandem amorphous holds great promise but is years away from the mythical 50 cent per watt panacea to global energy problems.

If the PV industry is to grow, it must reach markets in developing nations. Unfortunately, those who need PV the most can least afford it. If PV production volume does not grow, prices will not fall. Barring a miracle technological breakthrough, mass production is the only significant factor to drive prices down.

Japanese PV companies may dominate the market by the end of the century. They are well-subsidized by their government and have successfully marketed amorphous PV consumer items worldwide. All they need to do is stabilize their product to stop degradation and increase the size of their modules. If they can do that, and cut prices in half, the U.S. PV industry will be seriously threatened. But it must be remembered that PV is not like the automobile, which was a well-established consumer item long before the Japanese began competing for U.S. auto dollars.

As of early 1987, amorphous PV has not been around long enough (or is well enough understood) for us to know how long the cells will last. Single-crystal cells have been in use for decades and will continue to perform indefinitely. Some amorphous modules now have a three-year warranty. That's just about how long the technology has been out in the field.

Surprisingly, when surveyed, PV users almost always say they are glad they didn't wait for promised lower prices. Price, tax credits, and major breakthrough rumors notwithstanding, PV has found its niche in the global energy resource picture. Compared to bringing in a power line or operating a generator, PV really shines. For remote or portable applications, PV is unsurpassed. As for reliability, its almost 30-year success in space plus more than two decades of use on earth are proof enough that photovoltaics is on its way to becoming a billion dollar industry.

THE BOILING FROG STORY

We are told that the ability to adapt to changing situations is good: adaptation means survival. People who can change with changing circumstances are called survivors. But there's another side to adaptability.

An experiment was made using a frog and a pan of water. The pan was placed on a stove and the water was heated. When the frog was dropped into the hot water, it instantly jumped out.

The same frog, same pan of water, and same stove were used in the second part of the experiment. This time the frog was placed in the room-temperature water. The frog stayed in the water. The heat was turned on and slowly the water temperature rose. The frog stayed in the water. The frog's natural ability to adapt helped it adjust to the rising temperature. In fact, the frog adapted so well it was boiled.

When oil prices skyrocketed, there was outrage. People were shocked when gasoline prices rose from 25 cents to 50 cents a gallon. They swore they would never pay $1.00 a gallon for the stuff. But pay they did as prices climbed toward $1.50. When the price of gas dropped back to around $1.00 a gallon again, they were pleased with the bargain.

At the same time, though less dramatically, the price of natural gas, propane, and, especially, utility electricity was steadily climbing. We didn't notice so much because the fact only hits us once a month in our regular bill at small increments and in small print. Not at all like the two-foot-high prices at the gas station.

We have adapted to a slowly changing condition: the rising cost of energy in a world where more and more people are rapidly depleting already diminished resources. We are about to reach the point where we will no longer have the ability to leap to safety.

_____ Chapter 2 _____

Living With PV

A PV-POWERED HOMESTEAD

The evolution of my PV system parallels the development of affordable equipment.

In 1972 I left the city to live in the Ozarks of northwest Arkansas. After renting and leasing for a few years, I moved atop a 2200-foot mountain, the nearest power line one mile away. I had the general idea of producing my own power to eliminate monthly bills and to gain greater control over my life. Now I was faced with the reality of the project.

Living in the second poorest county in the second poorest state in the U.S. was a handicap. It took money to develop a homestead and still more money to set up a farm. In 1972 the per capita income of my county was under $2000 per year. I was locally employed as a sawmill hand for a while, then worked as a VISTA volunteer doing weatherization work.

But I had a goal and it seemed possible to achieve it. Others had done so before. I read in _Mother Earth News_ about people starting on a shoestring. Helen and Scott Nearing had made a similar back-to-the-land start over 50 years earlier. My ability to scrounge and make do with used or recycled things helped a lot.

I began seeking out others who had done what I wanted to do. A few newcomers to the region had set up windchargers. The local people, hillbillies who knew how to survive in this poor region, were also a source of information. The consensus was that my home was a good wind site and that eventually I would install a windcharger.

But money was tight and other needs more pressing, like home and barn building and just day-to-day living. For shelter, I quickly erected a pole-frame cabin with (secondhand) south-facing windows. The sun provided 50% of the daytime winter heat, but wood was the main heat source. I used a propane stove and refrigerator and kerosene lamps.

FIGURE 2.1: Some remote PV installations are really remote. Colorado Mountain College used llamas to transport a solar electric system to one mountain site. (Photo: D. Stewart)

After drilling a dry hole instead of a well, I was forced to haul water. Eventually I installed a 500-gallon cistern to catch rainwater, and used a hand pump. The tank, too small even for my modest needs, convinced me that a sure cistern is better than a chancy well.

Even if I had dug a good well, I would still have had the problem of getting the water out of the hole.

I built and tested several low-cost (under $50) solar water heaters, but none could withstand the quick freezes typical of the region, so I opted for a used 30-gallon propane water heater (which cost $17). I installed the water heater but kept the pilot light unlit. Spark igniting the water heater and gas cook stove was inconvenient but really saved fuel. In my first year I used less than 130 gallons of propane to cook, heat water, and refrigerate.

My electrical system started with a couple of 16-watt fluorescent lamps. Power came from the truck battery which was charged by daily trips to and from work. Battery power worked well except over long winter weekends: power consumption would deep-discharge the battery and the truck would not start on Monday morning. Pushing or gravity-powered starting was necessary. Using a battery designed for starting, not for deep-cycling, was bad economics anyway.

A friend had a windcharger he wasn't using and so loaned it to me. Before I got it wired up, its blades, in an attempt at self-destruction, gyroscoped into the tower in gusty winds. That mechanical failure and my studies about the reliability of solar cells convinced me to save and buy a small array.

By Christmas, as a result of the combination of windcharger failure, encouragement from Edd Jeffords of the Ozark Institute, a seemingly good deal on second-quality modules, and my desire to be a PV pioneer, I was watching reports of snowstorm power outages on a PV-powered tv. Having just a little electricity while thousands of homes were blacked out convinced me that PV was the way to go.

Wiring in the cabin was simple, consisting of secondhand outlets and scavenged wire and plugs. I had no regulator and used a truck starter battery. Three small 10-watt solar modules were all that I could afford. The whole thing was pretty primitive, but it worked. Wire was Romex 12/2 run the shortest possible distance or doubled up to keep resistance to a minimum. Standard outlets and plugs were marked with a dab of red paint to indicate polarity.

In the beginning I didn't have much sun power. The three 10-watt modules designed for boat deck mounting were mounted instead on a piece of plywood which could be tilted to adjust for the seasonal changes in sun angle.

I bought a 12-volt recreational vehicle demand pump and installed it beneath the cabin, tapping a pipe into the cistern for running water, under pressure, in the cabin. The pump drew 4.6 amperes. For music, I used a 12-volt am/fm cassette stereo with a homemade preamp and an old turntable retrofitted with an external 12-volt motor and belt drive.

Production from the PV array in June 1979 matched my modest consumption of 3 kilowatt hours. In winter, more lights were used. (The region receives only 45% of possible sunshine in winter due to cloud cover.) Winter PV production was only one-fourth consumption, but that was no problem because I had upgraded the battery bank and had a back-up generator.

I bought used deep-cycle batteries at salvage prices that seemed in like-new condition. (They were originally owned by the telephone company.) Six heavy 2-volt cells were hooked up in series for over 400 ampere hours of storage.

Because consumption was greater than winter PV production, and since it is essential to keep batteries topped off or fully charged to insure long life, I assembled a standby generator. I put together a 3.5-horsepower horizontal-shaft gasoline engine from an old roto-tiller and a used truck generator and regulator. A flex coupling was used to connect the two shafts together. The whole jerry-rigged thing was bolted to a piece of truck frame and put in a doghouse-type shelter 50 feet from the cabin.

Heavy wire purchased by the pound from a salvage yard was run along the ground from the back-up generator to the batteries in the cabin. I adjusted the regulator for a maximum 30-ampere charge. In 18 months I used only 8 gallons of gasoline. The charger would run for six hours on 0.75 gallon of gasoline for 180 ampere hours of charge.

By this time I was familiar with the power system and life was more comfortable. (I had installed a shower, but still used an outhouse.) And I was totally committed to PV power and was saving to buy more modules.

By joining with Stephen Cook of Jasper, Arkansas, and including others in the deal, we were able to get a good price on solar modules. I purchased four 33-watt ARCO modules which quadrupled my production. This made me 100% solar eliminating the need for the back-up generator. I also purchased a regulator. It was a big step, but I took it because the 40% federal tax credits were available at the time.

About a year later, I moved to five acres in a nearby valley and built a cabin. PV systems are portable and everything was moved in one trip. The array was removed from the roof and disassembled. The modules were wrapped in blankets and rode in the cab of my truck. The batteries, which weighed over 800 pounds, were secured with rope in the bed of truck to prevent spillage. I even took the wiring and fixtures to reuse them at my new home. While it took me a few days to get everything hooked up again, the move and installation could have been done in one day.

At the new cabin, I fastened the four modules to the roof on a fixed stand-off mount since the roof tilt was optimum. The batteries were placed on a pallet in the crawlspace beneath the cabin. Wiring from the array came through the cabin to the regulator and meters and onto the batteries. Lights and outlets were wired in a few hours.

The original 30-watt array was set up on the ground to charge an auxiliary 12-volt battery for portable power. This portable power pack (just a battery and carrying strap) and a DC pump came in handy for pumping water from a nearby shallow well. I bought a 550-watt inverter and was able to operate a small drill and saber saw. The inverter was also used to power a sewing machine and vacuum cleaner. When I got my first computer, the inverter powered it too.

By summer 1981 I had organized other PV module purchases. Hundreds of people were getting started with photovoltaics. It was happening all over. I bought four more solar modules, additional batteries and a Zomeworks tracker. With plenty of power, I began working full-time at my office/cabin selling equipment and answering questions from people all over the world. The book you are now reading was started during that period.

BACK TO THE CITY

Big changes were to occur. The realization that energy prices were destined to continue rising and the improvements to PV systems had created a groundswell of interest in this gentle technology. I had planned to move, so I made the relocation an opportunity to work in PV.

FIGURE 2.2: Joel Davidson's Pettigrew, Arkansas, cabin. This small PV system used only four 33-watt solar modules and a 420 ampere hour battery bank located in the crawlspace to power lights, tv, stereo, radio, fans, computer, appliances, and tools.

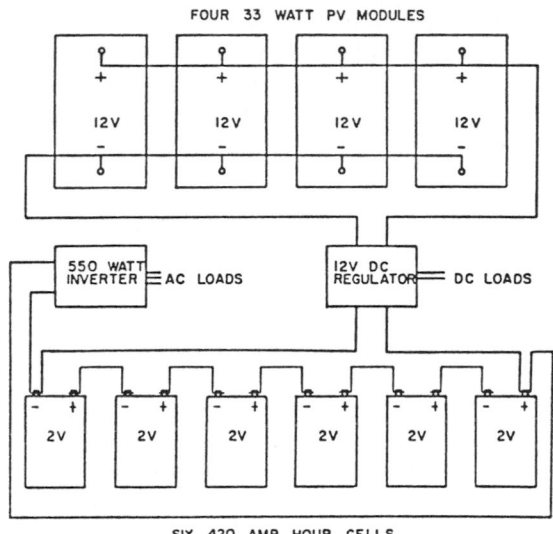

FIGURE 2.3: Block diagram of author's Arkansas cabin PV power system.

When I returned to the city, I brought my PV system with me. I sold the batteries to a friend and traded the tracker for computer parts, but I was not ready to part with the modules.

It was a shame to let my PV array sit in a box. I couldn't put them on the apartment building where I was living at the time, so I mounted them where I worked and set up a solar-powered office.

The office system evolved from a 12-volt set-up to an all AC configuration. At first, lights were 12-volt DC. The standard recessed fluorescent office fixtures were re-wired with 12-volt ballasts. The stereo was 12 volts, as was the tv monitor for the computer. The computer itself was powered by an inverter. Four Trojan deep-cycle, golfcart-type batteries were used for over 400 ampere hours of storage. A Specialty Concepts (SCI) regulator completed the system.

My now eight solar modules were powering my office 100%. The difference in solar input from Arkansas to southern California actually gave me an excess of power. It was for this reason that I went to a fixed 45% mount instead of a tracker. The array consisted of four 33-watt modules, three 28-watt modules, and one 35-watt module.

Production of over one kilowatt hour per day matched consumption. While the battery bank should be twice the size, performance was not affected. Any deficit in power was made up on weekends when the office was closed.

The next change in my travelling and evolving PV system occurred when Heart inverters became available. These high-efficiency inverters made all-AC operation practical. Their low no-load current draw and 85%-plus efficiency convinced me that it was time to show city people what PV could do for them.

I re-wired the solar array to 24 volts. No modules were added. The 12-volt regulator was upgraded to 24 volts by SCI. The series/parallel 12-volt battery bank was re-wired to a 24-volt series set-up. A 2500-watt Heart inverter operating a 24-volt DC input was the reason for the change-over. The job was done in one morning and a third of the wiring was eliminated.

The original ballasts for the lights were re-installed. Standard outlets were wired in and the room was now standard 120 volt AC. While there was a 10% to 15% efficiency loss in the change-over, it was worth it. The convenience of using standard appliances and office equipment was important.

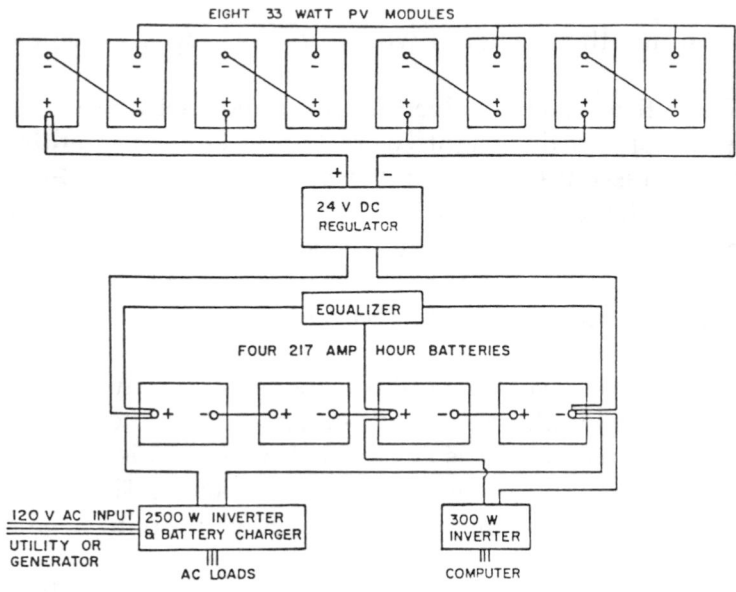

FIGURE 2.4: Block diagram of author's California office PV power system.

The office system was particularly unique because I had the opportunity to test the practicality of the standby system described on page 279. Although I produced all the power I needed, the standby system made it possible to power greater loads. When consumption exceeded production, the office automatically switched over to utility power. This could just as well have been a switch-over to a back-up generator at a location away from power lines.

I did not need this feature in my 100% PV-powered office. However, I wanted to show people living with utility power how they could start with a small PV system and gradually wean themselves off the grid. By having the ability to switch from PV to utility back-up, you can begin using PV power now. Adding to the array increases your energy independence.

ENERGY CONSERVATION

Living with PV can be as unique as life at sea on a sailboat or as ordinary as your usual day-to-day activities. You may have already been in a PV-powered office or home and, unless you saw the array, did not even know that the sun was providing the electricity.

It cannot be stressed too often—PV systems work well *if you are energy-conscious.* You must be willing to keep an open mind and find creative ways to use and *not use* electricity.

Energy conservation does not mean "doing without." During the oil embargo there was much talk about "freezing in the dark." Well, that didn't happen. People learned to tighten their energy belts and continue to live as before.

Energy conservation is not only dollar smart, it is wise. The energy you produce and save is like money in the bank for yourself and the entire planet. Conservation also means less pollution. We can no longer "foul the nest and move west." There is no place to hide from waste and pollution.

It has been shown that the lifestyle of the energy-conservative PV user earning an average income can be as full of comfort—even luxury—as the wealthy. How is that possible? Very simple. When you are producing your own power, you are working for and investing in yourself instead of the utility company. The money stays in your pocket.

The Cost of Utility Power

We are just beginning to understand the real cost of energy. The accounting must go beyond present and projected monetary costs to include environmental and societal impacts that will be borne by future generations.

While the days of skyrocketing electric bills may be behind us for a while, prices are still increasing. From August 1983 to October 1984, for example, Californians experienced an average electric rate increase of over 11.9%. The rest of the country more or less followed suit. Some areas were "spared" and only saw an 8% increase. Other areas were hit harder, most noticeably where utilities were passing on nuclear power plant operating costs, closedowns, and cost overruns.

Even the lucky ones are not that lucky. A modest 8% annual

increase in the electric bill means that it will double in ten years. At that rate a $50 per month electric bill will be $200 per month in twenty years. It might be a good idea to put something away now to pay for electricity when you retire—which is exactly what using PV is, an investment in your energy future. When you consider what electricity has cost you, is costing you, and will cost you, a PV power system becomes a reasonable investment.

Conservation Considerations

The first step you must take when deciding to set up a PV power system is to stop using luxury items such as electric can openers and blow dryers. For some people these devices are important, and they must pay for using them. But for most of us, our lives would be pleasantly simplified if we let our fingers do the working.

A number of new appliances have been designed for energy efficiency. Some, including air conditioners and refrigerators, have government-mandated efficiency tags so you can intelligently compare different brands. It will always pay in the long run to purchase the most energy-efficient model, even if the alternatives are less expensive. Some appliances made in the 1950s are actually more energy-efficient and reliable than those manufactured today. If you are looking for a used refrigerator, the 20-year-old manual defrost type could give you better service than a 5-year-old fancy chrome and avocado side-by-side.

On the other hand, many electrical devices in common use were designed for the days of subsidized "cheap" energy. Take a good look at the electrical equipment in your home. Are you using an old tv or stereo? Check the identification plate (usually on the back) and see if it uses more than 100 watts. If so, it is probably a relic from the days of low-cost grid power. Check all your devices and appliances. It's those cookers and heaters that really eat up the watts. The domestic electric water heater is one of the more wasteful appliances, unless it has a timer to shut it off when not in use. Air conditioning (which is really air refrigeration) and cooking with electricity (except brief periods of microwave use) may one day be outlawed unless the user either produces power or is willing to pay a penalty charge for high consumption. Home heating with electricity is so expensive that it is second only to air conditioning in power consumption in the temperate zones.

Table 2.1: The Rising Cost of Electricity

Year	Elec kWh	Heating Oil Gallon	Natural Gas Therm	Gasoline Gallon
1970	2.22	18.5	10.9	35.7
1973	2.54	22.8	12.9	38.8
1975	3.51	37.7	17.1	56.7
1978	4.31	49.4	25.6	62.6
1980	5.36	97.8	36.8	119.1
1982	6.86	118.6	51.7	122.2
1983	7.18	91.6	59.9	122.5
1984	7.56	91.6	60.6	119.8
1985	8.03	97.8	62.8	112.2
1986	8.10	71.1	56.8	74.5

Incandescent lights (the ones with filaments that heat) really waste electricity. A good way to check if a light is efficient is to touch it when it is on. Those old bulbs in your home are better heaters than lights. Fluorescent lamps are more energy-efficient, and the new ones do not have the noticeable 60-cycle flicker. A 20-watt fluorescent bulb will produce the same illumination as a 40-watt incandescent lamp—and that's real energy savings.

Just as important as converting to more efficient devices and appliances is the need to change energy consumption patterns. If you leave lights on in unoccupied rooms or use more lights than necessary, you will have to develop better energy habits before changing over to low-cost home-generated electricity. Even if you do not produce your own energy, it is wise to become aware of your energy usage and conserve. Most homeowners could cut their electrical consumption in half without suffering. The children of the future will appreciate your conservation efforts.

In Europe, where high utility bills have been a way of life for decades, smaller, efficient appliances are standard. Well-designed, unobtrusive blenders, hair dryers, and dishwashers seem appropriate in the smaller apartments that are the European norm, and usually consume less than half the wattage of their American counterparts. Scandinavians, Germans, and Swiss enjoy a higher standard of living than we do, while using one-half the energy per capita.

This is a good time to plug one of the most valuable books I have read since installing my own PV system. Michael Hackleman, a long-time user of wind and solar power, and a tinkerer who knows how important conservation is to the future, has written a great book called *Better Use of . . . Your Electric Lights, Home Appliances, Shop Tools—Everything That Uses Electricity* (available from **aatec**). For years, I have been telling people to stop buying books, and to take that money and get out there and "do it!" But this book changed my mind. Now I say, "Read Michael's book first." It tells you how to use the electrical equipment you want to use properly, maximizing the potential of your home power system. This is particularly important for DC power systems.

HOW THE PV HOUSE WORKS

At sunrise, the house utilizes either battery-stored PV power, power from the utility company, or power from a generator (grid/gen). Grid/gen is available either directly or indirectly via grid/gen-charged batteries. If the loads for the previous day were within the production/storage limits of the PV system, the occupants will get PV power from the battery bank. If the previous day's solar production was depleted (consumption exceeded solar production), the battery voltage then dropped below a preset point triggering grid/gen power and activating the standby battery charger. The occupants would then have access to grid/gen power until battery recharging was completed. At that point the system would revert to the autonomous mode.

During a sunny day, PV charges the battery bank and the home operates on solar energy. This begins as soon as the array voltage is greater than the battery voltage. During a cloudy day, the house will operate on stored power. Some solar power in the form of a trickle charge will enter the battery bank.

If the daytime loads exceed battery/PV capacity, the inverter will automatically switch to grid/gen power activating the standby battery charger. This occurs when the battery voltage drops below a preset point.

If the daytime loads are small and the batteries are fully charged, the voltage regulator, depending on the model, will automatically either stop the charge, reduce the charge to a trickle

FIGURE 2.5: This urban solar electric system provides part of this modern home's power requirements. Sixteen 35-watt PV modules flushmounted for a low-profile charge two separate battery banks through two 30-ampere charge controllers. A 2000-watt inverter provides AC power for winter lighting and summer air conditioning. Designed and installed by Greg Johanson, Solar Electrical Systems.

charge, or shunt the excess power to a secondary load (like a water heater element) or any combination of these and other features. This occurs when the battery voltage reaches the charge cut-out point.

In the evening when voltage from the PV array drops below the battery voltage, the regulator will disconnect the array from the battery bank preventing reverse current discharge. The house will then be on battery power with utility or generator back-up.

In the event of a utility power failure or generator breakdown, the system will access solar/battery power. In this mode, the PV system must provide all of the power. Upon return of grid/gen power, the system will automatically access this source should it be needed.

MODES OF OPERATION

A PV system can have several modes of operation. In the previous example we have seen:

1. Sun Charge Mode
   The sun charges the batteries. When the batteries are fully charged, solar electric production can be used to trickle

charge the batteries. Once the batteries are fully charged, power can also be diverted to a secondary battery bank or auxiliary load such as a fan or water pump.
2. Battery Mode
When the sun is not shining, the batteries provide power.
3. Grid/Gen Mode
When the sun is not shining and the battery bank is depleted, power from the utility or a standby generator can be used.
4. Emergency Mode
When utility or a standby generator has failed, all power will come from PV and the battery bank.

The following scenario is the possible sequence of events that can occur during the daily operation of a PV system with a 2500-watt inverter (24-volt DC input) that provides part of the power for a home. This system has automatic transfer and a standby battery charger connected to the utility company. By expanding the solar array, the home can be 100% PV-powered.

6 am   The battery bank is at 24.7 volts DC. The sun has not risen and array voltage is zero. The house is on battery power.

8 am   The sun is now shining on the array and array voltage is 25 volts DC. The batteries are being charged by the sun.

1 pm   It is a sunny day and battery voltage has reached 29.6 volts DC. Battery charging is complete. The system regulator has switched to the trickle charge mode and the batteries are receiving a float or trickle charge.

6 pm   The sun has set and the battery voltage is 26 volts DC.

10 pm   Due to heavy loads such as laundry washing and other evening activities, battery voltage has dropped to 24.1 volts DC. The automatic transfer switches the home to utility power and the standby battery charger comes on to recharge the battery bank. (In a home away from the utility lines, a generator can be used.)

2 am   The battery charger has recharged the batteries to 28.5 volts DC. Standby battery charging is complete and the home is transferred back to autonomous operation awaiting the morning sunshine.

Now that we have a general idea of how a PV home works, let's get on to the specifics.

Chapter 3

Conservation and System Design

DESIGNING FOR CONSERVATION

When you size a solar electric system, the first thing to consider is the load. The load is what will be powered, when, and for how long. Knowing the load is important because you must match solar production to consumption to insure that you have enough power to do the job.

Buying power from a utility does not require consideration of the load unless you are a major consumer, such as a manufacturing plant. However, it is a good idea to know your load anyway because your electric bill is based on it.

Producing your own power puts a completely different perspective on electrical consumption. Not only must you match production to consumption, you should also try to scale down your loads. Why? First of all, you will no longer be renting the use of utility electricity. You will be buying and operating your own power plant. By knowing your load, you can make your solar electric system as small and as efficient as possible with no over-capacity and no wasted dollars.

You also examine the load to determine the mix of equipment to be powered and when they will be used. Consumption patterns for an office will differ from those of a home or workshop. The combination of appliances used by a single person will differ from that of a couple with teenagers.

By examining the mix of equipment, you can size your inverter. The inverter must be able to handle the largest combined AC load to be used at any one time. While a 1200-watt high-surge inverter can power a refrigerator or a washing machine, it will not suffice if the well pump comes on at the same time. Therefore, doing the wash requires an inverter to carry the washing machine and well pump. Should the automatic switch for the refrigerator come on while the laundry is being done, the inverter must be capable of handling that surge and load as well.

This examination of your load requirements gives you the unique opportunity to design for efficiency. By examining the tasks to be performed, you can determine the most practical and efficient method of performing those tasks.

The example systems in this book were selected to help you think through good design. PV system designers look at the loads and select appliances and equipment based on the efficient use of power.

Let's look at kitchen lighting, for example. Most of us are familiar with the 100-watt incandescent bulb. For years that was the way kitchens were lit. "Modern" kitchens, a few years back, had two or more of these bulbs in globes or fixtures. Now we see newer homes with fluorescent lighting in the kitchen.

To efficiently light the kitchen, the first thing to do is to replace that 100-watt bulb with a 40-watt fluorescent lamp which will produce about the same amount of light (lumens) as the incandescent bulb. Not only have you cut the lighting load in half, you have also reduced the cooling load. How? An incandescent lamp is a better heater than a light. Most of the incandescent's production is in the infrared or heat spectrum. In a kitchen you do not need a light that is a heater, unless you live in a cave and want fire and light.

There is an additional step to take in kitchen lighting design.

You can re-examine the task you are performing and design for the task. Lighting consists of area illumination and task lighting. You know what it is like to try to work in your own shadow. That bulb in the ceiling may light up the room, but you end up in the dark as you work at the stove, sink, or counter. The solution is to place small 6- to 20-watt lights strategically above workplaces.

Finally, consider using the lowest-cost source of light—the sun. Natural lighting through windows and skylights is free and available everywhere. Don't cover up windows with curtains, blinds, or shades that can't be opened. Open them during the day. Not only do you get free lighting, you also get some heat in winter. In fact, the average home in the United States with a good southern exposure can get up to 50% of its daytime winter heating requirements through its south-facing windows. I'm not talking about a solar-designed house, just a regular house with regular windows.

So what have we done to the kitchen lighting? First, we have re-introduced daylighting. Next, unwanted heat from incandescent bulbs has been reduced. Third, through task lighting, lighting requirements have been cut. Fourth, area lighting power requirements have been cut in half. Fifth, the quality of lighting is improved.

It doesn't matter if you are not yet in your PV dream house. You can start now by switching over to fluorescents and good lighting design. If you are in your PV home, you can scale down your PV power plant or use the excess power once wasted on inefficient lighting for something else.

The possibilities are endless and exciting. Conserving does not mean doing with less: It means using what you have more wisely.

AN EXAMPLE IN CONSERVATION DESIGN

Bill and Jackie Perleberg of Golden, Colorado, have spent a lot of time thinking about energy and conservation, and they have put those thoughts into action.

Of all the people I know who believe in PV and live by their beliefs, Bill and Jackie are unique. They live in a modern home with all the conveniences anyone could want or need. Their home is still connected to the utility company, but they are almost independent, using a surprisingly small solar electric system. How are they able to live so comfortably with so little home-grown power?

Bill is a teacher and he learns a lot from his students. During the oil embargo, he asked them what could be done with a fuel-efficient car compared to a gas hog. He reports that 93% of the answers were something similar to, I'd be able to travel twice as far on the same amount of money. Only 7% spoke of being able to travel their usual daily routes for half the present cost.

Perhaps things have changed in the past ten years, but the point is we are basically consumers and measure our success by the amount we consume. To many, conservation means deprivation. Not so. Conservation means having something left to use later: that something can be energy, food, clothing, or money. Wouldn't it be nice to cut your utility expenses in half and use the savings to take a real vacation? That's what conservation can mean.

The Perleberg's home heating goal was to use a fuel others had not considered, possibly a waste or by-product. They tried wood, but that was a lot of work. They considered coal. Finally, they tried waste oil.

Did you ever wonder what happens to used engine oil? Some is recycled, but there are not many refineries. Most is dumped. The Perlebergs heat their home with this efficient, smokeless and odorless, high BTU fuel. While they don't like using a petroleum product, they will have a steady source of fuel as long as petroleum lubricates our engines.

For aesthetic fireplace fires, Bill and Jackie use warehouse pallets. Free for the hauling, they are seasoned, easy to handle, and require less processing than regular firewood. The nails which end up in the ashes are fished out with a magnet, netting Bill at least $15 per season as recycled scrap steel.

The Perleberg home was designed with electricity as the primary heat source. The 25-kilowatt electric furnace was tested but is never used. Waste oil is their primary fuel, with wood second, and electricity as back-up. As Bill points out, other locations may offer other primary fuel alternatives: corn cobs, sawdust, cow chips, or methane, for example. Think creatively.

Bill modified his furnace by putting switches on each of the five 5-kW heating elements, so he can control the number of kilowatts used. Why bother? Well, he uses the electric furnace duct system to

FIGURE 3.1: Bill and Jackie Perleberg's home located in the foothills above Golden, Colorado. Two 8-module passive solar trackers and a windcharger (not in photo) provide half the power for this comfortable home. (Courtesy Appropriate Systems, Inc.)

circulate heat via the built-in fan. That way his back-up heat source is always ready, if needed.

Water heating is also uniquely handled by careful thought and design. Bill built a solar water heater and reports that it works fine. He also pulls waste heat from his waste oil heater flue (can you beat that!) with none of the creosote problems associated with wood burning. Flue gases are diverted through a used commercial water heater tank minus original burners. Dampers in the flue pipe control how much flue gas is used to heat water. Solar-heated water is run through coils on the outside of this same preheat tank. The preheated water is then passed on to his conventional electric water heater.

Two other modifications were made to this conventional tank. First, more insulation. Everyone can benefit from that. If you put this book down right now and go out and buy a roll of 6-inch fiberglass insulation and wrap your water heater, you will save both the cost of this book and the cost of the insulation in less than one year.

Bill went further. He replaced the two 5.5-kW water heater elements with two 2.5-kW ones. Still plenty of hot water. Again downsizing to 1-kW elements, the only change was slower recovery. He finally ended up with 750-watt elements and no shortage of hot water. Granted they don't wash five loads of laundry a day, but spreading this chore over the week fits their lifestyle fine. They also switch their water heater on and off at the circuit breaker, heating water when appliances are off. All that adds up to lower consumption.

And they didn't stop there. Occasionally, the 16 tracker-mounted PV modules and 200-watt windcharger produce an excess of electricity. Bill uses a Charge-a-Stat (see the chapter 7) to divert this excess electricity to three 12-volt 500-watt water heater elements. Bill is so sold on this practical use of excess PV/wind power, that he sells the elements to fellow alternate energy system enthusiasts.

The house, I must remind you was originally all-electric. The conventional 240-volt AC electric range came with four cooking elements: two small 1325-watt and two larger 2350-watt. After these wattages were noted, half the range top was wired for 120 volt AC. The small element now draws 360 watts and the larger 620 watts. The difference in cooking at the lower voltage is 3 to 5 minutes. This keeps their overall demand down while cooking and using other electrical appliances.

They didn't like the high demand load from the conventional 3400-watt oven and Jackie likes to bake, so they installed a propane gas oven from a wrecked RV. They use only about 50 pounds of propane a year.

Mostly, they cook with microwave using a mechanical power selector and mechanical timer which Bill reports is more accurate. His old, but reliable, BEST inverter output could affect the solid-state timer of other microwave ovens. If they are not in a hurry, they cook at the lowest setting which adds about 10 minutes cooking time.

They also have a toaster/broiler rated at 800 watts (the label on your toaster probably says 1200 watts or more), an 800-watt iron (smaller and lighter than the standard), and a low-wattage 150-watt crockpot. Their electric fry pan is 1250 watts, but Bill points out that it cycles often and so doesn't push their demand up much. Label reading on appliances is a habit with them.

Yes, they do have a dishwasher. They disconnected the high-

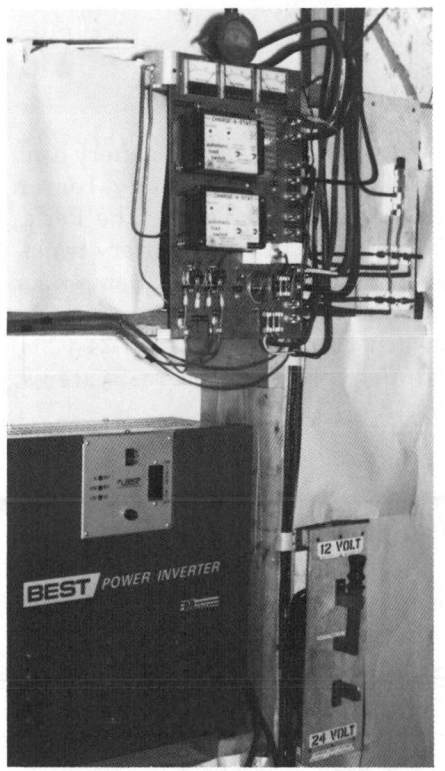

FIGURE 3.2: The control panel for the Perleberg system has two Charge-a-Stats, one for solar and one for wind. Large 100-amp knife switch to the right of inverter is used to switch two 12-volt battery banks to 24 volts for inverter operation. (Courtesy Appropriate Systems, Inc.)

draw sanitary elements and monitor the water temperature with a gauge on their water heater tank. They wash dishes at 140°F. The motor in the dishwasher draws only 700 watts.

Clothes dryers use a lot of energy, but are hard to do without if you live in a climate where solar clothes drying is not always possible. To ease the situation, Bill and Jackie wear permanent press clothes that can be put in a dryer and tumbled on "Air only." Bill got an old gas dryer, removed the burners and ran a 5-inch diameter duct from his waste oil heater to the dryer. Free heat. The dryer motor uses only 550 watts.

When buying appliances, Bill and Jackie take along a hand-held watt meter—a fascinating experience for the salespeople and something all energy-conscious shoppers should consider.

Why all the attention to watts? Because this is where "demand" comes in. And if you have an inverter in your PV system or if you are on utility power with demand billing, demand is important. Demand

FIGURE 3.3: Part of the battery bank used to store electricity for an energy-efficient home. The Perleberg's home has four battery banks, each with ten 6-volt 210 amp-hour golfcart-type batteries that can be switched from 12 volts to 24 volts. (Courtesy Appropriate Systems, Inc.)

billing, unlike declining block rate, means the lower your peak demand, the lower your rate. Put another way, the more you draw from the utility in kilowatts during a (usually) 15-minute period, the higher your rate. Not all utility companies offer demand billing. Most offer "declining block" which means if you use a lot of electricity, the last chunk of power will be sold to you at a lower cost than the first couple hundred kilowatt hours.

If you use an inverter, demand is very important because you can downsize your inverter by keeping your combined demands below a certain point. Let's say you have a home with a microwave oven, garbage disposal, and refrigerator in the kitchen. The refrigerator comes on automatically. However, if you run the garbage disposal and the microwave at the same time, you may need a 5000-watt inverter. If you do not use the garbage disposal when microwaving, you can use a 2500-watt inverter. You just saved yourself over $1000 in equipment.

Bill and Jackie have a 200-watt super-efficient Whirlpool 17 cubic foot refrigerator. Their well pump is a 0.75-hp 240-volt AC unit which draws about 1000 watts. These cycling combined loads determine their consumption of 1 kW for their monthly demand rate. The remaining appliances draw less than 1 kW each and are seldom used at the same time.

What else do the Perlebergs use? They have most modern appliances: hair setter, plenty of home lighting, power tools (except for a welder), etc. Their entertainment equipment is 12 volt DC and works as well or better than 120 volt AC. They sound pretty suburban. Except they fight the impulse to buy what is not needed. Everything you plug in, Bill says, is a parasite to your income.

In 1977 when the Perlebergs built their home, they could have gone with more propane and regular electric billing like their neighbors. However, they saw that conservation would be both rewarding in spirit and practice. At the time, federal and state energy credits for conservation and renewable energy systems were available. The lower demand rate billing from the utility company was also available. These economic incentives made the change-over attractive. With the savings over the years, they have been able to live very well.

What price do the Perlebergs pay for all this? Bill, who is naturally handy and likes to try new ideas, enjoys what he does. It is more than just a rewarding hobby. The Perlebergs are able to live by their beliefs and be an example to others. In the beginning there were minor lifestyle changes. Once appliances were labeled with wattage ratings, the rest was simple. They leave a 3 x 5 card on the range to show that "Laundry is in Progress." Sometimes it was inconvenient, but problems were solved and a few more cards made, such as, "Power Tools in Use," "Dishwasher On" and "Water Heater On."

No big thing, but the utility company thought it was. They were suspicious. They changed the kilowatt-hour meter several times during the first year as the Perleberg's demand dropped lower and lower. Then they began to place two seals on the demand reset. Finally, a district manager and a meter repairman came to their home unannounced, pulled the meter from its socket and ran load tests. The meter was checked for accuracy and magnetism. When Bill wanted to know what this was all about, he was told that he couldn't heat water for what he was paying, much less heat the house or dry clothes. So he gave them a tour. They left him alone after that.

As mentioned, the Perlebergs have a combined PV and wind system. They use both AC and DC and have both 12- and 24-volt DC circuits. The house was built for AC, with conduit for later running DC. Low-voltage wiring was part of the original plan.

Bill reports that the passive solar trackers are a bit slow in cold weather, not tipping back to the east until 10 am. His battery area is kept at approximately 60°F (15°C) and though there is good air movement, the doors are left open during equalization charges. The batteries were a bit sulfated when he got them so he has set charging cut-out voltage to 14.8 volts DC. The windcharger tower and the PV trackers are grounded with ground rods.

The Perlebergs purchase half their power from the utility company; their bills are under $20 per month. PV and wind make up the balance, about half and half. PV production is about 60 kWh per month. 16 each 35-watt modules = 560 peak watts x 4 peak hours of sun = 2240 x 30% tracker increase = 2912 watt hours per day at 70% combined inverter/battery efficiency = 2038.4 x 30 days = 61.1 kWh per month.

To go from 12 volts to 24 volts, Bill manually switches at the battery bank and then turns on the inverter. The inverter has a demand start and has worked well. Bill put AC watt-hour meters on several appliances to check usage. This breaks down monthly to refrigerator: 85 kWh; well pump: 8.33 kWh; hot water: 22.8 kWh. They save about $1200 per year through conservation alone.

Further savings may be realized in the future. Bill and Jackie have been using a 10 cubic foot LP refrigerator for about a year. The refrigerator also has 120-volt AC back-up. It runs for about 120 days on a 100-pound bottle of propane. The refrigerator and oven combined consume a bottle of propane in 105 days.

With their electrical consumption getting down to 85 kWh per month, a DC refrigerator (like the Sun Frost) would be considered if it were less costly. They are also considering purchasing a 5-hp stirling engine and running it with the surplus heat from their waste oil heater. The engine would run a DC generator to charge the battery bank during the heating season when solar battery charging is at its lowest production.

It is only a matter of time before time-of-day utility demand rate billing will be instituted in their area. But that will have little affect on the Perlebergs because "We will simply leave the grid during penalty times and return to the grid at low-cost times—or maybe not at all." Through conservation and the use of alternative energy systems, Bill and Jackie have the freedom of choice.

APPLIANCES AND THE PV HOME

In 1929 when home refrigerators first appeared, they were novel devices that made ice. Homeowners could now store food just as the Navy and commercial businesses had for years. It was amazing and the refrigerator was given a special place in the home. Now it is commonplace.

In 1939 television began to appear. Prior to that, home entertainment was the radio or do-it-yourself. Now we have tv, stereo, VCR, home movies, and a host of attachments that would make Edison smile. We take it all for granted.

Air conditioning, the bane of the low-power energy-efficient home, first came on the scene in 1952. Before that, people actually lived and worked without it. How soon we forget what life before air conditioning was like.

We live surrounded by wonders of science and engineering and consider them commonplace. To appreciate these marvels of modern technology and really understand our dependence on electricity, I suggest you go outside and turn off the main circuit breaker to your home. See how long you can survive without electricity and the conveniences it powers.

While for us just the thought of living in a world without electricity is awesome, most people on the planet do live without it. Some can remember when rural areas of the United States were first electrified. Now most of the U.S. is grid-connected.

The solar electric home goes one step further. Not only does it contain modern electric devices, it produces the power to operate them.

And when you become your own power plant manager, you think about what you are powering. You notice the quality of the lighting in a room. You listen to the sound of the motor in refrigerator compressors and washing machines. You wonder if the interference on the television is from your equipment or from the tv station.

While politicians debated "power to the people," some took that phrase literally. In doing so they gained new respect for electrical devices, and also learned how to break their dependence on the utility company monopoly. They have become truly "powerful."

Lighting

To review what we discussed earlier about lighting, incandescents are better heaters than light producers. Fluorescents produce twice the lumens, or light, as incandescents. In some places like California where energy conservation is the law, incandescents can only be used as "specialty lighting." New construction must use area lighting that produces 15 lumens per watt or better.

Most incandescents produce about 10 to 15 lumens per watt; fluorescents produce 25 or more lumens per watt. Panasonic manufactures a screw-based fluorescent that produces 45 lumens per watt. Philips has a similar but more expensive version of this bulb that produces 60 lumens per watt. For comparison, a standard 60-watt incandescent puts out 900 lumens or 15 lumens per watt. Mitsubishi's 15-watt bulb compares to a conventional 60-watt bulb while cutting 75% from lighting costs. If every home in America installed four Mitsubishi bulbs, the savings in electricity would be greater than all the power produced by the six largest nuclear power plants.

Use fluorescents wherever you can. If you don't like the light they produce, check out the different types available. There is an acceptable fluorescent light for every use. Artists use full-spectrum tubes which produce the equivalent of daylight for that "north skylight" light. There's even a mellow glowing tube to use above the bathroom mirror so you don't look like a zombie.

Refrigerators

Whereas once we went to the garden or the pantry or the corner store for fresh food, now we shop once a week and pack it in the refrigerator. There is an energy price to pay for that convenience. In the average home without air conditioning and where gas is used to cook and heat, the refrigerator consumes up to 20% of the electricity.

According to an Energy Commission report, if Californians used energy-efficient refrigerators that would save more energy than all the other conservation programs combined. The total savings would be $10 billion dollars. Buying a new energy-efficient refrigerator *is* cost-effective for everyone. The $100 or so higher price tag has a payback of two to four years. And with the estimated refrigerator lifetime of 15 to 20 years, the savings really add up.

In the energy-conservative home as much as 50% of the electricity is consumed by the refrigerator. What does this mean to the solar electric homeowner? It means that half the solar array is devoted to one appliance. It means that half the cost of the solar electric system is used to operate the frig.

There's got to be a better way—and there is. Far be it for me to suggest that you give up your refrigerator. I lived without one for a while and found it inconvenient.

A solution used by most 100% solar electric homes is the gas refrigerator. Old Servels are finding new homes. These relics are nothing to joke about. They work great, have no moving parts, and actually cost less to operate than their electric counterparts. Although they are only slightly more efficient than electric refrigerators, the gas to power them costs less per energy unit (BTU) than electricity. It may not always be that way, but that's how it is now.

Since the PV-powered home will probably have gas cooking, water heating, and space heating, all or in part, it is logical to use gas to refrigerate. Old gas refrigerators can cost as much as the fancy, new frost-free, electric energy hogs. If you can't find a used one, you can buy a new gas refrigerator. To find your local supplier, contact the gas company.

By going to a gas refrigerator, you have cut your electric energy budget nearly in half. If you want electric refrigeration, that's still possible, and perhaps ultimately more practical. There are modern, energy-efficient, no-nonsense refrigerator/freezers that use less than 4 kilowatt hours per day. Granted, that's a lot of power, but there are things you can do to help. You can place your refrigerator in a spot that does not get above 70°F (21°C), and that has lots of ventilation to keep the working parts cool. The addition of more insulation can also make a big difference.

One couple living in a PV/wind-powered home added insulation to their 25 cubic foot side-by-side refrigerator/freezer. They put 4 inches of styrofoam on the top and sides and 12 inches of fiberglass on the back. This reduced the duty cycle to 20%. Normally, the machine would come on for 5 minutes and be off for 5 minutes. Now it is on for 5 minutes and off for 20 minutes. It holds the cold in better and uses less electricity. In fact, it works so well, a small fan (PV-powered, of course) was installed to circulate the cold air inside.

This same method is used in supermarkets to prevent cold air from stratifying at the bottom of the cold compartment.

Robert Sardinsky put a PV system in his home in Maine. He insulated a 12-volt DC Dometic refrigerator to reduce the duty cycle. He even put insulation on the doors. Then he covered the insulation with wood paneling and trim. Now he has one of the most attractive and energy-efficient refrigerators around. It looks like fine cabinet work.

Dr. Larry Schussler of Sun Frost has also made refrigeration more practical for PV people. His super-insulated 12- and 24-volt DC refrigerators are the most energy-efficient in the world. The box is custom-made and well-insulated. The compressors (Danfoss BD2.5) are small and very efficient.

Sanyo has been making a very energy-efficient small refrigerator for sale in Japan. We will begin seeing this model and others in the U.S. as electricity becomes more costly. Presently, the three most energy-efficient single-door, manual defrost 10 to 12 cubic foot refrigerators are the Absocold (HM103S7W42) at 400 kWh/yr, the Kenmore (86311*0) at 439 kWh/yr, and the Sanyo (SR1020–2) at 439 kWh/yr. Woods and Panasonic make 14 and 16 cubic foot chest freezers that use less than 500 kWh/yr.

There is an excellent booklet available for $2 from the American Council for an Energy Efficient Economy (1001 Connecticut Avenue NW, Washington, DC 20036) entitled *The Most Energy Efficient Appliances*. Updated regularly, it lists appliances and their power consumption.

Some Tips About Refrigerators

Whatever refrigerator you buy, there are some common sense rules to guide its use.

1. Both refrigerators and freezers operate more efficiently when full but not over-crowded. It takes more energy to keep air cold than it does to keep foods cold. Be sure to allow enough space between items for air to circulate freely.
2. Locate your refrigerator away from the direct flow of warm air from the cooking range, heating ducts, or direct sunlight.

FIGURE 3.4: Robert Sardinsky's 12-volt DC super-insulated refrigerator. The attractive mahogany paneling covers two inches of rigid insulation board added to all sides of the stock unit's two-inch insulated shell, transforming an extremely efficient operation into a beautiful package. (Photo: Robert Sardinsky, Rising Sun Enterprises)

3. Set the temperature to medium and keep a thermometer in the refrigerator to be sure that you are operating at 34 to 37°F (1 to 3°C). Separate freezers are kept at 0 to 5°F (-18 to -15°C). Refrigerator/freezers with one outside door should be 15 to 20°F (-9 to -7°C) in the freezer compartment.

4. Turn off the automatic defroster. It is wasting energy by heating up the cold refrigerator. Defrost often manually and do not allow frost build-up of more than 1/4-inch.

5. Clean the condenser coils a couple times a year. Dust build-up causes your refrigerator to work harder.

6. Uncovered liquids cause frost to form. Keep containers sealed.

7. Allow hot food to cool slightly before placing in the refrigerator. You can leave most foods at room temperature for up to two hours. Thaw frozen foods in the refrigerator, not on the counter. If you put frozen foods in the refrigerator overnight, they should be ready to use by the next evening.

8. Keep the door closed. When you do open the refrigerator, keep it open for as brief a time as possible. Your refrigerator will not lose its cool with the door closed.

Laundry

When it comes to laundry, the only real difference between a PV home and a grid-connected home is how the chore is planned. Wash full loads only. If that means going back to the traditional once a week washday, so be it. That day should be when the sun is shining so you can also take advantage of solar drying.

Where outdoor drying is not practical, temporary lines can be used indoors. When I was a child in Philadelphia, we hung our laundry in the basement in winter. Delicate things were hung in the bathroom on a folding wooden rack. I still remember how the air in the house smelled on wash day—clean, fresh and moist, even though the house was closed up for winter.

The modern clothes dryer has just about replaced the energy-efficient clothes line. Electric dryers are out of the question in the PV home. Gas dryers work well and can be operated with natural gas or propane (remember how Bill Perleberg modified his?). Although I now have a gas dryer, most of the laundry is hung on the line. Machine dryers seem to shrink blue jeans shorter every washing. Solar-dried jeans may be a little stiff at first, but they sure last longer. Solar drying saves energy and clothes and makes bright things brighter. Sounds like an ad for solar dryers and it is.

Doing the wash with cold water saves energy. About 90% of the energy used in washing goes to heating the water. The motor and controls are pretty efficient. If you think you need hot water to get things clean, try a prewash and cold water detergents. For sanitary purposes (diapers and bedding from ill people), use hot water. Always use full loads and don't overwash clothes.

Dishwashers

The results of a Consumers' Union study says that dishwashers use less hot water than washing by hand and rinsing under constantly flowing hot water. Most dishwashers use 10 to 15 gallons of water per load. Now I wash dishes by hand and, by rinsing in a pan of hot water, can use as little as 5 gallons to clean up after a big turkey dinner. But then I've had a lot of practice.

Small Appliances

Small appliances consume very little energy. However, modern DC to AC inverters with no-load low-power modes can be triggered "on" by as little as a 7-watt load, or less. That means, if you leave even an electric clock on the inverter, it may no longer remain in its quiet, low-load mode and may, in fact, be wasting energy. Be sure to follow inverter recommendations for low-power loads.

Entertainment Equipment

Modern solid-state radios, televisions, video cassette recorders, and so on are very energy-efficient. Most operate internally on 12 volt DC or less and have a transformer and rectifier built-in to change AC to DC. If you are using a DC PV system, consider using DC entertainment equipment. Some of the best sound systems are made for cars. Sound from the higher-quality 12-volt DC units is usually superior to 120-volt AC radio/cassette players which cost twice as much.

Air Conditioning

Air conditioning has really spoiled us. However, it's a fairly recent innovation, and people did live without it. If you need air conditioning for health reasons, read no further. I can't help you. If you are in good health and have an open mind, read on.

The first thing to do if you now have an air conditioner is to turn it off. If you can live without one, do. If you can't, find out why. Is it the heat, the humidity, or lack of natural breezes through your home? Just why do you need refrigerated air?

If you use the most energy-efficient air conditioner available to cool a fair-sized room for eight hours, it would consume over four kilowatt hours of electricity or the output of eighteen 55-watt solar modules on a sunny summer's day in Houston.

Another way to look at the amount of energy air conditioners use is to examine one at different settings. The normal setting is 78°F (25°C) Any air conditioner will cost 46% more to operate at 72°F (22°C). If you reduce the temperature to 70°F (21°C), that same air conditioner will cost you 61% more than at 78°F. Now you begin to see why 78°F has become the government standard. Be

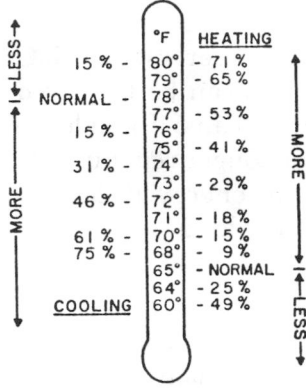

FIGURE 3.5: The relation of temperature to heating and cooling costs. If your normal thermostat setting is 65°F (18°C), it costs 41% more using any type of fuel to heat your home to 75°F (24°C). Air conditioning to 72°F (22°C) from the normal 78 (25°C) setting will cost 46% more.

sure the air conditioners in public buildings are set at 78 because it's your tax money they are spending.

Why does it cost so much to air condition (or refrigerate)? The refrigeration process is very energy-intensive. It takes a lot of energy to make something cold. In fact, it takes lots of energy to change the temperature of anything. There is an important factor involved here called "Delta T," which we will examine in a moment.

What if you have to have air conditioning? Most likely, you'll have to pay for it—whether in PV or utility power. There are a few options, though. You might be able to use evaporative cooling, if the average humidity where you live is low. Check with your old-timer neighbors and learn what they do. If no one in your locale is using evaporative cooling, then the humidity is probably too high.

Another way to stay cool is to let the breezes blow through your house. The home pictured in Figures 3.6 and 3.7, located in the humid Florida Keys, does just that. Adequate shading with large overhanging eaves keeps sunshine out; ceiling fans help to move air. The government services listed on page 19 will provide you with basic information on natural cooling.

Swimming Pools

I am often asked how PV can help reduce the cost of operating a swimming pool. Pools require maintenance: the water must be treated and filtered and, in most cases, heated. All this takes time and money.

If the pool is covered and cleaned often by hand, and used less often, pumping can be cut almost in half. Of course, that requires

FIGURES 3.6 & 3.7: This spacious and airy home in the Florida Keys has many energy-efficient features. Warm air is drawn up and out by ceiling fans. Rainwater fills a 15,000-gallon cistern. The solar electric system has three circuits: 115 volts AC for major appliances, 24 volts DC for all motor loads including a well pump, and 12 volts DC for lighting. The 16 solar modules charge 24 deep-cycle 220 amp-hour batteries. Designed and built by Solar Technologies of the Florida Keys, Inc., Key West, FL.

close monitoring of the pH and algae and bacteria growth to keep the pool safe. There is still the cost of heating and for that there are only two solutions: heat the pool with solar energy and use a pool enclosure, or turn the heater off completely. (Few pool owners, however, are willing to give up that clean, warm dip.)

There is a price for luxury. In fact, people without pools "subsidize" swimming pool operating costs by sharing the electric rate. Swimming pools are heated at the same electric rate as running a refrigerator or powering a hospital, not at a special "luxury" rate. We all pay the environmental impact of electrical production to power non-essentials.

APPLIANCE POWER CONSUMPTION

The following is a partial list of equipment and appliances and their typical wattage rating. Check your appliances for the exact numbers.

Equipment	Wattage	Equipment		Wattage
baby food warmer	165	radio/record player		109
blender	300	television (solid-state)		
can opener	100	black and white		45
coffee maker	1200	color		145
coffee mill	150	clock		2
corn popper (oil)	575	floor waxer/cleaner		305
freezer 15 cu ft		garage door opener		350
(manual defrost)	341	1/4-inch drill		287
juicer	125	circular saw		1150
microwave oven	1450	jig saw		1380
hand mixer	100	table saw		1380
refrig/freezer		sander		287
14 cu ft		sewing machine		75
(manual defrost)	330	sump pump		85
toaster	1100	vacuum cleaner		650
yogurt maker	25	air compressor		3000
air conditioner		battery charger		
5K BTU/hr	710	40 amps		900
dehumidifier	257	90 amps		1300
electric blanket	150	electric motors	start	run
attic fan	370	1/6 hp	500	300
circulating fan	88	1/4 hp	750	350
humidifier	177	1/2 hp	1500	600
waterbed heater	375	3/4 hp	2000	750
blow dryer	1000	1 hp	2700	900
hair curlers	350	2 hp	5500	1800
shaver	15	note: motor wattage increased with		
clothes washer	512	applied loads		
radio	71			

HOME HEATING

I have heated my homes with gas and with wood. The few times I have used electricity, I found it too expensive, costing well over twice as much as gas. For ten years I cut my own wood and used, first, an old cast iron heater, and then a less romantic but more efficient air-tight wood stove. Wood heating has come a long way since the oil embargo and energy crisis.

Gas heating is practical in the PV-powered home. A well-insulated house will use very little fuel. Super-insulated homes—with R-30 walls and R-50 ceiling, and air infiltration almost zero—are growing in popularity in direct proportion to rising fuel costs. If this jargon means nothing to you, learn about it. You could save hundreds of dollars per year by applying new energy conservation techniques to your home.

DELTA T

Delta T is an important concept. The Greek symbol delta (Δ) is used to mean difference. Delta T means difference in temperature. One of the most important things you can learn about heat (or cold) gain or loss is: *The greater the Delta T, the greater the heat loss.* This means that the transition from cold to hot or from hot to cold takes energy. The greater the difference between the hot and cold, the more energy used. Since energy and money equate to the same thing, the greater the Delta T, the more the cost.

How does this work? If it is 70°F outside your home and 70°F inside, then the heater is not working and you are using no energy. If it is 50° outside and you want 70° inside, you will be using some heat energy to warm the home. If it is 0°F outside and you want 70°F inside, then your heater will really be working. The Delta T between 70 and 70 is 0; between 70 and 50 is 20; between 70 and 0 is 70. Any Delta T of 20° or more is a lot. If you can feel the difference with your hand, lots of energy is being used.

Heating water takes energy. Let's say you have a gas water heater. The house is kept at 70°F. Do you need an insulating jacket for the water heater? Yes, you do. Why? Because the Delta T is probably 50° (120 - 70 = 50) and a lot of heat is being lost through the inadequate insulation. Even the best insulated water heaters can be improved with a 6-inch wrap of fiberglass insulation. Don't buy skimpy little water heater insulation kits. Put lots of heat-retaining material between the heater and the outside air even if the heater is indoors.

The same applies to your home. Insulate. After you stop air infiltration, insulation is cheaper than any other energy conservation measure. Of course, solar water heating for the PV home goes without mention. If you can get half or all of your hot water with a solar water heater, install one. In all cases, solar water heating is the only way to go with new construction.

This section on heating and cooling is included to start your brain working in new ways. Look at the whole picture. Don't take water heating and space heating and cooling as separate tasks. Think of them as a heating engineer would. You can even use fancy words like environmental control. But whatever you do, cut your energy budget through conservation and good design that addresses your real needs.

Speaking of insulation, rather than heat the room where I am writing this, I retain my body heat with layers of clothing. When I leave to do something more physical, I will shed some layers. A nice ride on my bicycle and I'll be down to shorts and a T shirt. Total cost for this flexible, well-engineered environmental control: zero.

GAS VS ELECTRIC

Why is there such an emphasis on gas appliances in a book about solar electricity? First, PV is not cheap: It requires a substantial investment to get started. Gas appliances will help keep the initial cost of your PV system down. Second, gas appliances are readily available and reasonable in price. You can buy and use them now, even if you are not yet in your PV dream home. Here is a list of average annual operating costs for appliances (1986 southern California):

Appliance	Gas	Electric	High-Eff. Gas
Space heating	$301	$ 657	$216
Water heating	138	352	105
Clothes drying	20	72	20
Cooking	35	73	35
Total	$494	$1154	$376

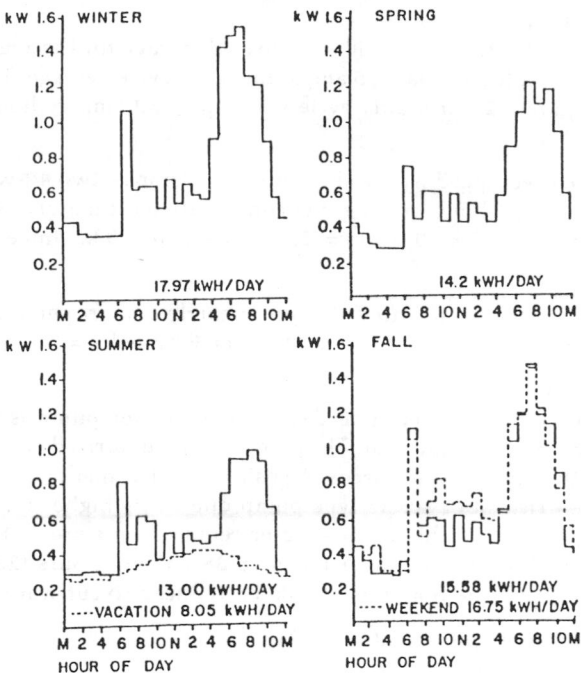

FIGURE 3.8: Seasonal Characteristics of Diversified Electrical LoadsDaily Profiles. These four charts show the pattern of energy use for a family of four living in an all-electric three-bedroom home using 5540 kWh/year. Note the use patterns based upon the hour of day and compare them to your patterns. An energy-conservative family will use less than half the energy to do the same tasks. (Courtesy DOE)

DOMESTIC PHOTOVOLTAIC POWER

This is another example of a solar electric home design. The design options at the end of this section are good examples of the "cut and try" reasoning necessary to design a properly balanced system.

[From "Solar Photovoltaic Application Seminar: Design Installation and Operation of Small, Stand-Alone Photovoltaic Power Systems," July 1980.]

General
The PV system will provide electricity for remotely located homes where hook-up to utility grid lines is not economically feasible.

Refrigeration
A 6-cubic foot refrigerator will provide cold storage for the average family; 12-volt DC motor, 60 watts, 5 amps, 25% duty cycle average. Total ampere hours required: 24 hr x duty cycle x 5 amps = 30 ampere hours.

Lighting
The home is equipped with four fluorescent lamps—two 40-watt and two 10-watt—used 4 hours per night (a conservative estimate). 2 x 40 watts ÷ 12 volts x 4 hr + 2 x 10 watts ÷ 12 volts x 4 hr = 33.3 ampere hours.

Television
Television usage is estimated at approximately 4 hours per day at 24 watts equaling 8 ampere hours (4 hr x 24 watts ÷ 12 volts = 8 ampere hours).

Cistern Water Pump
A permanent magnet positive displacement water pump is used, which draws 5 amps under full load. The pump fills a pressurized tank, which then supplies the home's water needs. A positive displacement pump was chosen over a centrifugal or screw-type pump due to its higher efficiency (80% versus 35 to 50%). The family's water requirements are 200 gal/day; the pump provides 9.4 gal/min, requiring 6.38 ampere hours (200 gal/day is pretty high and the flow rate should be reduced to cut water usage.)

Total Charge Requirements Per Day
77.7 ampere hours per day

System Component Sizing
Array—220 peak watts or 13.75 peak amps
Battery—680 ampere hours at 12 volts DC
Voltage regulator to prevent excessive outgassing due to overcharging

Notes
1. A manual switch isolates the load for maintenance purposes.
2. The system is fused to prevent battery drain if a short circuit develops. Standard bus fuses (car fuses) are acceptable and readily available.
3. A voltage meter with its switches located in the battery storage box isolates and monitors the battery or array voltage.
4. An ammeter with switches located in the battery storage box monitors total array output current or total load current.
5. Electrical storage is provided by low self-discharge lead-acid batteries.
6. Venting or recombiner caps for the batteries prevent dangerous hydrogen gas build-up.
7. Batteries may undergo a maximum 60% depth of discharge.
8. Disconnect switches are not required between the PV panels in the array due to low system voltage (12-volt DC).
9. An opaque cover should be used during maintenance to cover and effectively "turn off" the array.

10. Due to the location of the battery storage in the home, special care must be taken to insure proper venting of gases and protection of the battery terminals against accidental shorting.

Design Options
(Note: July 1980 DOE Prices)
1. Using 11 each PV panels (12 volt DC, 20 peak watts @ $11.25/peak watt = $225) seasonal storage is not needed, tilt angle used is latitude plus 15°.
2. Using 10 panels, seasonal battery storage needed is 128.5 ampere hours at latitude plus 15° tilt.
3. Tilting 10 panels at the degrees latitude increases seasonal storage to 402.2 ampere hours.
4. Optimal choice: between one extra panel or 128.5 additional ampere hours of battery storage. (Battery cost is $170/kWh @ 500 hour rate @ 60% depth of discharge = $437.)
5. Difference is $37 - $225 = $212.
6. Percent difference is 212/225 = 94.27%.
7. Conclusion: choose the additional panel.

STAND-ALONE VS UTILITY INTERCONNECT

A General Electric Corporation PV study published in 1979 concluded that self-sufficient stand-alone PV systems were less preferable than utility-connected PV systems. The assumptions for this conclusion may be wrong. In fact, a goal of the study was to encourage widespread use of utility-tied PV home power systems. This has not happened. The report's comments on stand-alone systems are worth reviewing in light of developments since 1979.

An important example concerns stand-alone systems. In this approach, the residence energy systems are designed for complete self-sufficiency. Besides the solar energy and storage elements, the system requires back-up generating capability as would be provided by a diesel generator. The diesel generator requires additional space, which probably will be limited in the 1986 time frame; it will require more maintenance than conventional equipment; it may result in additional noise during operation which may affect neighbors; and it may present environmental problems.

The impact of a stand-alone system on the power conditioning equipment can be extensive involving rather complicated means for sequencing loads because of the system's limited peak power capability. Such restrictions would almost certainly involve computer software for load programming. Although this approach can be satisfied with available technology, its use would involve the need for new devices and would impose severe constraints on residential living patterns. There is no doubt that such systems can be designed. However, as encountered during preliminary studies by Lincoln Laboratories, problems of energy management for stand-alone systems can be formidable. Because of such complications, the almost certainty of higher costs, and the limited applications of this approach, it (stand-alone) was dropped from further considerations in the study.

Fortunately, others did not drop PV stand-alone systems. Perhaps, the researchers did not know any better. Perhaps, it was not profitable for them to seriously examine stand-alone PV. For whatever reasons, they were not only wrong on several counts, but appear prejudiced. It may be possible that such conclusions, regardless of how well-intentioned, have slowed the acceptance of stand-alone PV.

Let's look at the conclusions and use them as touchstones for practical PV system design. Is a back-up generator required? Thousands of PV users today would answer no. If a PV system is sized to handle the home's loads and those loads are managed properly, there is no need for a back-up generator. If the PV system is undersized or several large loads are used at the same time, a back-up power source will be needed. But must it be diesel?

Most remotely located PV homes use propane for cooking, refrigeration, clothes drying, and heating. It seems logical that if a back-up generator must be used, propane would be the fuel of choice. Not only does it run cleaner, but it can be set up to run quieter, too. In those instances where homes do need a generator, space for the generator site is hardly a factor. Maintenance, noise, and environmental problems must be considered regardless of the fuel.

In urban or suburban PV systems with utility power lines to the home (as in this study), why not use utility power for back-up? It is hard to imagine that the study investigators could overlook utility back-up with the thousands of existing commercial and industrial uninterruptable power systems (UPS) in use. These UPS use utility power as a primary source and battery storage as back-up. It doesn't take much imagination to see that priorities can be reversed and utility power can become back-up. However, it took actual PV users to implement this practical approach to back-up power for stand-alone PV.

On the subject of power conditioning equipment, solid-state

inverters were available and in use at the time of the study. Sequencing the various loads by computer was also being done in industry and business. It may have been natural for the people doing the study to feel the need for "complicated means for sequencing loads" and "computer software" to manage loads, but why these devices would "impose severe constraints on residential living" is hard to imagine. Why not let the best computers in the house manage the loads: the brains of the occupants. Thus, constraints would be a matter of choice.

What's so hard about remembering not to run the clothes washer, dishwasher, garbage disposal, and table saw at the same time? What family even lives that way? Managing loads to operate within the limits of the PV system and the inverter can be easy. If a person forgets and runs too many things at one time, either a circuit breaker trips as in a utility-serviced home or voltage transfer occurs and the grid (or generator) provides back-up to compensate. If you overload the average home, all you get is a blown fuse or tripped circuit breaker.

While energy management may be complicated for some people, this study is wrong on other counts. First, the high cost of electricity in any form has created a whole array of energy-efficient appliances which help lessen the loads in modern homes. Second, high energy costs have made us all energy managers, willing or not. Third, the average person will put out a little extra effort to manage loads if it is to his or her advantage.

Our conclusions are very different from the 1979 study. Stand-alone PV systems can be less complicated, cost less, and have relatively unlimited application in areas where utility lines already exist. The average family can put in a PV system with voltage transfer, and be on the way to energy independence. Once installed, the system can be expanded to carry more of the house's electrical loads until it is either totally self-sufficient or to the point where most of the electric bill is eliminated.

As a footnote, a Westinghouse study in 1980 of utility intertie PV systems concluded that:

> Based on the equipment cost and sellback rate assumptions made in this study, it is a general conclusion that on-site electrical storage is essential to widespread application of on-site photovoltaic total energy residential systems.

Perhaps the validity of battery storage is in the eyes of the beholder.

Located in a farming community in central California, just a short drive from the Sierra Mountains and giant Redwood trees, is a home that gets most of its power from the sun. In the yard next to the satellite dish are twenty-four 55-watt solar modules mounted on passive solar trackers. Power is stored in deep-cycle batteries and used through a 2500-watt inverter. The family of four has all the modern conveniences: dishwasher, 19 cubic foot refrigerator/freezer, 21 cubic foot freezer, automatic washer, plus summer comfort using evaporative cooling and occasionally air conditioning. For nine months, this energy-efficient home is 100% PV-powered, averaging 8 kWh per day solar production. Winter production is 4 kWh per day and utility power provides back-up. With one more tracker array and an expanded battery bank, this family can say goodbye to the power company and hello to complete energy independence.

Chapter 4
Making the Decision

If you are in the process of making the decision of whether to use PV to power your home, you need to know both sides of the story.

ADVANTAGES

1. There is a one-time cash outlay to purchase solar cells.
2. There is no monthly utility bill.
3. Users are not affected by electricity price increases or inflation.
4. Modules are reliable, sturdy, and lightweight.
5. They can be used wherever the sun shines.
6. There are no moving parts to wear out or break.
7. Modular system design can be augmented as money permits and needs require.
8. DC appliances are compatible with recreational vehicle equipment.
9. New solid-state high-efficiency inverters make the use of AC appliances practical.
10. Battery technology has long been proven reliable.

11. Modules can be used in conjunction with commercial electricity, generators, wind, or hydropower.
12. Users are not affected by commercial power outages.
13. Autonomous systems do not pollute at point of production or use.

DISADVANTAGES

1. The initial system cost is high.
2. Electricity is not produced at night.
3. Very cloudy weather significantly reduces power production.
4. Storage batteries must be serviced.
5. An inverter must be used to power standard AC appliances.
6. The power output per dollar invested is low.
7. A back-up generator or other power source may be needed to maintain batteries if the system is undersized.
8. The manufacture of solar cells may produce some environmental pollution.
9. Small solar arrays need seasonal adjustment to maximize production.

PV AND THE ENVIRONMENT

In keeping with energy-conservative consciousness, we need to consider the impact of our actions on others. So we need to examine the environmental risks in the manufacture of photovoltaics. One study states that "although the uncertainty range of public health risk estimates is large, possible public health effects imposed by the inhalation of cadmium and silicon emitted throughout the photovoltaic energy cycle are small." Further, it says "the largest public health risk of this system appears to be related to the release of air pollutants associated with fuel combustion and process emissions in the material supply cycles."

In the actual production of photovoltaic equipment the risks for occupational accidents "when compared with industry averages appear to be small." Workers, however, can be exposed to toxic chemicals in the production of PV modules. There is reason to be concerned about this exposure. Fortunately, the care taken in producing PV along with production automation and legal safety guidelines for workers has helped establish an excellent record for the industry.

As for production process emissions, more work is needed to reduce or eliminate this airborne source of pollution. Although the pollution stops once the module is made, increased module production means increased production emissions. If clean room conditions are extended to all parts of the production process, not only will the quality of the modules improve, but the emissions will be contained.

What does all this mean to the end user? No one would want their use of PV to result in a production worker's ill health. Therefore, it is important to consider more than just price when buying equipment. Be sure that the manufacturer has a good safety record. Do not support slipshod workmanship or unsafe working conditions.

SAFETY CONSIDERATIONS

PV systems are remarkably safe, and that's because PV users adhere strictly to the safety codes. They are aware of the dangers inherent in electricity, and they watch for them. Wires of the right size are installed according to the National Electrical Code®* (NEC®) and local ordinances, eliminating electrical hazard. By following battery manufacturer's recommendations, the storage bank will be safe. Safety systems are installed in grid-connect, protecting unwary utility line workers. It is easy to have a safe PV system and a safe solar society—but you have to be an active participant.

INSURANCE

Your photovoltaic system is an investment and should be adequately covered by insurance. You should consider two forms of coverage: standard homeowner's and liability. It may be possible to simply add the equipment onto your present homeowner's policy.

Coverage should include protection against hail, wind, falling trees or branches, fire, lightning, vandalism and theft. If someone

---

*National Electrical Code® and NEC® are registered trademarks of the National Fire Protection Association, Inc., Quincy, MA 02269.

else installs your system, be sure you are covered against improper installation. Your storage batteries and electrical equipment may be unfamiliar to your insurance agent. Suggest that the agent speak to your PV supplier.

Insurance to cover liability claims for personal injury, burns, electrical problems, and any utility interconnection should be considered. Finally make sure your homeowner's policy protects in case a strong wind blows the array into a neighbor's home. (Though obviously, in that case, the best insurance is to mount and secure the array properly in the first place.)

GETTING STARTED: A SCENARIO

Many PV people started on their path to energy independence with a gas generator. I've seen this transition hundreds of times. First a generator and some breakdowns and some repairs, followed by some more breakdowns. Caught in the costly dilemma of replacing the generator (a noisy, expensive high-maintenance, and unpredictable device they don't particularly like anyway), they begin to look for an alternative. So they buy a small PV system, usually under 500 peak watts. The PV system is used at first to power a few 12-volt DC lights and a stereo. Next a small, inexpensive inverter is added to run a sewing machine or blender or vacuum cleaner. By this time the family has a pretty good idea of what electrical equipment and appliances they want to use. Then it's just a matter of saving to buy the PV they need.

However, they have wasted time and money and gone through the frustrating process of trial-and-error. They now own a generator that is hard to sell and seldom used because it is worn out. They have a bunch of 12-volt lights and gadgets and an undersized inverter. Should they change over to standard 120-volt AC appliances? Probably yes, but what to do with what they already have?

If they had thought things out in the beginning, they could have avoided a lot of extra expense and had a better PV system for about the same cost. That's what this book is all about. It will help you make long-range plans and either buy your entire PV system now or just those parts you need to get started. As things progress, you can add more modules and batteries without having to rebuild the whole system.

If I were to do it all over again, I would keep the acquisition of 12-volt DC appliances to a minimum. I would have wired for AC, standard house wiring. I would buy the battery bank, regulator, and inverter I would ultimately need, and buy PV modules as I could afford them.

This planned method of putting in a PV system has a couple of real economic advantages. It insures that my batteries will last a long time since I won't be deep-cycling them as often as a small "starter-kit" battery bank. Plus I won't end up with a bunch of outgrown equipment collecting dust.

If you do decide to start with a generator and plan to eventually replace it with PV, an added benefit of proper system planning and design is that the generator will probably be run less. Its value will be higher and you can use that money to add those final modules to your already adequate PV power system.

THE COST OF GETTING STARTED

While the economics listed here are based on comparing PV to the cost of bringing in a power line to a home site, there are many other factors to consider. The urban or suburban homeowner should be particularly concerned about rising utility rates. More importantly, we must think about the cost of environmental pollution and health risks, nuclear waste storage, and an international energy supply network that requires standing armies to defend. It is not within the scope of this how-to book to examine "bottom line" PV. Bottom-line mentality is almost always short-sighted. If your monthly utility bills total less than the cost of a PV system, try factoring in the inheritance value of the PV array and its power production to your children. Simply stated, short-term economics for non-renewable energy production seem profitable. Long-term economics for PV are excellent.

While it's nice to build a PV home from scratch, it's not necessary. By simply adding a PV array to an existing AC home, having the proper interface equipment, and using energy-efficient appliances wisely, people can begin using PV today.

Just how simple is it? PV modules can be mounted on the same type of array racks used for solar water heaters, or onto homemade frames. Inverters of all sizes are off-the-shelf-items. Interface equipment such as automatic and manual switches and circuit breakers used to link emergency power systems to hospitals, banks and so on are available. The list of energy-efficient appliances continues to grow.

If you live one-half mile or more from utility power lines and are energy-conservative, you can install a PV system to provide electricity for your home. Your PV system will pay for itself immediately compared to the cost of bringing in a power line (as mine did). You can own a clean and efficient power system and never pay an electric bill again. Does this sound too good to be true? Where's the catch?

A SOLAR SALES PITCH

Let's take a closer look at the claims being made. Just what do they really say?

1. If you live a half a mile or more from utility lines
2. and are energy-conservative
3. you can install a PV system
4. provide electricity for your home
5. pay for itself immediately
6. own a clean, efficient power system
7. never pay an electric bill again.

1. If you live half a mile or more from utility lines ...

While economic and political factors will cause statistical variations for different parts of the world, these numbers hold relatively true. Let's begin by looking at the cost of utility power to get an idea of how expensive electricity really is.

As of 1987, the cost to have power lines installed to a new residence ranges from as low as $4 per foot to over $16 per foot. Some utilities run short lines for free. Free, that is, if you have an all-electric house and commit yourself to a high enough level of consumption.

Recently, a person building a new home and needing a power line was quoted $32 per foot by Southern California Edison Company. Two years ago the same utility requested a $3000 deposit to begin installation of a one-half mile utility line to a home in San

Bernardino County near Los Angeles. The total cost was over $53,000. That's over $20 per foot for the "privilege" of paying an electric bill. Worse still is the fact that monthly bills in that locale have been increasing at a rate of over 10% per year. Paying over $50,000 dollars to get a monthly bill that doubles every nine years or so sounds, and is, incredible.

But it's the same story all over the world—except where government rural electrification programs force everyone to pay indirectly for power lines through taxation—the cost of extending electric service is very high. One PV-powered homeowner in Colorado can see where the underground power line passes a few hundred feet from the living room window. In 1983 the cost to extend service to his home was quoted at $7000.

In the United States the average cost to have a utility line extended to a new home is $7 per foot. That's what we will use in our example. Contact the utility company in your area to find out how much they charge, what deposits are needed, and what restrictions or requirements regarding electrical use and home occupancy are imposed. Also find out how long you will have to wait to have a line installed.

Even conservation can cause our electric bills to go up. How can that be? Haven't we been told by the utility companies and the government to conserve and save? Well, yes, and higher prices have forced us to conserve and use less electricity, too. However, the cost of operating non-renewable power plants continues to climb. Even hydroelectric power production costs are rising. Use less power and the price per kilowatt hour goes up.

When energy conservation results in utility over-capacity, rate payers are placed in an awkward position. Some utility companies in the northwestern United States (where there is an abundance of hydroelectric power) have even asked their customers to use more power to keep rates low. That sounds crazy but it's true. It doesn't matter though since the rates eventually go up anyway.

2. and are energy conservative...

Which brings us to the next point: energy conservation. Much of this book is about energy conservation and for good reason. In a world with increasing population and finite resources, conservation is essential if we want a decent future for the inhabitants of this planet.

On a practical level, conservation is wise. The tips in this book may help you reduce your energy consumption to the point where you decide that PV is not for you simply because your electric bill is almost nothing. That's ok, too.

Americans use six times as much energy as people living in India. Some may reject that comparison because of the extreme differences in lifestyles. However, the fact that most western European nations use one-half the energy per capita that Americans use is a sobering thought.

Since the energy crisis became evident, more people have become serious about conservation. But to many others, conservation conjures up images of The Great Depression. They feel they have worked hard to get where they are and they want to enjoy it. There's nothing wrong with that, but consider that wise conservation methods leave you, and everyone else, more to enjoy.

Conservation and energy efficiency go hand in hand. For example, if you use an automatic washing machine and an electric water heater there is a simple way to reduce your utility electric bill by over $100 per year. Just switch to cold water and cold water detergent for most of your laundry. For the rest, use warm water. Now that doesn't sound like Depression misery.

Note that energy conservation is an important factor in the "sales claim" calculation. Without energy conservation, the cost of energy can be very high, not to mention environmental costs or our secure future. With wise conservation, power consumption is reduced and the size, and cost, of your PV power system is, in turn, reduced.

3. you can install a PV system...

You can install a PV system. Those words divide us into two groups. The smaller group is the do-it-yourselfers, DIY in their jargon. The rest of us are the unskilled, all-thumbs, "Wish I'd never started this," "Oh oh, some assembly required" victims of circumstance.

Our complex technological world requires specialization. To an extent, that is good. However, how many times have you called upon a specialist and paid high fees for some simple task? Granted, we sometimes hire others for convenience, to save time, and to get the benefit of their training and experience. In fact, I encourage novices to seek out the expert. But I also encourage them to first learn something about the work to be done. The job may be beyond your capabilities, but if you prepare you will know enough to wisely select the person to hire and to assess that person's work.

What does installing a PV system mean? It means selecting the right equipment. It means connecting the equipment properly. It means having an understanding of the equipment. If you are handy and have a basic tool box, PV should prove no more difficult than any other new task. If you can do a little construction work, have tuned a car or installed a stereo system, you have the skills.

If you are all thumbs, you can install a PV system the same way you service your car—by hiring someone to do it. However, read this book carefully so you will know what you are getting. A well-designed "turn-key" PV installation is a pleasure. Even the DIY would profit by bringing in an expert at some phase of the project, if only to check that everything is being done correctly.

4. provide electricity for your home...

PV has been powering electrical devices for over three decades since its development at Bell Laboratories in the early 1950s. And it can power your home. You may have seen photographs of the first PV-powered repeater installed near Americus, Georgia, in 1954. Everyone has seen news photos of solar panels on space satellites. As a reliable electrical generator, PV is unsurpassed.

Electricity produced by photovoltaics is direct current (DC) power. Most homes use alternating current (AC) power. Small cabins, boats, and recreational vehicles using DC can easily be converted to solar. AC applications require an additional piece of equipment—an inverter which changes the DC power to AC. Inverter applications are explained later in this book.

To provide PV power for your home you will need to first look at the equipment you wish to power, and the work you want that equipment to perform. There are a couple reasons for this. First, you must match production to consumption to insure that you have enough power. Second, the size of your inverter and other system components is based upon the work you want done. There may be better ways of performing the same task. This brings us again to the recurring theme of conservation. If you can perform the same task using less energy, your power costs will be less.

For example, say your car gets 20 miles per gallon, you drive 10,000 miles per year, and gas costs $1.50 per gallon. By driving a car that gets 30 mpg, you can save $375 per year. Same task, less energy, less money.

The same goes for tasks at home. You could simply add a PV power system to your existing home. (That was the way emergency generators were sized in the days of cheap fossil fuel.) Or you could install PV in an energy-efficient home and save on the size of the system.

5. pay for itself immediately...

To pay for itself immediately, the PV system must cost less than the cost of one-half mile of utility power line. Again, this depends on the amount of electricity you use. If you have an energy-conservative home, your PV system will cost less.

6. own a clean, efficient power system...

Are PV power systems clean and efficient? If you are lucky enough to live near a large PV power system such as the utility-sized megawatt plants located near Hesperia, San Luis Obispo, or Sacramento, California, pay them a visit. The Sacramento Municipal Utility District (SMUD) PV power plant is particularly worth seeing as it is right next to the Rancho Seco nuclear power plant, an interesting juxtaposition.

Or visit a recreational vehicle equipment dealer that sells PV and gas generators. Crank up a generator to power a light bulb. Then fasten a PV module to the same light and put the module in the sun. What more can I say?

7. never pay an electric bill again.

Once a month millions of people wish that their electric bills would go away. For those who fail to pay on time, eventually it does. Over 15,000 Americans, however—those who use PV—do not get a monthly electric bill.

Not all of us are ready for a total commitment. Some may want to obtain only a portion of their power from the sun. Some may want to have an emergency or back-up power system that gets its energy from the sun. Still others may want to slowly make the transition to PV, supplementing part of their utility power with PV or vice versa. The last group envisions a gradual end to their dependence on utility power. By matching the expansion of their PV system with utility rate increases, they hope to keep the wolf from the door. They plan to be totally energy self-sufficient by the time they retire. They want fixed expenses to match their fixed income.

COST COMPARISON:
PV VS BRINGING IN A POWER LINE

The system shown in Figure 4.1 has been in continuous operation since its installation in October 1981 without a back-up generator. The battery bank has been enlarged and upgraded to true deep-cycle batteries.

The prices listed are based upon comparable equipment sold in 1986.

12 each 35-watt solar modules (420 watts) replaced by	
10 each 50-watt modules (500 watts)	$4,000.00
3 each manually adjusted pole mounts replaced by	
1 each 12 module tracking mount	700.00
1 each regulator	300.00
24 each batteries (31.2 kWh total storage)	1,800.00
1 each 2500-watt inverter	2,200.00
misc. wiring and hardware	500.00
Subtotal	9,500.00

Now let's make it more difficult for the solar system. Add sales tax and professional installation fees:

6% sales tax	570.00
installation charges	1,500.00
Total	$11,570.00

How does this compare to the cost of running a power line one-half mile to the home site? At $7 per foot, one-half mile (2640 feet) will cost $18,480. Even at $5 per foot, the solar electric system is a

FIGURE 4.1: This home's photovoltaic system supplies an annual average of 1680 watt hours per day. It powers lights, color tv, radio, vacuum cleaner, toaster, blender, washer, power tools, and water pumped from a 180-foot well. Propane powers a refrigerator, water heater, dryer and stove. Located in Piute Mountains, California. (Courtesy ARCO Solar, Inc., and Greg Johanson, Solar Electric Systems)

FIGURE 4.2: Four 45-watt solar modules power Richard Tucker's home in Lyons, Colorado. Wired for 24 volts DC, the home also has a 2500-watt inverter. Battery storage totals 480 ampere hours. The system was installed during construction to power tools. Installer: Dr. Mark McCray. (Courtesy Rocky Mountain Solar Electric)

better deal, not to mention the monthly bills. If you did this installation yourself, you would be installing a solar electric power plant cheaper than the utility company could install a power line at $4 a foot. Photovoltaics can be less expensive than utility service to a new home site.

Reports presented by government and industry in the early 1970s predicted that PV would be economical in 1986 at 50 cents per peak array watt. And so 50 cents and 1986 became the knee-jerk response used by the media and PV researchers. No one seemed to think out the implications of that prediction.

Although there was always the hope of a "breakthrough," at no time did it really appear that PV would ever be that cheap. In fact, some companies tried to fulfill the predictions of lower prices by selling below cost. Hoping that relatively low prices would increase

FIGURE 4.3: It would have cost Jack and Nadine Pettry $66,000 to have utility power brought to their home in Colorado. Instead, they chose to purchase fourteen 35-watt modules on seasonally adjustable racks, 740 ampere hours of battery storage, and a 2500-watt inverter at far less cost. Their home has DC fluorescent and incandescent lights, solar-heated water, a submersible pump, and more. This 1500 square foot energy-efficient home has a solar greenhouse and high-mass Russian fireplace, and uses propane for back-up heat, cooking, and refrigeration. Installer: Dr. Mark McCray. (Courtesy Rocky Mountain Solar Electric)

the market and bring about widespread public use of PV, they forged ahead only to go out of business.

Waiting for breakthroughs may be all right for armchair voyagers. Hoping for a better tomorrow is better than having no hope at all, but taking action and creating the kind of life you want to live holds more than the immediate advantage of getting you off your duff.

Industry is responsive to consumer desires. If enough people buy and use PV now, business is going to see a market and try to profit from it. Think not? Compare today's computers with those sold just a few years ago. When people demanded fuel-efficient cars, industry responded. Likewise, we have a responsibility to express our desire for solar energy and independence. Only then will we see low-cost PV.

FIGURE 4.4: John and Joanne Willis's home in the Tortolita Mountains, Arizona, is $33,000 away from utility power. Thanks to tax credits, they were able to get PV power for under $4000. Resulting back-up generator fuel and maintenance savings mean this system will pay for itself in less than five years. This PV system powers everything except a spa run directly from the generator. Batteries store enough power for five sunless days. (Bill Cirrito, Electrasun Energy Systems, Tucson, AZ)

Chapter 5

Sizing a PV System

Practical use of solar power is a blend of science and art. Sizing a PV system can be reduced to numbers. However, before attempting the math, it's a good idea to examine the general principles behind a properly sized system. The variations of climate, the unique characteristics of the equipment selected, and the different ways people use electricity combine to produce a range of system possibilities. The most appropriate system is the one specifically designed for your application.

The science of PV sizing is the numbers. Power requirements, climate data, equipment specifications, and user-specific parameters are input. The output is the number of modules, size of wire, storage capacity required, control and power conditioning equipment. It all adds up.

The art of sizing a PV system involves the human element. You can operate a PV system "by the numbers," but to get the most out of your equipment, you have to feel the way it operates. When you drive a standard transmission vehicle, there are recommended speeds at which to shift. However, shifting by the numbers does not always work. By experience, you learn when to shift. You get the feel for driving a standard transmission by listening and observing. So it is with a PV system. By observing equipment performance, you can fine-tune system operation. The important goal at the time of sizing is to do some fine-tuning before you use the equipment.

The steps in sizing process are:

1. Determine your power requirements
2. Match PV production for your climate to your power requirements
3. Match battery storage for your climate

Each step in the sizing process is interrelated and each step determines the equipment to be used. Thus, the load ultimately affects the operation and the cost of your PV system.

The first general principle is conservation. With unlimited resources, you can produce an unlimited amount of energy. However, most of us have limitations. And even if we didn't, the finite resources of the planet must be shared.

Therefore, you should design a PV system that maximizes the use of equipment, performs the work you want done, wastes little, enhances the environment, is long-lasting, and is within your budget.

A good environmental yardstick for measuring your PV system is whether you can live with the power plant you are designing. Any technology that is a bad neighbor should be thrown out of the neighborhood.

DETERMINE YOUR POWER REQUIREMENTS

There are three main reasons for reviewing your power requirements. First, you must match production to consumption. This requires that you know how much you consume. To know how much energy is enough, you have to quantify your power usage.

For example, you are planning a dinner party. By counting the number of people and noting whether they are adults or children and whether they have any dietary restrictions, you avoid running short and disappointing your guests, or wasting food (and money) by making too much.

There's another important reason to look at your power

requirements. Close examination can help you determine whether you are doing what needs to be done in the best possible way. Each time you reduce power consumption, you are rewarded by requiring a smaller PV system that costs less money. This alone should provide plenty of motivation. In addition, think of all the resources you will not be using up.

Still another reason to examine your power requirements is to group loads by common characteristics. List all small or regular daily loads such as lighting, toaster, stereo. Separately list loads you use less frequently—laundry equipment and shop tools. Last, make a list of your large power loads such as well pumps, shop tools, freezers, etc. These large loads will determine how large an inverter you must use. If you restrict the size of large loads, the size of your PV system will be smaller and cost less. By doing so, you can use a smaller inverter for the smaller AC loads. A good example is using a 1/2-hp well pump or other motor load instead of a 3/4-hp motor load.

If you can get by with a 2500-watt inverter, go for it. Two words of caution though: Don't skimp on what you need and don't undersize or marginally size your equipment. That's false economy. If you need lots of water for your garden and trees, overworking a small inverter and pump to its early death is a bad bargain. Also, full-load duty-cycling a marginally sized inverter will result in its early demise.

A good way to downsize your inverter is to stagger the use of equipment that requires a lot of power. If you run the garbage disposal, the water pump and wash laundry while someone is using the table saw (the refrigerator is always on), you have added at least $1000 to the cost of your inverter for the "convenience" of doing everything at once.

The Three Lists Method

An easy way to determine your power requirements is to use the "Three Lists Method." This method makes it possible to start small, taking care of essential loads while planning for the systematic expansion of your PV system.

The first list is the "Actual List," an energy audit of your present electrical power consumption. Even if you plan to move from the city to the country, it is important to make this list. You may be moving from one location to another, but you will be the same person with the same basic needs and the same ways of satisfying those needs. Besides, it's easier to list what you actually have now to get the knack of system sizing.

Simply go from room to room and write down every electrical device. List all the kitchen appliances and don't forget the bathroom and the garage or shop. Include well pumps and the washing machine. What about the refrigerator?

Write down the amount of power each device draws. You can find that information—in watts or amperes—on the identification tag or label. Also note the voltage, as some things come only in 120 or 240 volts AC. If any appliance does not have a power information label, use the charts in the energy conservation chapter and the DC Appliances, Devices, and Loads section (page 98). Note which appliances you will use when you set up your PV system.

Beside each listing, write the number of minutes or hours it is used each day or each week. Be accurate. You may be surprised at the numbers. Multiply the minutes or hours of use for each device times its wattage (amperes times volts equals watts) and put that in the next column.

Total the amounts in the watt hours column. Multiply that total by 30 days and divide by 1000. Your final figure should match your electric bill's total monthly kilowatt hours. If you are not even close, check your math and see if you have missed anything like the water heater or air conditioner.

When finished you will have completed an energy audit of your electrical loads. The audit is a profile of who you are electrically. You have taken the first major step to becoming your own power plant manager.

Next make your "Dream List." Now that you understand how to do an energy audit, this list should be easy—though you need a little mental preparation. Sit in your favorite chair and close your eyes. Imagine that it's five years from now. You are in your dream home. Your PV system works fine.

Take a mental walk through each room. List all the lights and appliances. Step outside to list power loads such as the well pump and shop tools. Be generous: remember, this is a dream list. Be sure to include everything from your Actual List that you will likely use later. If you are planning the purchase of a new automatic washer, include it on the list, noting its actual power requirements. Your

Dream List can also be used to help you shop for energy-efficient appliances. Include the number of hours you will use each device. Complete this list as you did the Actual List.

The last list is simple. It's the "Bare Bones List." Again, sit yourself down and close your eyes. It's five years in the future, but things haven't gone so well. You are living in a PV-powered home, albeit somewhat scaled-down from original intentions. The Bare Bones List is the least you will live with, remembering you are not on this earth to punish yourself. It includes essentials and defines the lower limits of the PV system you are going to size. Proceed as with the previous lists and total up the kilowatt hours per day.

Why three lists? Few people seeking to become energy independent can do so all at once. For most of us it is a gradual process because of time and money limitations. With these three lists, you are able to make the transition from your present power consumption (Actual List) to future consumption and production (Dream List). The safety net of the Bare Bones List helps set priorities and indicates what needs to be done first.

With your three lists you now know your power requirements (and a little more about your personal values). Are there any electrical tasks that can be done more efficiently? Perhaps some things can be done without electricity. Sometimes it costs less to use a gas refrigerator than an electric refrigerator, especially if you plan to heat and cook with gas. Some people use their transition to energy independence as an opportunity to use renewables such as wood for heating. Now is a good time to learn about new techniques which save energy. Did you know that while a microwave-convection oven may use more electricity when operating, it actually reduces total annual power consumption because it is so efficient? Make a point to learn more about energy and how to use it wisely.

MATCHING PRODUCTION TO CONSUMPTION

Now that you know your power requirements, you need to determine how much power is needed to meet them. This is pure math based on the output of the PV modules and the efficiency of the power conditioning and distribution equipment you select. There are guidelines to help simplify the system and reduce costs.

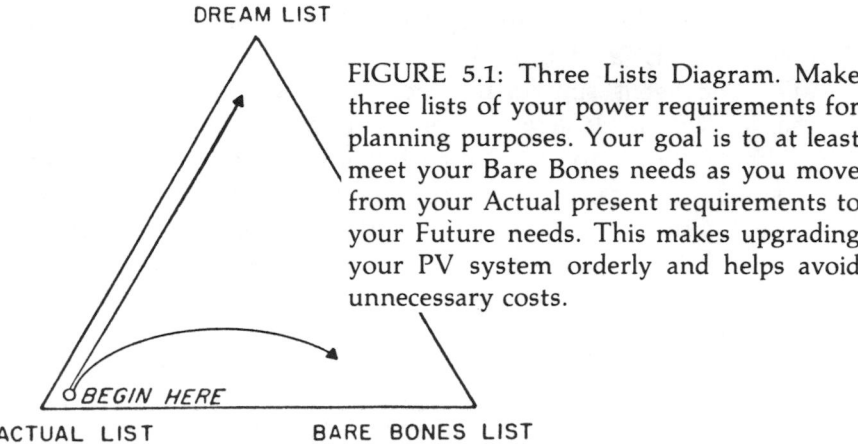

FIGURE 5.1: Three Lists Diagram. Make three lists of your power requirements for planning purposes. Your goal is to at least meet your Bare Bones needs as you move from your Actual present requirements to your Future needs. This makes upgrading your PV system orderly and helps avoid unnecessary costs.

For example, if you live in a unique microclimate that differs from the general locations in this book (such as a place that gets significantly more rain than a location 50 miles away), you must allow for that difference. Coastal fog versus inland sun or increased solar radiation at higher elevations can make a significant difference.

The time of year and the time of day you use power can make a difference in the size and configuration of your system. You may live in a hot climate and need cooling fans or you may have irrigation requirements. These seasonal loads can be partly met by using a solar tracker or by manually adjusting the array.

In the temperate zones there is more sun in summer than in winter. Tracking the winter sun in the United States increases PV production only 10% or 15%; however, the 40% to 50% increase in production by using a tracker in summer can make all the difference in determining the size of your array.

BATTERY STORAGE

Battery storage sizing is also a mathematical process. However, location makes a big difference in battery bank size. If you live in a cold climate, it is worth the extra effort to locate your batteries where they will stay at 70°F (21°C) to reduce battery bank size. The same principle in reverse holds true in hot climates. Keeping your batteries cool can prolong their life. One way to do this in the tropics is to put the batteries under the house or in a partially buried battery compartment.

Power requirements also affect battery storage. Your battery bank must be able to handle power surges associated with motor equipment switched on and off. A good rule of thumb is to size your battery bank ampere hour capacity to at least four times your largest surge load. For example, if your have a 100-ampere surge load, the battery bank should be at least 400 ampere hours (20 hour rate) in size. Surge rating is often listed on equipment. If a motor's surge rating is not known, figure three to five times operating current for the surge.

POWER CONDITIONING AND REGULATION

The type of regulator you use with your PV system depends on the system size and its special features. If you have DC loads, you might want a low voltage disconnect to protect your batteries from being deeply discharged and possibly damaged. Perhaps you have excess PV production for part of the year. A load diversion option may be appropriate. Meters for monitoring your system voltage and charging current are also offered by most regulator manufacturers. Automatic generator starts and temperature compensation are also available. Consult with your supplier for recommendations.

As previously mentioned, the size of your inverter is determined by the largest load or combined loads, and when and how they are used.

WIRE SIZING

While wire size is discussed in depth later, it's a good idea to start thinking about it now. Some people prefer to use DC appliances to avoid the cost of an inverter and its associated power loss. However, DC wiring must be large enough to handle the load. If undersized, it can rob you of power. Keep your array and battery bank as physically close to each other as possible. An inverter for AC loads permits use of standard interior house wire sizes.

Though wire seems a small item costwise, it adds up. If you site your PV array at a distance from the house, it is best to locate the batteries nearby and use AC. But before making any final decisions, price out the possible wiring schemes.

SIZING—AN EXAMPLE

Now that we have looked at the basic sizing considerations, let's size a system. While the people in this example are fictitious, their power requirements are similar to those of many people I have actually worked with. Although our example system here is in a rural setting, there is no reason why people living in the city could not put in a similar set-up and partially or completely unplug from the utility company. Several hundred city and suburban homeowners started using PV because they planned to eventually move to the country. Some did and took their PV system with them. Others still live in the city with reduced electric bills. One family of four in Los Angeles has cut their 1200-kWh per month electric bill to less than one-fifth that amount. Another energy-conservative family of four gets all their power from the sun nine months of the year. If the example seems either too austere or overly consumptive, that's fine. Each of us has our own ideas of what is essential and what is too much.

Our example system is used by a couple who moved from the city to the country. They both have jobs that do not require commuting. When they made the move, they became part-time self-employed doing the work they did in the city. In addition, they are developing new income-earning skills. They like to garden and grow a lot of their vegetables.

They have been buying energy-conservative, sturdy appliances for the past few years in preparation for the move. The water well on their property is 180 feet deep. They have built a large storage building which will eventually be the barn and workshop, once the house is built. For now, they live in a trailer. The property is two miles from the nearest power lines. They knew this when they bought it, but were not deterred. It is a beautiful piece of land with fine trees, an excellent house site, clean water, and the right price. The road is good and the nearest neighbors are well out of sight.

They use propane for cooking and back-up heat. Wood from the property is their main heat source. They have a propane water heater. A gas clothes dryer is used in winter and on rainy days. They will use a propane refrigerator for the first few years while saving for an energy-efficient electric refrigerator/freezer, and additional PV modules.

The appliances they brought with them are being put to immediate use. They don't mind living in the cramped trailer, but they're not going to rough it with all their appliances and tools packed away. Let's look at their electrical power requirements.

Load Device	How Many	Watts	Hours Per Day Used	Days Per Week Used	Daily Watt Hours
Living Room:					
lights (reading)	2	22	2	7	88
light (area)	1	40	2	7	80
television	1	100	4	7	400
radio/stereo	1	100	2	5	142.9
VCR	1	100	2	2	57.1
Kitchen:					
lights	2	40	3	7	240
lights	2	20	0.5	7	20
microwave oven	1	1450	0.5	3	310.7
toaster	1	800	2 minutes	7	26.7
mixer	1	150	5 minutes	2	3.6
juicer	1	125	30 seconds	7	1
food processor	1	450	10 seconds	7	1.25
vent fan	1	88	0.5	2	12.6
Bath:					
lights	2	40	1	7	80
Bedroom:					
light (room)	1	40	1	7	40
light (reading)	2	20	1	3	17.1
Appliances:					
washing machine	1	512	2	1	146.3
dryer (motor only)	1	300	1	1	42.9
iron	1	1000	0.5	1	71.4
vacuum cleaner	1	600	0.25	2	42.9
sewing machine	1	75	0.5	1	5.4
clocks (rechargeable battery)					
Equipment:					
water pump	1	1050	2	7	2100

(continued next page)

Tools:
(During construction, tools are powered by the generator providing battery charging for nighttime shop tool use. Once the house is built, hours of tool use are unknown.)

Future Needs:
(for system planning purposes at 50% duty cycle)

refrig/freezer	1	360	12	7	4320

To calculate the daily power consumption of each load device, use the formula:

Quantity x Watts x Hours x 7 (for one week) ÷ 7 =
Watt Hours per Day Consumption

The total power consumption for the home is 3929.15 or 3930 watt hours per day. This figure does not include tool use or the future electric refrigerator/freezer.

Let's analyze the power consumption by type of load. It is obvious that water pumping is the heaviest load requirement. Using a standard 1/2-hp submersible pump and pressure system, the household will get 600 gallons per day. The pump draws 8.75 amperes at 120 volts or 1050 watts. It pumps five gallons per minute into a 30-gallon pressure tank. Five gallons per minute equals 300 gallons per hour. Could the couple get by with less water? Yes. With only two 10-minute showers per day and a water-saving showerhead that delivers a good spray at 2 gallons per minute, showers total 40 gallons. At the wash basin they might use an additional 2 to 5 gallons. The toilet could be a water-saving unit requiring only 2 gallons per flush. There are toilets that use even less.

Dishes are washed in two basins, one for soapy water and one for rinse. (A much better method than continually running rinse water.) The garden is gravity watered by a spring-fed pond. With proper planning and water conservation, they can cut their daily average water consumption by half or more.

Lighting loads are the next greatest requirement, totaling 565.1 (rounded to 566) watt hours per day. This is an average because lighting usage varies greatly throughout the year. In winter, more time is spent indoors; the shorter days mean artificial lighting is required in morning and early evening. In summer, of course, lighting needs are reduced.

To conserve, some people get rid of kitchen appliances and gadgets. Let's see if that will help to reduce the total load. if we eliminate the toaster, mixer, juicer, and food processor, the daily load is reduced 32.55 watt hours. Hardly worth the inconvenience, since this couple already owns and uses these things.

Perhaps the microwave should go. With rising propane costs and possible fuel shortages, phasing it out may mean future cooking on a wood stove. Since a move to the country often means a significant reduction in income, buying enough PV to power the microwave now may be the only time it will be affordable for several years. I'd say keep the microwave.

Water pumping and lighting total over two-thirds the power requirements for this household. In most cases, reduced water consumption made possible by conservation would cut these combined loads to one-half the total kilowatt hours. However, we will use the initial figures for our system sizing example.

Preliminary Module Sizing

Once you know the total daily power requirements, you can perform the preliminary system sizing—an aid in the ultimate selection of the power production equipment needed.

There are three sources for solar radiation data in Appendix B: the ampere hour charts, the average number of peak sun hours maps and, finally, the specific location charts.

A fourth source of information is your equipment supplier. However, do your homework first and be a knowledgeable buyer. Use your supplier to double-check your figures. You can also use your figures to double-check your supplier. While most suppliers are ethical, there are some who will sell you what they have, rather than what you need. If there is a big difference between your figures and the supplier's, ask for an explanation. If the explanation makes sense (i.e., is based on more accurate local climate data), use it. If not, buy elsewhere.

My 32-acre mountaintop property in Arkansas had three distinct microclimates. The north slope was cool to cold. The top of the mountain was windy with greater temperature swings. The south slope was semi-arid and hot in summer. Vegetation, soil moisture, and solar radiation differed greatly on this small parcel. Local climate can differ as much as 10% from the charts and yearly weather may differ even more. But don't let this discourage you from seeking accuracy in sizing. The validity of climatic averages has been proven over the past decade in actual solar electric homes

Our example's power requirements are 3930 kWh/day. The location is 40° north, 106° west near Grandby, Colorado. Using the ampere hour chart in Appendix B you see that Grandby has an average 12.1 ampere hours per day of solar production using a 35-watt 12-volt module through a battery bank with wire loss factored in. If you change that figure to watt hours, you have 145.2 watt hours per day (12.1 x 12 volts = 145.2). Change that figure to show the output of a 55-watt solar module, since that is what will be used in the system. 145.2 ÷ 35 watts x 55 watts = 228.17 watt hours.

Since the home uses AC appliances, you must factor in inverter loss. Using a solid-state inverter operating at 90% efficiency, usable electrical production from one 55-watt module is 205.35 watt hours per day (228.17 x 0.9 = 205.35). It must be remembered that this is based on yearly figures with the module tilt fixed at 55° south. Seasonal mount adjustments and tracking will increase annual production an average of 30%. For now use the fixed mounting data.

By dividing the requirements by module production, we find that this couple will need approximately 19 modules to meet their power requirements. If they reduce their water consumption by one half, they can get by with 14 modules. The obvious lesson here is that conservation can significantly reduce the size and cost of a solar electric system.

Checking Your Figures

Now you will, more accurately, determine the size of the array using the maps on page 339 and the chart on page 355. We see from the map that the location averages 5 peak sun hours per day. The chart shows that nearby Denver has 5.7 peak sun hours for a latitude tilt angle. That is the number we will use.

You again select the 55-watt module. This module will produce 313.5 watt hours per day in our location. Now factor losses. Wire is sized according to *National Electrical Code* regulations, battery losses are typical, and the high-efficiency inverter used has an average 90% rating. Module losses due to temperature are not a major factor as the voltage from the 36-cell module selected will meet battery charge voltage needs at all times. The losses are: wire losses 2%; battery losses 20%; and inverter losses 10%.

The formula for converting raw module output is:

55 (module wattage rating) x 5.7 (peak sun hours) x 0.98 (wire losses) x 0.8 (battery efficiency) x 0.9 (inverter efficiency) = Available Power per Module

For our example we have 55 x 5.7 x 0.98 x 0.8 x 0.9 = 221.2 useful watt hours per day from a 55-watt solar module. The daily load (3930 watt hours) divided by available module power output is 17.76, that is, 18 each 55-watt solar modules to do the job.

If you want to work the sizing figures in reverse from watt hour daily load to number of solar modules, the formula is:

Watt hour per day load x 1.02 (wire losses) x 1.25 (battery losses) x 1.11 (inverter losses) ÷ 5.7 (peak sun hours) = 55 (module rating)

Module Operating Temperature

Module temperature and reflected sunlight will affect the output of a solar array. As the temperature of a solar cell increases, its efficiency decreases. This can significantly affect system performance in summer when using modules with less than 33 cells. Using 77°F (25°C) as the base temperature, there is typically a 12% voltage drop at 116°F (47°C). Therefore, a 36-cell module rated 17.6 volts at 77°F will put out 15.5 volts when heated by the sun. A 30-cell module rated 14.5 volts will only put out 12.7 volts (which will not "top off" a battery) at the higher temperature. This temperature drop is factored into the sizing calculation for modules with 33 or more cells.

Average reflectance is also factored into the simplified sizing calculation. Reflected light from snow will increase array output by 10% or more as lower temperatures also increase cell efficiency. Conversely, a solar array on a dark roof or in the middle of a black asphalt parking lot will have a slightly lower than average output.

Battery Sizing

Next we'll size the battery bank. The Battery Storage Requirements Map in Appendix B shows that a system located near Denver will need 4.8 days of battery autonomy. Use the battery autonomy formula (Appendix B) to determine actual battery bank capacity.

FIGURES 5.2 & 5.3: PV power installation on an Indian reservation in the southwest. In winter lower cell operating temperatures and sunlight reflected off the snow increase array output, partly offsetting reduced output due to cloud cover and shorter days. (Photos: ARCO Solar, Inc.)

This formula was developed for use with true deep-cycle batteries kept in a protected area to avoid temperature swings. If the building is unheated, the battery compartment should be insulated. Normal charging will produce some heat. However, use a high-low thermometer (available for under $20) to find out what temperature swings actually do occur. Use the temperature correction factor from Table 5.1 to adjust for swings.

Our example couple, in using 3930 watt hours per day, will need 47.16 kilowatt hours of battery storage for 100% autonomy

Table 5.1: Temperature Correction Factors

Temperature °C	Correction Factor	Temperature °F
−10	1.10	+14
−15	1.55	+ 5
−20	2.05	− 4
−25	2.75	−13
−30	3.50	−22
−35	4.25	−31

(3930 x 4.8 x 2.5 = 47,160/1000 = 47.16). If they use a golfcart-type battery like the Trojan T-105 (217 ampere hours x 6 volts = 1302 watt hours), they will need 36.2 batteries. Before completing battery sizing, we need to know the system primary (DC) voltage. That voltage is determined by the size and type of inverter used.

Inverter Sizing

The largest load in the home is the water pump. This pump was specifically selected to (1) serve the needs of the occupants, (2) be readily available and serviceable, (3) have a low initial cost, and (4) most importantly as far as the solar electric system is concerned, allow for a medium-sized inverter. The 1/2-hp pump motor will work well with a 2500-watt inverter. If the couple selected a 3/4-hp pump, they might have had to get a larger, more expensive inverter.

The largest combined load will be the water pump and washing machine. All loads will be managed to limit the largest combined or simultaneous load from exceeding these two devices. While it may be possible to set up a gravity water storage system instead of pumped water for the washing machine, this may not be practical or convenient.

When the laundry is being done, the water pump will cycle on and off. The combined laundry and water pump surge should be within the inverter's surge capability. Be sure that the inverter you select can handle surges at least four times its rating.

The inverter selected has a DC input voltage of 24 volts. This voltage determines the configuration of the solar electric system. Standard 12-volt nominal solar modules must be used in groups of two for 24 volts. Eighteen 55-watt modules (990 watts) will work fine. Twenty-two or twenty-four each 35-watt modules could be used instead. Twenty-four each 35-watt modules (840 watts) will produce surplus power; 22 modules (770 watts) may fall short of winter daily needs.

Since a generator will be used during construction, it might be a good idea to consider an inverter with a battery charging option. A separate battery charger will also work. Be sure to check the quality and efficiency of the battery charger to avoid running the generator any more than necessary. Also important is the type of switching (automatic or manual) used to transfer between PV/inverter to generator/battery charger.

Regulator Sizing and Selection

The solar array must produce 3930 watt hours per day to meet this household's power requirements. It must be configured for 24 volts to meet the input voltage requirements of the inverter. If 18 each 55-watt (17.6 volts, 3.13 amperes) modules are used, the total current from the solar array will be 28.2 amperes (18 ÷ 2 x 3.13 = 28.17). In other words, a nominal 24-volt solar array using the 55-watt module described will consist of 9 parallel 2-module series strings.

The regulator must be capable of handling 28.2 amperes; 30-ampere 24-volt regulators are available. Be sure the regulator can handle the full array load current. Some regulators must be sized based on the short-circuit current of the array, which is higher than load current. In any case, always connect the array *after* the battery bank load is wired to the regulator. Never disconnect the battery bank without first disconnecting the array.

Since this couple has a generator to handle construction equipment loads, they might want to use the generator as back-up power. In this case, the automatic generator start option offered by some regulator manufacturers might be a good idea. The initial cost for such options is low and retrofitting is often not practical. Other regulator options to consider are monitoring meters, automatic disconnects, and low and high voltage alarms. Monitoring meters make it easy to know your PV system status. Alarms, particularly for low voltage, can alert the forgetful user that it is time to conserve energy or turn on the auxiliary generator. Automatic low-voltage disconnects will shut off loads that are inadvertently left on should the load consumption exceed solar production. While it is nice to have some of these features, be sure they are really necessary to you.

The Art of System Sizing

You have now designed a solar electric system. It consists of:

Production: 18 each 55-watt solar modules
Storage: 36.2 each Trojan batteries (rounded to 40)
Regulation: 1 each 30-ampere regulator
Conditioning: 1 each 2500-watt inverter

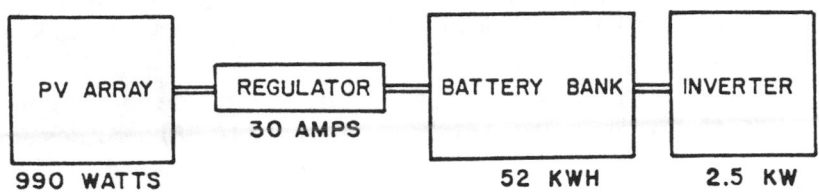

FIGURE 5.4: Block diagram showing rating and capacity of PV system components.

But you have to do something about the battery bank, since you can't have two-tenths of a battery. The system requires 24 volts. The battery selected is 6 volts. Therefore, 4 batteries in series are needed to get 24 volts. Nine parallel strings of 4 each T-105 batteries in series will give you 46.872 kilowatt hours storage. You are still shy 260.4 watt hours. That's not much compared to total battery bank capacity. However, if the batteries are to be stored in a cold location, it may be a good idea to buy 40 batteries.

Of course, you could select another battery. But you do not have to select a battery bank that totals up to exactly what you have calculated. Slightly more battery storage is a good idea if you can get a good price on your battery bank. Any good deep-cycle battery, bought locally from a reputable dealer, should do the job. If you downsize the battery bank, you will be sacrificing both autonomy and battery longevity. Smaller battery banks may cost less, but will cycle more often on the same load. More and deeper cycling shortens battery life.

Can we downsize the solar array? Yes. By putting the modules on a tracker, you will increase winter production by 10% to 15%. Summer production will be increased by 30% or more. As winter lighting loads are an important concern, downsizing should be limited to no less than winter loads.

For the example system, a 55-watt module on a tracker will produce approximately 272 watts in winter. (See the map on page 342.) Winter insolation (halfway between the 4 and 5 line) is 4.5 peak hours (55 x 1.1 [10%] x 4.5 = 272.25). Thus, 14.44 modules will do the job. As the system requires modules in groups of two in series, the homeowner must decide whether to cut consumption and squeeze by with 14 modules or go to 16 modules. In either case, the tracker has paid for itself by knocking 2 or 4 modules off the system.

If the water requirements can be cut in half, the couple could use a tracking solar array of only 12 (actually 10.58) modules and 36 (actually 34.56) batteries. Do the calculations yourself. Play around with the numbers and see if you can squeeze more from this solar system. Plug in some of your own power requirement figures and see what happens.

SIZING YOUR OWN PV SYSTEM

By using this sizing method, you can design your own PV power system. Once you know the system location, the power requirements and the characteristics of the equipment that will be used, you can calculate any PV system.

While the numbers are still fresh in your mind, grab your calculator and design a PV system to suit your needs. If you are energy-conservative, you will be amazed at how few modules and batteries are needed. If you have large power requirements, do not despair. Here's some information to sleep on before you juggle the figures again.

Budget System Sizing

Many people find that the system they "need" is beyond their budget. Rather than give up on energy independence, look at your budget. It should include both the initial cash outlay and what will be saved over the next two, three, or more years.

The couple from Grandby, Colorado, has $7000 to put into their solar electric system. If they really push it and allow for savings gained by not using the generator as often, they can pull together $8500.

Now we need equipment costs. Using 1987 figures, a 55-watt solar module sells for approximately $400. Batteries are $60 each. Eight-module trackers are about $600. The regulator is about $300 and the inverter approximately $2000. Therefore, the 16-module system with two trackers and 40 batteries will cost about $13,000 with wiring and miscellaneous hardware included.

That's almost twice their budget. What can be done? Well, if we look at our costs, we see a relationship between system cost and module cost. A very rough rule of thumb is that half the cost of a solar electric system is in modules.

With $7000 they can afford about 8 or 9 solar modules ($7000 ÷ 2 ÷ $400 = 8.75), which will meet half their power needs. They might consider buying an 8-module system now and adding on later, if they are willing to use the generator a few hours a day. During house construction, the generator can serve double-duty powering tools and charging batteries.

If the system is cut to eight modules, the cost has been reduced to $8500 or so. At this point, however, they might be tempted to downsize the battery bank: by reducing it to half, they can come in very close to budget. Unfortunately, it's not that simple. If daily power needs are drawn from a battery bank of half the size, the batteries will be cycled more often and probably more deeply. That means batteries will age faster and need replacement sooner. It also means greater reliance on the generator.

It is false economy to reduce the battery bank. Forty batteries will last at least six, and more likely ten, years. Twenty batteries for the same load may last five or more years, but are more likely to last three to five years. If it seems likely that their income will increase within five years, then economizing now with 20 batteries may be ok. If the money is not going to be available for 40 new batteries in five years, I recommend trying to negotiate a better battery price. Forty batteries is more than a pallet load. Perhaps direct purchase from a large battery distributor would save money.

What's the real cost for the downsized budget system? Running the generator regularly means fuel, service (such as oil changes, repairs, parts), and a shorter generator life. It may also mean not having enough power if the generator fails. And fail it will. Parts may be miles or days away. Repairs can be costly and time-consuming.

In review, system sizing is based on real needs based on daily average power requirements. Climate data is used to predict average solar array performance. Battery storage is based on the number of days of autonomy required for your climate. The regulator must be able to handle the current from the solar array. The inverter must be sized for the largest single and combined loads. The wiring must be sized to carry the current.

It takes planning and ingenuity to get started in PV on a limited budget, but thousands of people have done it. So can you.

DC PV SYSTEMS

DC Appliances, Devices, and Loads

The words appliances, devices, and loads are interchangeable as we use them here. This section will describe some of the hundreds of loads possible with a PV power system. (A look at the index of any J.C. Whitney catalog will widen the possibilities.)

All the items below are 12 volt DC. If inverters and low-wattage standard 120-volt AC equipment were included, this list would be much longer.

air compressors	electric door locks	polishers/sanders
air conditioners	electric oil changers	radios
alarm timers	electric window lifts	refrigerators
amplifiers	fans	relays
battery analyzers	flashers	revolving lights
battery testers	float, tank	roof fans
blowers	fluorescent lamps	sirens
bug killers	hair curlers	spark plug cleaners
burglar alarms	hair dryers	spotlights
CB equipment	heaters	stereo tape players
cigarette lighters	hoists	stereo equalizers
circuit breakers	horns	stereo amplifiers
clocks	intercoms	trouble lights
coffee makers	lights, all types	vacuum cleaners
defroster blowers	lights, quartz	water heaters
digital clocks	lights-on alarms	water pumps
door jamb switches	motors	winches

FIGURE 5.5: A 40-kW PV array and battery bank provide all of the power to this full-sized home located in New York's Hudson River Valley. The house also features direct solar gain, super insulation, heat mirror glazing, and active domestic water and space heating. (Photo: Solar Design Associates)

Listed below is the power consumption of some of the more common 12-volt DC loads. Use this list when determining your power requirements and sizing your system.

Lights
Fluorescent 12-volt DC
    16-watt (two 8-watt bulbs) 1.4 amps
    30-watt (two 15-watt bulbs) 1.9 amps
    22-watt circular 1.4 amps
Incandescent 12-volt DC type
    15-watt standard base 1.3 amps
    50-watt standard base 4.2 amps
    (screw base and automotive type rated on bulb or package)
Quartz-iodine, metal halide or tungsten-halogen
    17-watt 1.5 amps
    34-watt 2.9 amps

Entertainment/Communications
12-inch black and white television (16-watt) 1.4 amps
13-inch color television 4.1 amps
am/fm car radio with tape player 0.5 amp
Turntable (converted to 12 volt DC) 0.5 amp
Radio/telephone
    receive 0.3 amp
    transmit 2.5 to 15 amps

CB radio
> receive 0.3 amp
> transmit 0.5 amp
Intercom 0.5 amp

Appliances
Coffee pot 11 amps
Popcorn popper 16 amps
Toaster 20 amps
(Mixer, blender, can opener, food processor, etc. see inverter section)
Washing machine (wringer-style with 12-volt DC 1/4-hp motor) 22 amps
Refrigerator
> 12-volt DC Koolatron (3 cubic foot) 2 amps
> Danfoss compressor BD 2.5 (up to 8 cubic foot) 3.7 to 6.3 amps
> Norcold model 8010 (9 cubic foot) 20 amps
Vacuum cleaner
> Sears 12-volt DC (108-watt) 9 amps
> J. C. Whitney hand-type 4.5 amps
> J. C. Whitney upright 20 amps
Fans
> 8-inch blade, oscillating 1.4 amps
> 6-inch vent fans 2 amps
> Converted 12-inch fan with Datsun three-speed blower motor 4.5 amps

Tools
Chain saw (Minibrute) 100 amps
Drill (3/8-inch) 12 amps
Air Compressor (60-watt) 5 amps
Polisher/sander 7 amps
Hoists up to 100 amps
Cordless tools with ni-cad power packs, as rated
Motors, as rated

Water Pumps
March 893 0.9 amp
Honeywell 1/4-hp with separate pump 11 amps
Jabsco (4 gallons per minute) 4.5 amps
J. C. Whitney non-automatic (2 gallons per minute) 1.5 amps

Miscellaneous
Water purifier (24-watt) 2 amps
Bug killer (12-volt DC) 1.2 amps
Travel iron 10 amps
Curling iron (40-watt) 3.5 amps
Digital alarm clock 0.3 ampere hours per day total (continuous duty)

Small Inverter Loads

There is a temptation with small PV systems to use all 12-volt DC equipment. However, you will find that not all equipment comes in DC models. Converting everything to 12 volt DC can be trying even for the handy. Just as your choice of appliances and equipment is based on personal needs, so too should the power you select fit your needs. In almost all cases it is less expensive to use 12-volt DC fluorescent and quartz lights than incandescent bulbs. The use of DC water pumps with a cistern water supply is more efficient than other methods. But trying to convert or make some devices fit a 12-volt DC PV system can be more costly than using existing AC equipment and an efficient solid-state inverter.

A 600-watt Heart or 550-watt Tripp-Lite inverter will allow you to use AC equipment. You can also use low-cost new or secondhand appliances that operate on 120-volt AC. Small power drills and saber saws work well. A zigzag sewing machine motor draws only 0.85 amp and gives much more versatility than a treadle machine. Many kitchen devices are used intermittently (blenders, mixers, can openers, electric knives), and can be powered by an inverter.

When selecting an inverter for the occasional 120-volt AC load, be sure to consider your load. Cheap AC motors often draw five times the running or rated operating current to start, so the inverter must be sized to carry that initial load. If you have a 120-volt AC grain mill, you might consider putting in a DC motor, as the original AC motor is probably an induction-type and will need a big inverter to power the start-up.

Try to match your devices to the inverter. An examination of the power requirements of a collection of 120-volt AC devices showed that most of the smaller ones were in the 2- to 3-amp range. It was less costly to replace the larger motors—such as those for pumps and large power tools—than to buy the big inverter they would require to power them.

Typical Loads

The following lists show the equipment used in actual homes. You may want to refer back to these examples when preparing your own system sizing lists.

Home 1

	ampere hours/day
Two 16-watt fluorescent lamps used 4 hours daily	1.4 x 2 x 4 = 11.2
Automotive am/fm/cassette used 4 hours daily	0.5 x 4 = 2
8-inch oscillating fan used 4 hours daily	1.5 x 4 = 6
	Total =19.2

Home 2

	ampere hours/day
Three 16-watt fluorescent lamps used 4 hours daily	1.4 x 3 x 4 = 14.4
Two 30-watt fluorescent lamps used 4 hours daily	2.5 x 2 x 4 = 13.6
Automotive am/fm/cassette used 4 hours daily	0.5 x 4 = 2
Turntable used 2 hours daily with radio as amplifier	0.5 + 0.5 x 2 = 2
12-inch black and white tv used 6 hours daily	1.4 x 6 = 8.4
Water pump used 1 hour daily	4.5 x 1 = 4.5
Inverter (550-watt, intermittent use)	
sewing machine 0.5 hour daily average	
(85 watts x 1.11 inverter eff. ÷ 12)	7.9 x 0.5 = 3.95
vacuum cleaner 0.5 hour weekly average	
(2.3 amps x 120 volts x 1.11 ÷ 12)	25.6 x 0.5 7 = 1.82
blender 0.5 hour weekly average	
(750 watts x 1.11 ÷ 12)	69.4 x 0.5 7 = 4.95
	Total = 55.62

DC Sizing Using Ampere Hours

The following examples of DC power systems show how to size small- and medium-sized PV systems using ampere hours. Sizing small PV power systems can be simplified by using the ampere hour method. Since solar module output is rated in amperes (under load) and battery capacity is typically rated in ampere hours (20-hour rating), using amps can be convenient.

Example 1

Let's begin with a small PV-powered cabin owned by nature-loving people who prefer simplicity. The cabin is a passive solar building heated with wood. The occupants use wood or propane for cooking and refrigeration. They have jobs outside the home and use only hand tools. The occupants have a collection of recorded tapes. Water is supplied by a gravity-fed spring.

Two living room lamps (20-watt fluorescent) used 4 hours per evening
One kitchen lamp (30-watt fluorescent) used 2 hours per day
One bedroom reading lamp (0.25 amp incandescent) used 1 hour per night, average
One automobile am/fm radio tape recorder (0.5-amp) used 4 hours per day, average

The total power requirement for this PV system is:

2 x 1.7 x 4 = 13.6 ampere hours per day ⎫
1 x 2.5 x 2 = 5.0 ampere hours per day   ⎬ lighting
1 x 0.25 x 1 = 0.25 ampere hours per day ⎭
1 x 0.5 x 4 = 2 ampere hours per day (radio/tape player)

Total ampere hours per day = 20.85.

Example 2

Our second sizing example is a rural home near Phoenix, Arizona, occupied by a family of four. It has three bedrooms, a solar greenhouse, and full indoor plumbing. This time we will make two separate load requirement lists—one for winter and one for summer—since the family's activities differ with the seasons. Spring and fall load requirements are an average of the two quantified seasons.

Lighting
Four each 30-watt fluorescent lamps 6 hours per night
Two each 30-watt fluorescent lamps 2 hours per night
Four each 0.75-amp reading lamps used 2 hours per night, average

Entertainment
One 12-inch black and white tv used 4 hours per day, average am/fm radio (0.5-amp) used 4 hours per day
Record turntable (0.5-amp) used with amplifier (0.5 amp) for 2 hours per day, average

Appliances
Washing machine (wringer modifier with 1/4-hp 12-volt DC motor)
Three one-half hour load three time per week
(3 x 22 amps x 0.5 hour x 3 times per week = 99 ampere hours per week) totaling (99 ÷ 7 days) = 14.2 average ampere hours per day

Water pump used in pressure system

4.5-amp draw for 1.5 hours per day, average

Digital alarm clock
0.3 ampere hour per day

120-volt AC load (powered by a 300-watt inverter):
Sewing machine (0.85 amps AC) for 0.5 hour per day
Vacuum cleaner (2.3 amps AC) for 1.0 hour week or 0.14 hour per day
Small power tools (average 2.5 amp AC) used 0.5 hour per day, average

Since there are more daylight hours in summer, lighting requirements are one-half winter needs, however, power tools are used three times as much. In addition, two 8-inch fans (1.4 amps) are used 4 hours per day on the average. All other loads remain the same.

The winter power requirement for this PV system is:

4 x 2.5 x 6 = 60.0 ampere hours per day ⎫
2 x 2.5 x 2 = 10.0 ampere hours per day ⎬ lighting
4 x 0.75 x 2 = 6 ampere hours per day ⎭
1 x 1.4 x 4 = 5.6 ampere hours per day (tv)
1 x 0.5 x 4 = 2 ampere hours per day (radio/tape player)
1 x (0.5 + 0.5) x 2 = 2 ampere hours per day (turntable)
14.2 ampere hours per day (washing machine)
1 x 4.5 x 1.5 = 6.8 ampere hours per day (water pump)
0.3 ampere hour per day (digital clock)
AC loads
    0.85 x 0.5 = 0.425
    2.3 x 0.14 = 0.33
    2.5 x 0.5 = 1.25
        2.005 x 10 (120 volts to 12 volts) x 1.11 (98% inverter eff.) = 22.26

Winter total ampere hours per day = 129.16

The summer power requirement for this PV system is:

4 x 2.5 x 3 = 30.0 ampere hours per day ⎫
2 x 2.5 x 1 = 5.0 ampere hours per day ⎬ lighting
4 x 0.75 x 1 = 3 ampere hours per day ⎭
9.6 ampere hours per day (entertainment—same as winter)
14.2 ampere hours (washing machine—same as winter)
6.8 ampere hours per day (water pump—same as winter)
0.3 ampere hour per day (digital clock—same as winter)
22.26 ampere hours per day (summer inverter load)
2 x 1.4 x 4 = 11.2 ampere hours per day (fans)

Summer total ampere hours per day = 102.36

As you can see, determining your power requirements is simply a process of listing each load or device, the power it consumes, and the number of hours it will be in use. We have converted watts to amperes to simplify the sizing of the PV array and the battery storage. Just use the Ohm's law chart (Appendix A) or your multitester to obtain estimated or actual current draws.

You will also note that we have increased some power loads. In the second example, the actual mathematical answer for the washing machine load is 14.1428 ampere hours per day. We rounded this off to 14.2 ampere hours per day to simplify the math. To allow for the possibility of wire resistance loss and variations in usage patterns, it is a good idea to avoid conservative estimates.

In the second example, notice the significant difference (26.8 ampere hours per day) between winter and summer daily power requirements. In general, power requirements are higher in winter than summer because of increased lighting requirements and indoor activities. Of course, this may not be true in your situation—everybody's different.

These examples are just guidelines to help you get started in determining your own power requirements. You may find that you have to adjust your figures to meet your budget. (In that case, it's not a bad idea to follow the Three Lists Method.) Whenever you make a list, be accurate and be honest. If you are now on the utility grid, list every item and compare your figures with the kilowatt figure on your electric bill (an accuracy check).

Sizing Your PV Array

After you have determined your power requirements, the next step is to size your PV array to meet them. The following procedure has been modified from ARCO Solar design data. The Multiplier Factors Map (Appendix B) is a convenient tool for sizing for sizing your PV array, and is based on the more detailed Peak Sun Hour maps. When using the numbers on these maps, be sure to pay attention to the units represented. A multiplier factor is just a number used for convenience to simplify the sizing process. The radiation maps list actual and average measurable units. Keep this difference straight so you won't be multiplying apples by oranges. The tables in Appendix B give information on a number of specific locations worldwide.

Step One. You have already determined your daily power or load requirement. Now you must divide your daily power requirement by 24 (hours) to get the *continuous load amps.*

Step Two. Multiply the continuous load amps by the multiplier factor for your PV array location. This will give you the power requirements for your PV system in *peak amps.*

Step Three. Divide the peak amps by the peak amps of the PV panel you propose to use. Use the manufacturer's peak power current (amperes) rating (@ 100 mW/cm^2 and 25°C). You now have the number of PV modules required for your 12-volt DC system. If the figure is a fraction, round it off to the next highest whole number. If your system is 24, 36, or 48 volts, then connect your 12-volt modules in series to get the proper voltage. A 24-volt system will require twice as many modules as a 12-volt system with the same peak ampere requirement.

Now let's see how this simplified method works. We will use Example 1—the small cabin that requires 20.85 total ampere hours per day.

Step One. Compute the continuous load amps.

$$20.85 \div 24 = 0.8687 \text{ (rounded to 0.87)}$$

Step Two. The cabin is located near Omaha, Nebraska. From the map we can see that the multiplier factor for this location is 7. Multiply the continuous load amps by the multiplier factor to get the peak amps requirement.

$$0.87 \times 7 = 6.09$$

Step Three. Divide the peak amps by the manufacturer's peak ampere rating for the PV modules to be used. In this example, we will use a solar module rated at 35 watts, 2.26 peak amps.

$$6.09 \div 2.26 = 2.69 \text{ PV modules}$$

Thus, this system will require three 35-watt PV modules. By rounding off the number of modules required to the next whole number, you insure that the system will have adequate power during winter, with a surplus of power in summer.

Now let's compute the size of the PV array for the larger home in Example 2. To insure that the home has adequate power year-round, we must use the winter total ampere hours per day (129.16). Using our formula and the Multiplier Factors Map, all we need to know is the peak ampere rating of the PV module to be used and the location of the home. In this case we will be using a 55-watt module rated at 3.35 amps peak power.

Step One. 129.16 ÷ 24 = 5.38 continuous load amps

Step Two. 5.38 x 5 (multiplier factor) = 26.9 amps

Step Three. 26.9 ÷ 3.35 = 8.03 or 9 PV modules

Sizing Your Battery Storage

When attempting to design the optimum PV system, you will find that there is a trade-off between the number of modules in your array and the size of the battery bank. There is also a relationship between array tilt and storage size. In general, a fixed array is tilted to boost winter output at the expense of summer output. This averages out the system's daily performance during the year and decreases the amount of battery storage needed to get through winter months. Such a design is optimum for steady, year-round power. If power needs vary significantly with the seasons, then the storage, tilt, and array size must be designed to accommodate such variations.

A fixed PV array will be tilted just like a fixed solar water heater—that is, latitude plus 15°. But sizing the battery storage is a complex trade-off based on a few simple principles. If there were no clouds, no seasons, and no limits to battery performance, you would need only enough storage to last through the night. But in the real world you need more.

First, there will be cloudy days when the array will not be able to fully recharge the batteries. You must have enough capacity to last through several overcast days in a row. Five days of autonomous storage is the minimum in the continental U.S., except for the sunny southwest.

Second, there are fewer hours of usable sunlight in winter and a battery's average state of charge could be lower during winter. Even at its predicted lowest, your storage must have at least as many ampere hours of "required" storage as it does autonomous storage. That means, if you need 200 ampere hours of storage to take you through cloudy periods, you will need at least 200 ampere hours additional required or reserve battery capacity.

Third, battery performance has its limits. Repeated cycling between charging and discharging will eventuallly shorten battery life. Both sulfation (growth of lead sulfate crystals on the battery's plates) and stress fatigue can be partially prevented by adding more capacity so that the depth of discharge never gets below 60% in winter. However, your batteries will get old and weaker over the years. They will last longer and work better if protected from extremes of heat and cold—especially cold. People have located their batteries in the house, in an enclosed crawlspace, and even in a well-drained underground chamber to take advantage of everything from the earth's thermal inertia to normal house heat. If you don't do this, you must increase your battery capacity.

All of these factors must be considered and they add up to a lot more than one night's storage. In the very sunniest locations, the battery capacity will be 7 to 10 times larger than the average daily load. The stormier climates can require 20 to 30 times the daily load. If your area has a history of weeks of cloudy and partly cloudy weather, seriously consider the trade-off between the cost of additional battery storage (twice as much as you might need in some areas) as compared to the regular cost of battery replacement every three to five years. Some battery banks have lasted five years and longer—and they were used telephone company batteries to begin with.

If your batteries will experience freezing temperatures regardless of preventive measures, the battery storage capacity should be increased. Table 5.1 will be helpful in allowing for this. Find the lowest extended time period temperature for your area and multiply your calculated storage capacity by the correction factor.

There are many methods for calculating the amount of battery storage you will need. A basic rule of thumb for approximately three days battery storage is to size the battery bank based on the daily total ampere hours per day requirement. The ratio is 10 to 1. For every 1 ampere hour daily load requirement, you will need 10 ampere hours of battery storage. The example cabin requires 20.85 ampere hours per day times 10 for a total of 208.5 ampere hours of battery storage.

Let's see how useful this rule of thumb is. In winter, the cabin's location can have up to seven cloudy days in a row. That means we need at least 145.95 usable ampere hours of storage in the battery bank. If we want our battery bank to last a long time, we must not deep-discharge the batteries too often or too deeply. It is recommended that you use no more than 40% of your battery storage capacity during discharging or cycling, remember you can experience a 30% loss of battery-rated storage due to internal battery resistance, self-discharge, and wire resistance in your system. Thus, the size of the battery storage needed for the example cabin should be at least 365 ampere hours. As 6-volt deep-cycle (golfcart) batteries usually come in the 200 ampere hour size, this system would need four such batteries. The battery configuration would be two sets of parallel batteries connected in series.

COLORADO MOUNTAIN COLLEGE SIZING EXAMPLE

Colorado Mountain College (CMC), in Glenwood Springs, Colorado, is unique among educational institutions. Seeing a growing need for qualified PV installers, CMC instituted a workshop/seminar program to provide practical information about solar electric systems. Their focus has been on stand-alone small-scale residential systems.

The PV program was developed as a segment of the school's regular eleven-month vocational solar curriculum. Program participants include electricians who want to expand into PV, solar dealers, and homeowners seeking do-it-yourself knowledge. In addition to regularly scheduled courses, CMC has occasional intensive hands-on workshops. The professional teaching staff at CMC also gives solar presentations and classes around the world.

The following example system is excerpted from the Colorado Mountain College Comprehensive Design/Installation Manual.

System Description

The CMC example system represents a simple stand-alone photovoltaic lighting application common to remote homes and vacation cabins. It includes the basic components of a safely installed system.

Electrical load information: one 20-watt fluorescent light; one 10-watt incandescent light; three 25-watt incandescent lights. All lights are 12 volts DC used 4 hours a day, 4 days a week. There are no AC loads.

Batteries are stored at room temperature. Battery efficiency is estimated at 80%. Five days of autonomy are required at 50% depth of discharge. Batteries are 12 volt at 100 ampere hours. The solar panels are pole-mounted at 55° to be used near Denver, Colorado (40° north latitude). The modules used are rated 12 volt DC (nominal), 2.23 amperes load current (2.4 short-circuit current). The controller or regulator current capacity is based on the maximum array output and comes with adjustable low-voltage disconnect and meters.

Wiring information: array to controller is 28 feet; controller to battery is 2 feet; each light is 10 feet from controller. The equipment ground requires a 10-foot wire from the panel to a ground rod.

PV System Sizing Worksheet

Electrical Load Estimation:

Loads	Volts	x	Amps	x	Qty	=	DC Watts	x	Hrs Used	x	Days Used	7 / Days	=	Watt Hours
Incn.Lt.	12	x	0.83	x	1	=	10	x	4	x	4	/ 7	=	22.86
Flor.Lt.	12	x	1.67	x	1	=	20	x	4	x	4	/ 7	=	45.71
Incn.Lt.	12	x	2.08	x	3	=	75	x	4	x	4	/ 7	=	171.43

Total Connected DC Watts = 105 Average Daily Load = 240

Battery Sizing:
240 daily watt/hours/12 volts = 20 ampere hours/day
20 x 5 days autonomy/ 0.5 (50%) discharge limit/ 110 ampere hour battery
2 each 12-volt batteries in parallel

Array Sizing:
20 ampere hour/day load / 0.8 battery efficiency / 5.6 peak sun hours =
4.46 array peak amps / 2.23 module amps = 2 modules in parallel

Controller Sizing:
2 modules x 2.4 short-circuit current = 4.8 maximum array amps
105 total connected DC watts/12 volts = 8.75 maximum load amps
Use a shunt controller rated at least 5 amps
with a load rating of at least 10 amps

System Wire Sizing:
4.8 maximum array amps 28 ft 1 ea. 10/2 AWG
8.75 maximum load amps 2 ft 2 ea 8 AWG
5 ea branch circuits 10 ft with 10/2 AWG
10 ft equipment ground #6 bare
5 ft electrical ground #6 bare

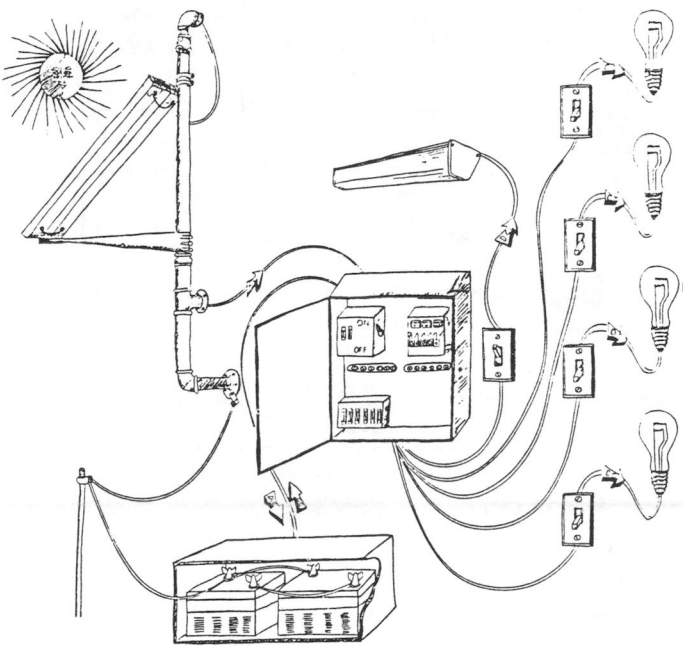

FIGURE 5.6: Cabin lighting system illustration. (Courtesy Colorado Mountain College)

A DOE STUDY ON DC PV

The following "DC versus AC Power Distribution" report is from a General Electric Company 1977 study (Conceptual Design and System Analysis of Photovoltaic Systems, Report No. ALO-3686–01). It should help you think through the inverter question.

Introduction

This analysis was undertaken to assess the merit of directly supplying certain residence loads with the DC power available from the solar array/battery power sources rather than inverting the power to AC with losses resulting from inverter inefficiency. In making this assessment, the question of regulation must also be borne in mind. Although the inversion of DC to AC carries with it a nominal penalty of 12% in inefficiency, relatively good output AC regulation can be achieved at the same time within nominal limits of ±5%. Regulating DC from an unregulated DC source (of which the array/battery combination is typical with a regulation range of 30%) also involves a penalty of about 12%. Thus,

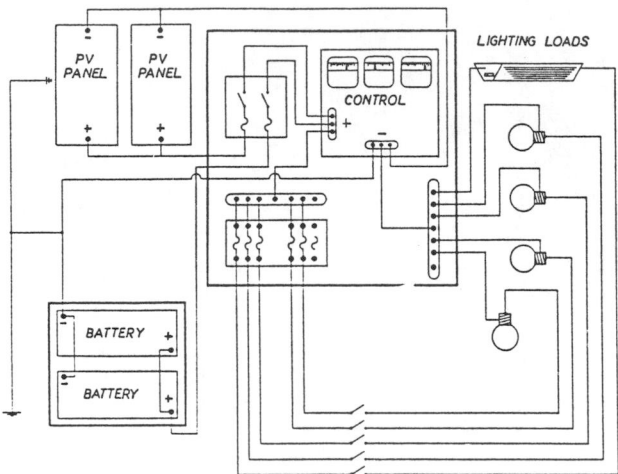

FIGURE 5.7: Cabin lighting system schematic. (Courtesy Colorado Mountain College)

power economy benefits would only result with using unregulated DC. An assessment of typical house loads is presented below concerning their operability with DC in general.

Classification of Appliances

Typical residential loads can be classified by their type of energy transformation. The four major categories are: resistance elements; universal motors; induction motors; and induction coupled loads. These resistance elements include simple heat producers such as electric range elements and incandescent elements, such as light bulbs. The universal motor is found in a great number of the portable appliances where the compact size and low cost are important and of primary concern. Where the low brush life or variable speed of the universal motor is objectionable, such as refrigerators and washers, the induction motor is used. In long life, low maintenance applications, such as time-keeping and sound reproduction, a synchronous induction motor is used. The fourth class of energy transformation utilizes induction coupling and includes such examples as fluorescent lamps, entertainment electronics power supplies, and microwave oven supplies.

Classification of Controls

Presently, residential loads are controlled by two types of control elements. The simple contact pair is the most prevalent and is found in both dry and mercury wetted forms. The second type of control element is the thyristor. Both unipolar (Silicon Controlled Rectifiers) and bipolar (Triacs) devices are found in limited quantities in typical residences.

The simple contact pair is found in every appliance or circuit to guarantee complete line isolation when the appliance circuit is turned off. An extension of this control is the simple bimetal temperature control found in appliances such as irons and skillets. These switches are required to cycle on the order of four or five times per minute in most applications. A further extension of this technique is in some universal motor speed controls where a miniature flyball governor is used to open and close contacts, thereby controlling the motor speed. Appliances and controls utilizing relays also rely on the contact pair. Many heating systems use relays for control and interlock functions.

The thyristor controls are presently found in lamp dimmers and motor speed controls where phase control techniques are employed to control the effective voltage delivered to the appliance or lighting load. A new application of these devices will be in appliance temperature controls where the thyristor, used in the zero crossing mode, can eliminate unwanted conducted and radiated electrical noise which is inherent in contact opening and closing.

Appliance Listing

The most common electric appliances are listed in Tables 5.2 through 5.9 pertaining to: Climate Control, Food Preparation, Food Preservation, Home Care, Home Entertainment, Laundry, Lighting, and Personal Care. Included in the listing are: Average Rated Wattage, Type of Energy Transformation Employed, Types of Common Controls, and Average Yearly Energy Usage.

The average rated wattage and average yearly energy usage figures have been compiled from information provided by the following sources: Niagara Mohawk Power Corporation, National Power Commission 1970 estimate, and CEQ 1973 estimate. The major types of energy transformation and control types have been marked on the tables with an X. The minor types have been marked with an M. For example, the first entry in the Climate Control List (Table 5.2) is the Air Cleaner. It has X marks on the induction motor and transformer coupled since there is usually an induction motor fan drive and in the case of electrostatic air cleaners, a high voltage transformer. These units usually have simple on-off switches. However, Room Air Conditioners are available with the three control types indicated. Most Radiant Heaters have no air circulation fans. However, some do use a small induction motor to cool the metallic surfaces of the unit and provide some convection heating. Therefore, there is an M in the Induction Motor column.

Due to the large range of appliance rating and diversity of suppliers and users, the average rating and yearly usage estimates cannot be extremely accurate. This listing is to be used as a guide to appliance energy transformation and control types.

Operation on DC Voltage

The operation of these household appliances will be discussed first with respect to the suitability of the basic energy transformation means to utilization of DC

Table 5.2: Climate Control

Appliance	Average Rated Load (k)	Energy Transformation				Control Type						Average Yearly Energy Usage (kWh)
		Resistance Element	Universal Motor	Induction Motor	Transformer Coupled	On-Off Switch	Push Buttons	Bimetal Relay	Phase Control	Magnetic Relay	Zero Crossing Switch	
Air Cleaner	50			X	X	X						216
Room Air Conditioner	800			X		X	X	X				860
Bed Covers	177	X				X	X					147
Dehumidifier	257			X		X		X				377
Fan—Attic	370			X		X						291
Fan—Bathroom	105			X		X						
Fan—Circulating	88			X		X	X					43
Fan—Window	200			X		X	X					
Furnace Fan	294			X		X		X		X		
Heater, Radiant	1250	X		M		X	X	X				176
Heat Pump	12500	X		X		X		X				
Heating Pad	65	X				X	X					16
Humidifier	177			X		X	X	X				163
Oil Burner	263			X	X	X		X		X		

Table 5.3: Food Preparation

Appliance	Average Rated Load (k)	Resistance Element	Universal Motor	Induction Motor	Transformer Coupled	On-Off Switch	Push Buttons	Bimetal Relay	Phase Control	Magnetic Relay	Zero Crossing Switch	Average Yearly Energy Usage (kWh)
Blender	386		X			X	X		X			15
Broiler	6700	X		M		X	X	X				100
	1400											
Can Opener	100		X	X		X						0.3
Carving Knife (Line)	092		X			X						0.8
Carving Knife (Batt)	092				X							0.8
Coffee Maker	894	X				X		X				106
Deep Fryer	1667	X				X		X				83
	1448											
Dishwasher	1250	X		X		X	X	X				363
	1202											
Egg Cooker	500	X				X		X				
Frying Pan	1250	X				X		X			M	186
	1196											
Sandwich Grill	1250	X				X		X				33
	1161											
Hot Plate	1257	X				X		X				90
Microwave Oven	1450			M	X	X		X				190
Range w/Oven	12200	X		M	M	X	X	X		X		1175
Range Self-Clean	12200	X		M	M	X	X	X		X		1205
Roaster	1333	X	M	M		X	X	X				205
Toaster	1250					X		X				39
	1146											
Trash Compactor	400		X	X		X						50
Waffle Iron	1116	X				X		X				22
Waste Disposal	445		X	X		X						30

Table 5.4: Food Preservation

Appliance	Average Rated Load (k)	Resistance Element	Universal Motor	Induction Motor	Transformer Coupled	On-Off Switch	Push Buttons	Bimetal Relay	Phase Control	Magnetic Relay	Zero Crossing Switch	Average Yearly Energy Usage (kWh)
Freezer (15 cu ft)	341			M		X		X		X		1195
Freezer (frost-free/15 cu ft)	440			M		X		X		X		1761
Refrigerator (12 cu ft)	238			M		X		X		X		728
	241											
Refrigerator (frost-free/12 cu ft)	333			M		X		X		X		1217
	321											
Refrigerator/Freezer (14 cu ft)	326			M		X		X		X		1137
Refrigerator/Freezer (frost-free/14 cu ft)	615			M		X		X		X		1829

Table 5.5: Home Entertainment

Appliance	Average Rated Load (k)	Resistance Element	Universal Motor	Induction Motor	Transformer Coupled	On-Off Switch	Push Buttons	Bimetal Relay	Phase Control	Magnetic Relay	Zero Crossing Switch	Average Yearly Energy Usage (kWh)
		Energy Transformation				Control Type						
Radio	71					X	X					86
Radio/Record Player	109					X	X					109
Television (B&W)	160											350
	55					X	X					120
Television (color)	300											660
	200					X	X					410
Slide Projector	300	X		X		X	X					
Movie Projector	600	X		X		X	X					

Table 5.6: Home Care

Appliance	Average Rated Load (k)	Resistance Element	Universal Motor	Induction Motor	Transformer Coupled	On-Off Switch	Push Buttons	Bimetal Relay	Phase Control	Magnetic Relay	Zero Crossing Switch	Average Yearly Energy Usage (kWh)
Clock	2			X								17
Floor Polisher	312											
	350		X			X						15
Germicidal Lamp	20	X			X	X						170
Sewing Machine	75		X			X				X		11
Vacuum Cleaner	630		X			X	X					46
Water Pump	454			X		X			X		X	
Circular Hand Saw	800		X			X						
Table Saw	800		X			X						
Drill (Hand)	250		X			X			X			
Soldering Iron	125	X			X	X						
Garage Door Opener				X	X	X					X	

Table 5.7: Laundry

Appliance	Average Rated Load (k)	Resistance Element	Universal Motor	Induction Motor	Transformer Coupled	On-Off Switch	Push Buttons	Bimetal Relay	Phase Control	Magnetic Relay	Zero Crossing Switch	Average Yearly Energy Usage (kWh)
Clothes Dryer	4856	X		X		X			X			993
Hand Iron	1000	X				X			X			144
Washing Machine (automatic)	500			X		X						103
	512											
Washing Machine (manual)	286			X		X						76
Water Heater	2500	X								X		4219
Quick Recovery	5000	X								X		4811

Table 5.8: Lighting

Appliance	Average Rated Load (k)	Resistance Element	Universal Motor	Induction Motor	Transformer Coupled	On-Off Switch	Push Buttons	Bimetal Relay	Phase Control	Magnetic Relay	Zero Crossing Switch	Average Yearly Energy Usage (kWh)
Incandescent (interior)	75	X				X	X	X	X			
Incandescent (exterior)	150	X				X		X				
Fluorescent	20				X	X						

Table 5.9: Personal Care

Appliance	Average Rated Load (k)	Resistance Element	Universal Motor	Induction Motor	Transformer Coupled	On-Off Switch	Push Buttons	Bimetal Relay	Phase Control	Magnetic Relay	Zero Crossing Switch	Average Yearly Energy Usage (kWh)
Hair Dryer	380	X	X	M		X	X					14
Heat Lamp	250	X				X	X					13
Curling Iron		X				X		X				
Heated Curlers		X				X		X				
Shaver	14		X		X	X						1.8
Sun Lamp	279	X				X						16
Toothbrush			X		X	X						0.5
Vibrator			X			X		X		M		2
Water Pic			X			X						

voltage. A second consideration will be the suitability of the various controls to operation on DC.

The residential loads with resistance element energy transformation can operate on DC voltages. Also, the universal motor can operate on DC voltages. However, the induction motors and induction coupled appliances rely on the periodic voltage reversal of AC to function. Therefore, these appliances cannot operate on DC voltage.

Of the resistance element loads, those most suitable for DC are the simple heat producers. These units follow a square law relationship between voltage and power transformed into heat. Many of these appliances have thermostatic controls such that the DC voltage might vary moderately ($\pm$10–20%) while the control would compensate for the input power variation. For example, an electric frying pan set for 350°F may only have power supplied to the resistance heater for fifteen seconds out of a minute which will raise the temperature of the pan and load to some upper limit, say 375°F. During the following 45 seconds, the temperature will decrease to some lower limit, say 325°F, at which time the bimetal controller applies voltage to the resistance element to start the cycle again. With a low input voltage, the cycle might be 25 seconds on, 35 seconds off, while the high input case would result in 10 seconds on, with 50 seconds off.

Due to the square law power conversion of these elements, the most significant variable in performance of these controlled resistance elements would be a low maximum output at low input voltages. Simple uncontrolled heaters, such as hot plates and radiant units also suffer from this poor regulation. That is, a 10% reduction in voltage would result in a 20% reduction in heat.

Incandescent lamps can operate on DC voltages. However, two phenomena must be considered. The first is the fact that the light output of an incandescent lamp follows the Stefan-Boltzman Law. This axiom implies that the light output is proportional to the fourth power of the operating voltage. Therefore, a 10% reduction in voltage will result in a 40% reduction in light out. The second phenomenon is metallurgical in nature. The incandescent tungsten filament in a lamp undergoes a grain boundary modification on DC operation. This phenomenon results in a reduced life for some lamps when operated on DC. Grain boundary modification is very temperature-sensitive. Lamps operated at and above 2800°K (the temperature range for most household lamps) show little of this characteristic during their life. Long-life lamps and very low-wattage types, such as night lights, operate at reduced filament temperatures (below 2700°K) which make them susceptible to this effect. Thus, the majority of the lamps used in residences will be unaffected by DC operation.

Universal motors do not rely on the periodic AC voltage reversal to operate. These units can operate from both AC and DC; hence, the name universal. However, most motors incorporated in present appliances have been optimized for AC operation since DC is not readily available. When the motors are operated on DC, the exhibit accelerated negative brush erosion, resulting in a 50% decrease in life. The speed of these motors is also dependent on the input voltage. A decrease of 10% in voltage would result in a 10% decrease in speed. In most blower applications, this would result in a 20% reduction in output.

The induction motor and transformer coupled devices rely on the periodic sinusoidal voltage reversals to operate. If these units are connected to a DC source, a heavy current will be drawn with resultant appliance failure.

If universal motors are excluded due to life considerations, the controlled resistance elements are the only present residential loads suitable for operation with DC voltage.

The contact pair as used in the typical residence has been optimized with respect to cost and function. To reduce the cost of contact material and mechanical actuators, the switches have been designed for AC operation only. The arc which is established as the contact pair opens is extinguished during the periodic current reversal. At the instant of zero current, the arc can be extinguished quite easily with inexpensive contact materials and small gap spacings. A DC switch must be designed to extinguish an established arc with both voltage and current available to sustain it. Therefore, more massive erosion-resistant contacts are needed with an increased spacing for DC operation.

For this reason, any appliance to be operated on DC would need an improved control means. Even the simple on-off switches of most appliances would need modifications to operate safely.

Conclusions

Unless severe performance compromises are to be tolerated, AC voltage appears most suitable for residential loads. Major load and control modification would be necessary to render present appliances suitable for DC operation. Even then, the DC voltage would have to be regulated to ±15% which is standard for most AC utility systems.

Here are some things to consider:

1. The use of unregulated PV in direct applications such as water pumping and fan operation shows economy of power. When the demand for water (irrigation) or air circulation (attic cooling) is greater, unregulated PV also has economy of design.

2. To use the charts for examining your own power requirements, compare your actual appliance wattage and number of hours used. The numbers used in the study are very high and reflect the sometimes absurd use of energy-inefficient appliances. For example, in the Food Preparation section Waste Disposal is listed at 30 kWh per year. A garbage disposal uses 445 watts or 7.41 watts per minute. Therefore, the study lists the garbage disposal daily operation at 11 minutes per day. Since most households with a garbage disposal use it two or three times a day for approximately 30 seconds each time, 4 kWh/yr would be a more realistic number.

3. It is important to note that switch contacts for DC operation must have larger contact surfaces, a larger contact gap, and possibly a capacitor to reduce sparking when the switch is turned off (contacts open).

4. Most universal motors are 120 volts AC/DC and are not suitable for 12-volt DC systems.

FIGURE 5.8: The year-round island home of Dick Buckheim and Susan Bailey in the Newfound Harbor Keys, Florida (30 miles east of Key West) is predominately PV-powered. A Heart inverter provides AC for lighting, ceiling fans, stereo, tv, power tools. The range and refrigerator are propane. Their solar system is completed by a solar hot water heater, sun-synchronous hot tub, and PV-powered cart for trans-island transport.

Chapter 6
Modules and Mounts

Designing a PV system is easier if you use the component approach. In the sizing process, consider the individual parts and how they will fit your particular needs. More importantly, when you buy your PV system, all the components may not be available from one supplier. Rather than substitute what is in stock, you should know and get exactly what you need.

The parts of a PV system are:

1. Solar Modules
2. Mount
3. Regulator
4. Batteries
5. Inverter
6. Wiring and Distribution Equipment

In this chapter, we will discuss modules, site selection, and mounts.

When I first started using PV in 1978, I had a very limited income. I was on my own when it came to selecting my first modules. I spent a lot of time reading manufacturer's literature and the impression I got was that PV modules were all about the same. Everyone's modules were well-built, using the finest materials under the most exacting conditions. It sounded great, but there was not enough solid information available for me to judge if what I was reading was true. There were few PV users who were not also selling modules; it was difficult to get an unbiased opinion.

So I bought what I thought was a good deal. The modules were hand-assembled and encapsulated in plastic acrylic designed to be mounted on boat decks. Now that sounded pretty rugged. In addition, the manufacturer just happened to have some good modules with scratches on them that he would let me have at reduced price. What a deal.

Well, that deal turned out to be a bust. The modules were not well-made. Solder interconnects between cells began to fail. Output was as rated, but I could expect more scratches on the encapsulant and eventual discoloration or yellowing which would adversely affect output. As the module heated up in the sun, the cells pulled away from each other. The superstrate was expanding and contracting (in heat and cold) one way, the cells bedded in silicon moved another way, and the wire interconnects still another. Pity the poor solder joint.

In 1980 I retired those modules along with the money spent on them and bought good modules.

SOLAR MODULES

Solar Cells and Arrays

Solar cells produce about 0.5 volt DC. The amount of current depends on the size of the cell: larger solar cells produce more current. Most cells now manufactured are between 0.100 amp and 4 amps produced by 4-inch cells. (Note: See chapter 1 for a brief overview of the history, physics, and manufacture of solar cells.)

To increase the voltage, solar cells are connected in series. To increase the current, they are connected in parallel. Load is the equipment being powered.

In a typical 12-volt module, 30 to 36 cells are connected in series to produce the voltage to charge a 12-volt battery. The total voltage of the module must exceed the battery voltage to "push" the charge into the battery. Most solar panels produce 14 to 17 volts. The battery then stores the electricity for later use. For example, a 33-cell module produces 16 volts at 3 amps or more open circuit. By connecting modules in parallel to create a PV array, the amount of current is increased. Three such modules will produce 9 amps at 16 volts.

Blocking diodes (Figure 6.4), electrical one-way gates which permit current to flow in only one direction, are used with some solar modules. The diode prevents the battery from being discharged back through the solar panel at night.

When several panels are connected you have an array. Any number of solar cells and batteries can be connected in both series and parallel to produce the needed voltage and current. Care must be taken to match cells and batteries because bad or low-power cells in either the PV modules or the batteries will diminish output.

Figure 6.5 shows three panels connected in parallel. A voltage regulator has been added between the battery and the array to prevent overcharging and to protect the battery.

To monitor the PV system, voltmeters and ammeters are installed (Figure 6.6). Voltmeters are connected parallel to the load. A fuse is added in the circuit to protect the load or equipment from damage due to a current surge.

What to Look For

In review, a solar module is made of cells connected together to produce the rated voltage and current. These interconnected cells are encapsulated to protect them from the environment, and this encapsulation may be more than one layer. Modules should have a frame or support for protection and strength, and some way of being connected to other modules or the battery bank.

The solar module that I prefer and use is generally accepted as the industry standard. Tried and proven in the field, these modules will last decades. The cells are redundantly interconnected which means they are soldered together with at least two flat wires. This double wiring will keep the module active even if one interconnect is damaged.

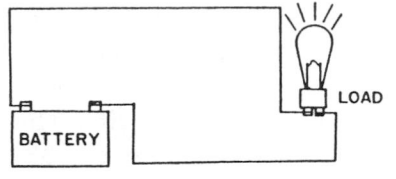

FIGURE 6.1a: A battery converts chemical energy into electricity.

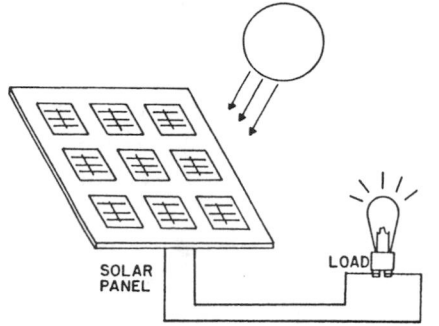

FIGURE 6.1b: A solar array converts light into electricity. Nothing internal gets used up or wears out.

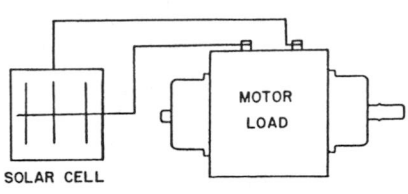

FIGURE 6.2: Small toy motors, like the one shown here, will operate on the output of a single solar cell. This simple device consists of a 4-inch solar cell, a short-length of two wire leads for positive and negative connections, and a small DC motor.

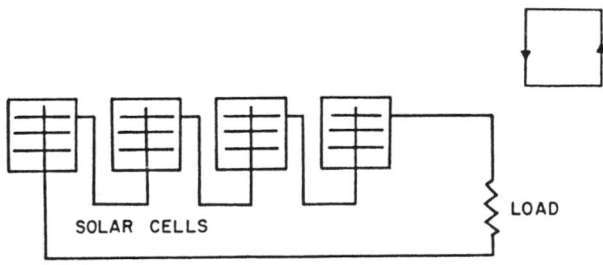

FIGURE 6.3a: Solar cells in series. The total voltage is the sum of the individual cell voltages, but the current is the same as that of a single cell.

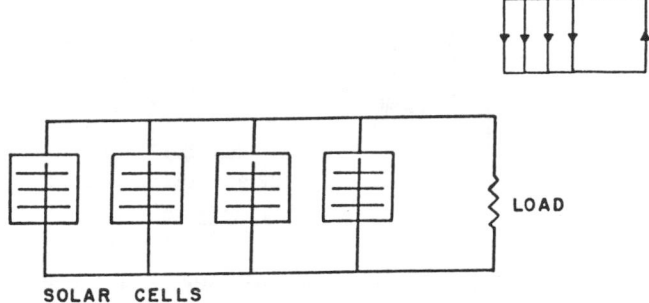

FIGURE 6.3b: Solar cells in parallel. Here the voltage would be that of one cell, but the current is the sum of the individual cell currents.

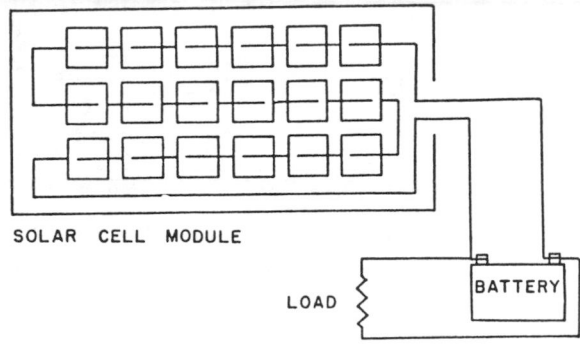

FIGURE 6.4: A solar module, constructed as a single panel with a built-in blocking diode, connected to both a load and a battery.

The two electrical interconnects between cells also provide a lower resistance path for electricity. The cell interconnects should be soldered along the length of the cell, and not at just one or two spots. The connected cells should then be hermetically sealed from the environment, absolutely airtight. Any moisture or air that gets to the cell interconnect or grid will cause corrosion and eventual failure. In a perfect world with a perfect hermetic seal, solar cells will produce power forever, or at least as long as the sun shines.

Some manufacturers use ethylene vinyl acetate (EVA) or some other plastic for cell encapsulation. Whatever the material, the

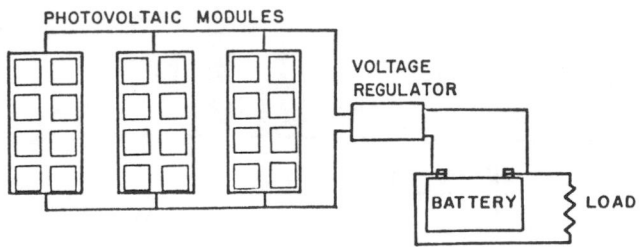

FIGURE 6.5: An array of three solar modules connected in parallel. Usually each module is a single panel, but some manufacturers make large panels of more than one module. These modules are independent and can be connected in series or parallel, depending on need.

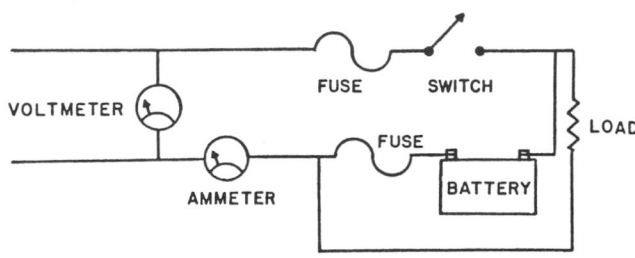

FIGURE 6.6: A voltmeter and an ammeter can be connected to a small solar electric system to monitor the performance of the array and the battery. Note fuse protection and disconnect switch.

encapsulate should be clear and stay clear. Once the cells are protected from air and water, they need to be protected from physical damage. That means a strong superstrate, or cover. You could use a bullet-proof plastic to protect the cells, but that's not such a good idea. First, it's expensive. Second, it is easily scratched. While it may be years before it clouds or yellows, I doubt if its transparence will last as long as the inert cell.

Another reason not to use bullet-proof plastic (and I mention this material as just one example of a plastic superstrate) is because it won't completely protect the cells if they are positioned right up against it. Should a bullet bounce off the superstrate, impact shock could fracture the cell. Such a point impact might cause cell failure.

It is also unwise to put a crystal solar cell under a flexible plastic material because the temptation to flex the module over a curved

FIGURE 6.7: Some people make their own modules. They either acquire surplus cells on their own or attend hands-on PV workshops to learn assembly. *Practical Photovoltaics* details step-by-step procedures for construction modules. Author Dr. Richard Komp leads several hands-on workshops each year. (Photo: David Ross Stevens)

surface and fracture the cells is too great. It's better to use a tempered glass cover and admit the cell's fragility. Glass is not such a bad choice: your home has glass windows. If vandals shoot up your solar array, they will probably do your windows too. Let's face it, there's not much of that stuff going on. If there were, protecting your solar cells would rank second to protecting your life.

So a good solar module has redundantly connected, hermetically sealed, glass-encapsulated cells. The glass front or superstrate also serves as an insulator. Should lightning strike your grounded array, the cells will continue to function.

A strong weatherproof substrate—the backing material of the module—is also important. Most module substrates are white to reflect sunlight and help keep them cool, though now modules are available in all colors. One manufacturer has used a yellow substrate. And why not? If the substrate material is going to yellow over the years, may as well make it yellow to start. A black substrate heats more rapidly than a white one, but black radiates heat almost as rapidly as it absorbs it, and so dumps the heat quickly.

A solar module should not be so large that it deflects or bends because glass does not bend much before it breaks. The size of the module is important for more than just reasons of strength. If the module is not large enough to be an economic unit, or if it is too small to be economically made and bought in quantity, then your costs will be higher. Several small specialty modules will cost more than a few standard-sized units. On the other hand, if the module is too large, shipping and handling costs go up.

Next we come to the frame. The solar module I prefer is a glass unit. The frame of this solar electric "window" should be strong and weatherproof. Modules need a frame, and it is a good idea to buy a module complete with its own. Aluminum, anodized aluminum, steel, and even wood are used for frames. Wood looks great but needs regular treatment to protect it from the weather. Steel can rust unless it is stainless, painted, or epoxy-coated. Aluminum, unless processed specifically for marine environments, will pit and corrode.

The frame should be strong enough to be used as the array support or part of the support. It should be easy to drill. It should look good. A well-made solar module should have a look of quality, the same quality appearance as a high-performance sports car.

Finally, we come to the interconnect box. Early solar modules had a couple of lead wires coming out of the encapsulant to which you soldered your plus and minus leads. This was not a good idea. Too often the leads broke at the point of entry into the encapsulant and you were left with the dreaded task of tearing into the module and soldering a repair add-on lead. This field-expedient repair would eventually break, too. Worse yet, you chanced breaking the hermetic seal.

Next, manufacturers began putting terminals on the back of the modules. This was better, but still vulnerable to damage. No matter how the terminals were reinforced, someone would figure out a way to knock them off. Soldering the terminals back on is relatively simple for an experienced technician, but it's best to prevent such damage. By securing your wire leads to the module frame instead of letting them flop around in the weather or be snagged, terminals are less likely to be pulled off.

For commercial installations, manufacturers began producing protective weatherproof terminal boxes. Some were metal, most were plastic. These boxes protect terminals from physical abuse and the weather, and provide an enclosure for bypass diodes. They also prevent accidental shorting of connections. The use of these boxes led, in part, to the Underwriters Laboratory (UL) listing of some modules. Interconnect boxes are a definite improvement.

If you have modules without interconnect boxes, consider adding them. Use plastic electrical junction boxes with covers, similar to those used for wall outlets and found at the hardware store. Cut the back off the box. Place it over your module terminals and seal/glue it to the back of the module with silicon. (Use tube silicon for sealing bathtubs.) The box has knock-outs through which to re-connect your array. Let the silicon dry for a few days before screwing on the box cover. It's that simple. Now you have fancy weatherproof interconnect boxes.

There are other types of modules available, but use your judgment when buying anything other than glass-encapsulated metal-framed modules. However, for special applications like boat deck mounting, curved surfaces or locations subject to damage by impact, you might consider specialty modules.

Some modules are mounted on metal and covered with a plastic coating. These are flexible. I hesitate to mention them because they are easy to abuse. Boat owners like them because they can be bent and curved to the boat deck and because they can be walked on. Bending increases the likelihood of cell fracture, however, and inevitably something is going to be dropped, causing a point of impact fracture to a cell. If you do use these modules on a boat, don't fasten them to the deck by screwing through the module. Even if there is space to screw without hitting a cell, don't do it. Punched holes in modules are the beginning of the end, as it breaks the hermetic seal.

There are other flexible modules available. Some come with stainless steel backs, some with plastic, some are small panels with canvas webbing holding them together. You pay a premium for flexibility. If you can do the job with a rigid module, forget the flexible units.

The amorphous modules now on the market have some unique characteristics and advantages. Being of monolithic construction, they usually have only two module wire-soldered interconnects. The cell interconnects are made during the deposition process and are integral to the cell/superstrate. Disadvantages to amorphous modules are efficiency and expected lifetime. At present, single-crystal technology is more advanced than amorphous technology. The typical single-crystal module is between 10 and 12% efficient in converting sunlight to electricity. Amorphous modules are about half as efficient.

Lifetime for the single-crystal module is measured in decades. Cells made in the early days of PV are still working. When asked, "How long will the panels last?" I try not to make unrealistic claims. Assuming that the installation of quality modules has been done well, is properly grounded, and not subject to physical damage, no one really knows how long they will last. Good single-crystal modules are warranted by some manufacturers for ten years. A warranty longer than ten years for anything is almost unheard of. The best amorphous cell warranty, as of this writing, is three years. Amorphous modules for outdoor use have only been in field testing for five years.

Commercial, industrial, agricultural, and military applications are generally based on a twenty-year design life (because that's how long most businesses keep long-life equipment "on the books"). However, more than one PV array in space has celebrated its twenty-fifth anniversary.

Jet Propulsion Laboratory's accelerated aging tests indicate that well-made modules will last for decades. Fifty years from now some PV modules will still be working "like new."

Module Characteristics

PV modules have physical and electrical characteristics which make certain modules better choices for certain applications.

The output voltage of a module is determined by the number of cells. The 33-cell module is the industry standard. However, 30- to 36-cell modules are also available.

The load voltage output of a 33-cell module is around 16 volts at 77°F (25°C). (We will not consider the open-circuit voltage of a module because that's not how modules are used.) A module is connected in a circuit under load. The number of modules and the size and state of charge of the battery bank or a load will determine system voltage. Temperature also determines voltage.

In general, solar modules will lose about 10% efficiency at normal operating temperature, typically 116°F (47°C). In full sun on a calm, warm day, a 33-cell module will actually put out about 14.5 volts. A 12-volt battery likes a charge of 14.3 to 14.8 volts, so this is adequate. This is why many residential-use modules contain 33 cells and are called nominal 12-volt modules.

However, in heat the 36-cell module performs better. You can always depend on sufficient voltage to drive a charge into your battery bank. You also have enough voltage for an occasional equalization charge. Even if you lose 15% of the rated power in extreme heat, you can keep your batteries charged.

Surprisingly, 36-cell solar modules also work well in locations with very little sun. The higher voltage means the module is conducting current to the batteries in low insolation conditions. Of course, regular cut-out voltage can limit the maximum charge.

The 30-cell module, popular among recreational vehicle owners because of its smaller size and lower cost, performs adequately most of the time. However, in the desert southwest, a 30-cell module will not fully charge a 12-volt battery. RVers usually drive enough to make up the difference in required charge voltage.

When selecting modules, look at operating voltage under load and at operating temperature. I don't recommend modules with fewer than 30 cells. Don't waste money on modules with more than 36 cells, either. There is a reason the 36-cell module has become the standard.

Module load current times load voltage is the module's load wattage rating. Manufacturers generally list the module wattage at 77°F (25°C), so it is important to look at voltage and current at actual operating temperature. A nominal 12-volt module should put out about 15.5 volts at 122°F (50°C). At that temperature, module current will be at or slightly above the 77°F rating.

Performance characteristics presented in a spec sheet's small IV curve chart mean very little. Supposedly, you can check the fill factor (indicated by the sharpness of the curve's knee), a measure of efficiency. However, the chart provides only a general picture of a typical module—not the test results of a specific module.

Other module characteristics are given—ground continuity (typically less than 1 ohm) and leakage current (typically less than

40 milliamps at 3000 volts DC), but the important one is load voltage and load current at normal operating temperature (NOT).

One of the most important things to consider is the warranty. If a manufacturer will back its product for ten years, you should feel confident, too. Of course, this assumes that the manufacturer will be around to honor the warranty. Buy good modules made by a reliable manufacturer and sold by a reputable supplier and someday your children and grandchildren will inherit your array.

Reading a Spec Sheet

Specifications for solar modules vary. Most spec sheets, such as the one on pages 131 to 132, provide the information you need to size your solar electric system.

General Information

This midsize power module has a range of applications. Most manufacturers offer modules in the 50-watt range. Metal-framed, glass-encapsulated, hermetically sealed modules should last decades and carry at least a ten-year warranty. Though not listed on the spec sheet, this Solec module does have a ten-year warranty.

Almost all ten-year warranty modules exceed the standards set by the U.S. government's Jet Propulsion Laboratory. Modules that pass JPL's Block V series of environmental and performance tests for thermal cycling and thermal shock, impact, wind load, humidity, freezing, corrosion, and electrical isolation should far outlast their warranty.

Some manufacturers boast the UL label. Underwriters Laboratory safety-tests products for a fee. The U.S., German and Japanese governments also perform extensive module testing. Most building inspectors will approve modules with a UL listing.

Features and Mechanical Specifications

Module size and weight are important. Information on frame thickness, mounting hole location, and the shape and size of interconnect boxes or terminals is also given. Oversized modules may incur additional shipping charges: check with your supplier on shipping weight and method.

S-53 Solar Electric Power Module

The SOLEC Utility Module is one of our most popular midsize PV modules. The S-53 consists of one series string of 36 four inch square cells which are encapsulated in a very cost effective module for large power applications.

The S-53 Module produces power efficiently and reliably for applications ranging from small scale battery charging to large scale utility systems, including pumping and lighting systems.

This midsize power module, like SOLEC's other standard modules, is designed to exceed all JPL Block V requirements. We used tempered low-iron glass as the front surface to provide for maximum transmission of light and impact resistance. The cells are laminated to the glass with layers of ethylene vinyl acetate (EVA).

The Module back surface is white Tedlar™, a substance which is not only an effective moisture barrier but also an excellent reflective surface. To give structural integrity to the module, a frame is added to the laminate and sealed with a special polyurethane polymer to provide rigid mechanical construction and effective moisture barriers for a maintenance-free operation.

The frame is made of extruded 6063-T6 anodized aluminum alloy. The module junction box is UV resistant and weather-proof with a moisture-tight wire strain reliever.

"Bringing the Sun Down to Earth."
Since 1976

FIGURE 6.8: Solec S-53 spec sheet.

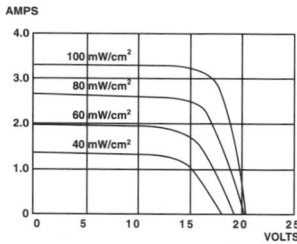

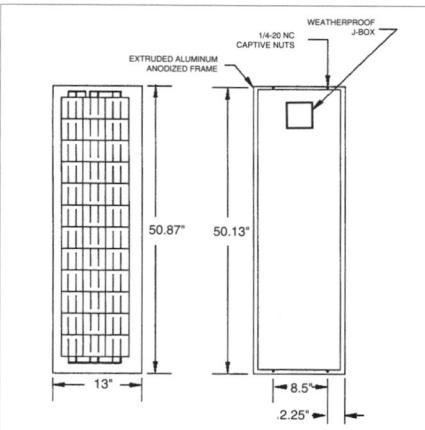

This graph demonstrates the effect on power output of varying light intensities at a cell temperature of 25°C.

This graph demonstrates how peak power and voltage are affected by changes in cell temperature.

Electrical Characteristics

S-53 electrical characteristics at solar intensity of 100mW/cm^2 and a cell temperature of 25°C (77°F).

Power (peak)	53 Watts
Voltage (peak)	17.10 Volts
Current (peak)	3.10 Amps
Voltage (open circuit)	20.30 Volts
Current (short circuit)	3.40 Amps
Cell size:	4 in (101.6 mm) square
Circuit:	1 strings of 36 cells in series

Physical Dimensions

Length	50.75 inch	1290 mm
Width	13.0 inch	330 mm
Thickness	1.5 inch	38 mm
Weight	13.2 lbs	6.0 kg

NOTE: Specifications given are examples of standard production modules and are subject to change without notice.

* The electrical characteristics represent nominal output. Actual production units may show a distribution of +/- 10%.

 **solec** International, Inc.

Solec International, Inc.,
12533 Chadron Avenue, Hawthorne, California 90250
TEL (310) 970-0065 • FAX (310) 970-1065

Performance Data

There are many ways to rate a solar module. Industry standards have not been established although most manufacturers use 1000 watts per square meter at 77°F (25°C) cell temperature and 1.5 air mass. This is slightly more than the sun's intensity in the Sahara Desert at high noon. A module's peak power rating is the highest combination of peak load current times the peak load voltage. Variations up to 10% are sometimes noted because of varying test standards and to allow for production tolerances. Few terrestrial applications require modules with more exacting tolerances.

Short-circuit current and open-circuit voltage are listed. This information is useful when sizing your regulator or testing a module straight out of the box. Even then, short- and open-circuit information is only useful if you have a calibrated reference solar cell. The human eye cannot discern light intensity in the same way the ear can judge sound intensity. Sunlight, bright to the eyes, may vary from 700 to 1500 watts per square meter intensity.

General IV (current times voltage) curves are just that—general. (Some manufacturers have been known to use the same curve on all their spec sheets.) Note the current output curve at different light levels. At about 40 watts per square centimeter, you hardly cast a shadow. Overcast is about 10 watts per square centimeter. This module has a very low threshold and may actually produce about 0.25 amps during a daytime rainstorm.

While not particularly useful, the example IV curve does give the different current and voltage outputs at three temperatures. As temperature increases, current increases slightly but voltage at the knee of the curve (peak power point) falls off about 10% to approximately 15 volts. In summer when the ambient air temperature is above 80°F (26°C), your module may reach 140°F (60°C) or more. A breeze helps to keep cell temperature down.

Some manufacturers offer modules with various combinations of electrical connections and module colors. Black modules may be aesthetically appealing, but they have a higher normal operating temperature and associated power loss.

Module voltage drops as temperatures rise, reducing actual power from this 53-watt module to as low as 45 delivered watts. Voltage drop begins at lower temperatures for black modules.

All solar cells have reduced output at higher temperatures. It is the nature of solid-state semiconductor material to lose efficiency in heat. Since modules are out in the hot sun, expect at least normal operating temperatures and associated voltage drops.

Normal operating temperature information is important. While 17.1 volts may seem too high for 12-volt charging, 15 volts is not. Thus, this 36-cell module will work well at high temperatures, but a 30-cell module may not provide enough voltage in summer to fully charge a 12-volt battery.

Another aspect of temperature rise and voltage drop is the effect reflectors and concentrators will have on the module. While it is possible to enhance solar input to Air Mass 2 without damaging most modules, it is not recommended unless you can keep temperatures down. Note that most manufacturer warranties are void if solar radiation striking the module has been enhanced by reflectors or concentrating devices.

When selecting a solar module, ask questions. All your questions should be answered to your satisfaction before you buy.

Table 6.1: Effect of Module Temperature on Power Production
As temperature increases, output voltage drops. All three modules are from the same manufacturer and rated 3 amperes. Nominal 12-volt system regulator cut-out is 13.7 volts. (Insolation: 1000 watts/square meter)

	77° F (25° C)	122° F (50° C)
36 Cells	44.0 W	43.8 W
33 Cells	43.5 W	40.3 W
30 Cells	41.8 W	30.7 W

SITE SELECTION

Obviously your PV array must be sited to receive full sunlight for the full day. However, if your home is surrounded by trees, don't start cutting them down. Spend some time sitting on the south side of the house and thinking. Consider the passive cooling and aesthetic aspects of the trees as opposed to harsh sun and stumps. Look at the sun path and note when shade covers your proposed location. It may be that some judicious trimming or slight relocation of the array will solve the problem.

Although the winter sun is lower in the sky and casts longer shadows, deciduous trees with open branch patterns (and without their leaves) may let enough sunlight through. But be prepared to sacrifice at least 10% of your solar production because of branch shading. In the south, where cooling is the greater problem, it is best to leave as much vegetation around the house as possible. In this case, you may have to locate the array away from the houseon the lawn, near the garden, or on a pole mount.

Selecting the site for your PV array can become overly complicated. Let's make it simple by first saying that your array should have full sunlight between the hours of 9:00 am through at least 3:00 pm. Early morning and late afternoon sunlight is diffused because of atmospheric moisture or haze and produces less power. For fixed mounts it is difficult to catch the late-summer sun after 4:00 pm without sacrificing ease in mounting.

Solar angles (altitude and azimuth) define the shade pattern or solar window: the area of the sky through which the sun appears to travel (see Figure 6.9). As mentioned above, the solar window must be free of shading during the optimum solar radiation collection hours of 9:00 am through 3:00 pm. The Energy Task Force of New York suggests a way to estimate the solar window. Stand where the collector is to be mounted and face true south. Point so that your finger and eye are horizontal. As shown in Figure 6.10, place one fist on top of the other (see the table to determine the exact number of times). Sight over the top of the fist at true south and 30° east and west (with adjustments in fist height) to determine shading effects. Any objects above your fists will cast a shadow on the collector; anything below will be below the lowest path of the winter sun.

It goes without saying that in the northern hemisphere your array should face south, as close to truth south as possible. Magnetic south, or compass south, is not the same as true south. Check Figure 6.11 or ask a local surveyor for the magnetic declination of your area. However, don't worry about this too much as your array can deviate as much as 15° from true south and still collect 90% of the sun's energy. Still, proper orientation is rewarded with maximum solar production.

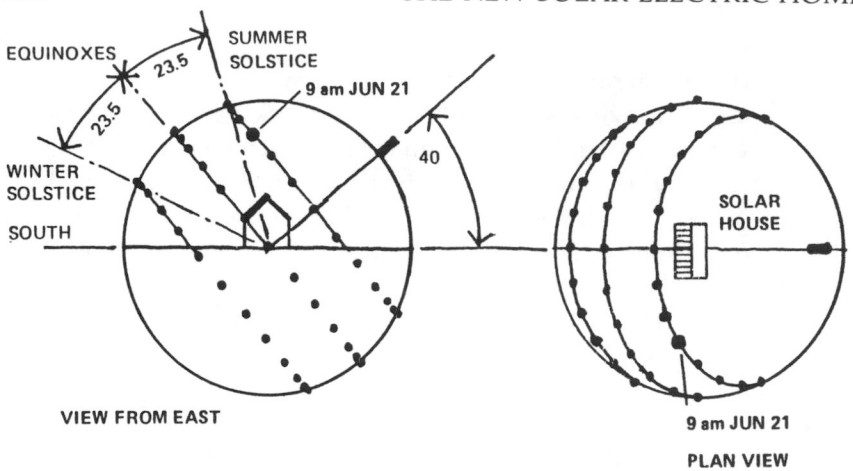

FIGURE 6.9a: If the sun's path were traced on a spherical ceiling, the view from the east and from the top would look like this. Each day's path describes a circle around the axis pointing toward the north, the axis here tilted from the horizon by a latitude angle of 40°. (Courtesy HUD)

FIGURE 6.9b: Another way to plot the same data is via a Mercator sky map which shows the sun's path on a rectangular grid of azimuth and elevation. Clear plastic guides are available from which you can make a working version of this diagram. (Courtesy HUD)

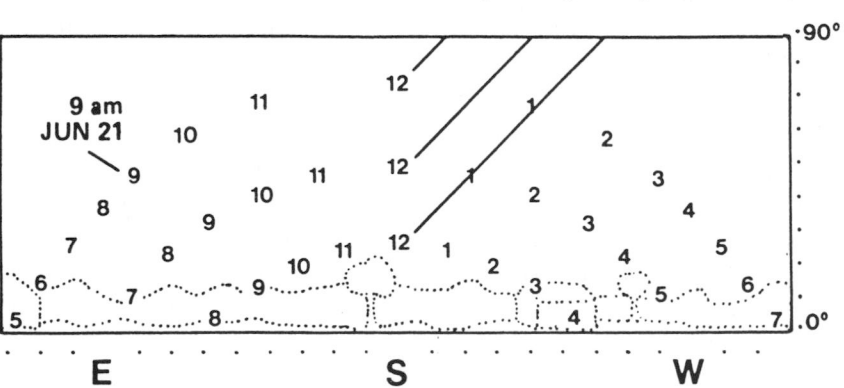

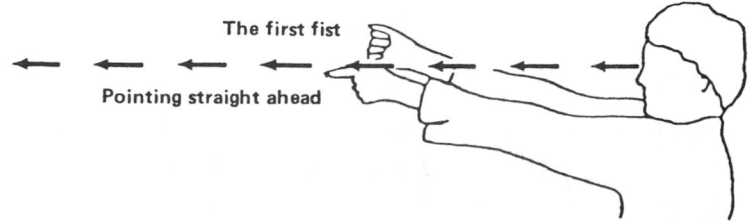

Latitude	12 O'Clock Position = 0° Bearing	11 O'Clock Position (East) and 1 O'Clock Position (West) = 30° Bearing Angle
28°N	4½ fists (47° Alt.)	3 fists (30° Alt.)
32°N	3½ fists (34° Alt.)	2½ fists (26° Alt.)
36°N	3 fists (30° Alt.)	2¼ fists (23° Alt.)
40°N	2½ fists (27° Alt.)	2 fists (20° Alt.)
44°N	2¼ fists (23° Alt.)	1½ fists (17° Alt.)
48°N	2 fists (20° Alt.)	1½ fists (14° Alt.)

FIGURE 6.10: Determining the solar window or shade pattern. (Courtesy The Energy Task Force of New York)

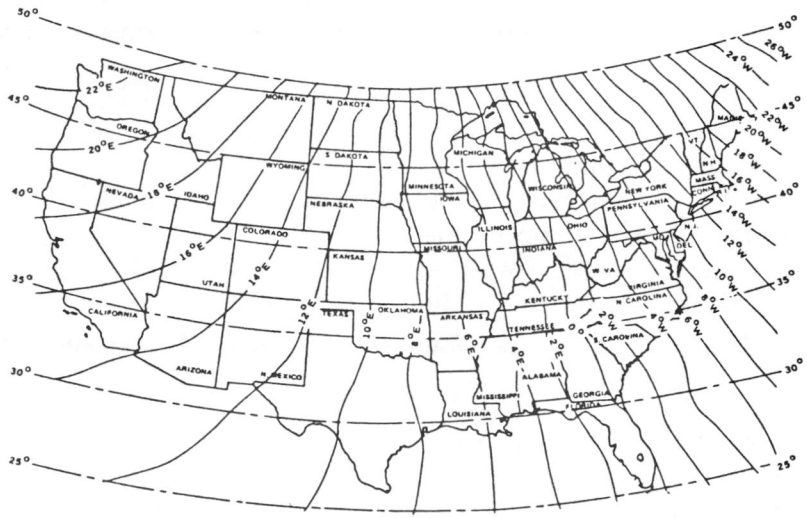

FIGURE 6.11: Map showing the difference between magnetic north and true north. (For example, central Texas magnetic north compass reading is 10° east of true north.)

MOUNTS

An Overview

The purpose of an array mount is to secure the modules to a particular location. To reduce wire losses and keep cost down, the ideal location is almost always as close as possible to where the power will be used. In some cases, mostly in utility-sized power fields, the modules are mounted where there is more sunshine, such as in the desert. Since the power is to be transmitted over miles of wire, location at point of use in this instance is secondary. However, residential PV systems usually combine production and point of use because it makes economic sense.

The most common mount used in residential applications is the stand-off roof mount. It's low in cost, easily built and installed, and often the most efficient use of materials. It is a good idea, however, to consider mounts in general, including some of the less common styles, to see how they might better serve your needs.

What are the general types of mounts? There are roof mounts which will be discussed shortly. There are also ground and pole mounts. Then, there are non-standard mounts. One non-standard type is the portable mount for small, lightweight modules used by campers and the military in field exercises. For example, a radio operator can hang a small module over his back and keep a small gel-cell or nickel-cadmium battery pack charged while hiking.

Another portable mount is simply a standard solar module with a carrying handle or strap. One farmer in Nebraska fastened a handle to the metal frame of a 35-watt module to carry it around and charge batteries on tractors and DC battery-driven motorized hay conveyors, and recharge electric fence batteries located on his property.

When I lived in Arkansas, I would pack a module and tv set on my horse. At a sunny patch in the woods, I'd set up for a little football watching. The 12-volt DC tv worked fine directly from the module. The module carrying strap was just a piece of baling twine.

Some PV modules on recreational vehicles are roof-mounted with wing nuts for easy removal. The driver can park in the shade and set up the array in the sun. Another movable mounting is a set of legs or folding sawhorse frame stored until needed.

Boat mounting PV modules can be a problem, as small boats do not offer much space. One way to get around this problem is to mount the modules off the transom or rail. There are kits available for rail mounting. Deck mounting of PV modules on boats is a common but questionable practice. Some modules are made with a plastic superstrate so they can be walked on, though this increases the likelihood of damage. A slippery module doesn't seem such a wise thing to have on deck.

A neat solution to limited deck space is to fasten modules together with standard door hinges. Four modules can be folded into a 1' x 4' x 8" thick bundle. This folding array can be pre-wired and, by using automotive-type quick disconnects, can be quickly put into use. When not in use, it can be stowed below deck.

Of course, when modules are stored away, they are not charging batteries. If you do use a portable mount, be sure that you are not letting your batteries go bad. Partially discharged batteries can be brought up to full charge only if they are connected to PV modules in the sun.

One recreational vehicle owner who wanted emergency lighting for his home has what might be called a mobile array. He parks his RV next to the house and uses a trailer-type connector to plug into his house. (The 12-volt DC circuits and fluorescent lights in the house are separate from the main house wiring.) Thus, his RV batteries and array are available for home use, too. Since he always has excess PV power, he has gotten into the habit of using the DC lighting instead of grid power at home.

Another portable mount is a trailer with a PV array (and sometimes batteries). Farmers with stock watering wells at various locations use trailer-mounted arrays and move them from well to well. The SUNRUNNER solar vehicle (see chapter 14) was transported on a trailer with batteries and inverter which could be pulled up to a house for emergency power.

Small ground- or pole-mounted arrays can provide portability. It is possible to back a pick-up truck to a four-module solar tracker mount, load it, and move it to another location. Thus, one array can serve two battery banks.

Temporary mounting is sometimes necessary, but it is important to make a temporary mount as strong as a permanent mount. An unsecured mount is dangerous, and it can also be costly should it fall or blow over.

I know some people who are using PV for water pumping. In a hurry to get the well in operation, they set up their modules temporarily by putting metal fence posts in the ground, stringing wire across the posts, and leaning their modules against the wire. A wind came from behind and blew the modules over. Luckily none broke. If you are planning a temporary test, lay the modules on the ground.

If a glass-encapsulated module should fall and the glass front break, it may still be serviceable. Test the output of the module. Be careful not to cut yourself or drive bits of glass into the cells. If you still have current and voltage, use a clear silicon or epoxy sealant suitable for outdoor use to re-seal the module. The goal is to keep air and water from the cells and interconnects. By running a bead of silicon around the perimeter of the module and hermetically sealing on another glass front, you can get many more years of service. The module won't look as nice, but the electricity will be just fine.

Whenever setting up an array mount, consider how you will ground the array frame for lightning and accident protection. Grounding is covered in the wiring chapter.

Chapter 11 contains more information on mounting.

Tracking Mounts

There are two types of tracking mounts, active and passive. Active trackers are motor-driven. Their one advantage is extreme accuracy and stability in high winds if clock-and-gear or chain driven. Their major disadvantage is that they use power to track.

If you are concentrating or using reflectors, a tracker is generally needed to keep the point of focus on the cells. However, I do not recommend concentrating or intensifying solar input to standard modules. The life of the module may be shortened because of heat and stress on the materials (solder interconnects, wire, substrate).

Manual tracking is also possible. If you are at the array site daily, you can tilt or point the array to the east in the morning, south at noon, and west in the afternoon. This will definitely increase the output of a small array. Do not forget to return the array to the true south position when you are no longer manually tracking.

Generally I recommend using a passive solar tracking mount. Even if your site does not allow for a ground-fastened pole-mounted passive tracker, consider attaching the tracker pole to your home or using a custom roof-mounted tracker. If your site has winds over 100 mph, however, a tracker may not be suitable. (See the Roof Mount section.)

One tried-and-true passive solar tracker is the Track Rack developed by Steve Baer of Zomeworks, a pioneer in the solar energy field. I first discovered his work when I was building geodesic domes in 1970. Steve had developed what he called zomes, zonal structures which expanded on the geometry of domes. Steve and Holly's zome home with drum-wall passive solar heating and movable insulation is an early shining example of design creativity.

The Track Rack is beautiful in its simplicity. The drive mechanism is two canisters of freon connected by a tube. Freon with its low boiling or gas point is used to shift weight from one side of the tracker to the other. In the morning, the sun shines on the canisters located at the extremes of the black frame. One side is shaded by a reflector/shade. The freon in that side is liquid. The freon refrigerant in other canister exposed to the sun, being hotter than the east canister, is gas. Gas is lighter than liquid and the tracker is weighted to the east. During the day the balance point of the tracker shifts as the amount of freon, liquid to gas, shifts. Accurate to within 5°, the tracker follows the sun arc across the sky 120°. When the tracker reaches the extreme west position, it stays there until morning. In the morning the sun is shielded from the east canister and heats the west canister from behind. In about 30 minutes the tracker automatically returns to the east to begin the day's tracking again. With no motors or pistons the tracker is simple, reliable, and energy-frugal. The increase in PV power production with this tracker averages 30% year-round over a fixed latitude mount. The improvement ranges from 10% to 20% in winter to 40% to 50% in summer. For an eight-module tracker that's like having 1.6 extra modules in winter and 3.2 more modules in summer or an average of 10.3 modules for the year. In all cases, the tracker more than pays for itself. Zomeworks guarantees its tracker for ten years. It uses specially made shock absorbers to dampen motion in high winds, and at about 30 mph will move out of the wind by itself.

In cloudy winter weather, the tracker will increase production about 10%. If you live in a cold, cloudy location like the northeastern U.S. and your primary loads are winter lighting, a tracker might not

FIGURE 6.12: Zomeworks' 24-module Track Rack. Installed by Pacific Energy, San Luis Obispo, CA. (Courtesy Zomeworks Corp.)

be suitable were it not for the ease in mounting modules (the tracker comes with pre-drilled holes). When ordering a Track Rack, or any mount, specify which modules you will be using to insure proper hole alignment.

As an extra benefit, the Zomeworks tracker can be tilted up and down from 15 to 45°. Making the altitude tilt adjustment is easy and worth the effort. Some people plan their twice-a-year system check and tilt adjust for March 15th and September 15th. They check their wiring and array, and perhaps clean the modules. The extra minute to adjust the tracker is rewarded by approximately 10% more yearly production.

If you require more power in summer anywhere in the temperate zones, a tracker is ideal. On a clear summer day, track-mounted modules at a fixed latitude mount will deliver up to 55% more energy. This means more water pumping or more fan ventilation.

The tracker, however, must have a clear sun window. Shading will interfere with its movement. If you have trees or a mountain ridge blocking the sun at the extreme east or west, the tracker may still be a good choice because it will help you grab as much PV power as possible at your less-than-perfect site.

The Zomeworks tracker is made to hold four, eight, twelve, and twenty-four module configurations. If you buy a Track Rack, consider the next size up from what you think you need. The added cost is insignificant and it's nice to have room to expand.

What are other methods of tracking? The large solar arrays built by ARCO Solar are examples. The world's first one megawatt power field employs two-axis tracking. These computer-aimed motorized large mounts follow the sun from east to west and seasonally adjust up and down. They also move to a flat position in winds over 30 mph to relieve wind loading. When flat, the modules still produce electricity.

The ARCO array field at the Carrisa Plains 6.5-megawatt power station is the world's largest PV production unit. It also uses two-axis tracking and has reflectors on the mounts to enhance solar collection. In general, this scheme is not cost-effective, due to loss of cell efficiency caused by heating plus the added cost of reflectors and mounting. The most cost-effective large tracking array is sited at the Sacramento nuclear power plant (see Figure 1.5). This two-megawatt PV power field uses horizontal single-axis tracking. The mount looks something like a handrail parallel to the ground and the modules move east to west. Such a mount is simple and low in cost. Even on a south-facing hillside (20 to 45°), this type of mount would be easy to install.

Concentrators and Reflectors

Almost every concentrating PV array scheme requires tracking adjustments which add to the cost and maintenance of the solar array.

Hybrid arrays appear attractive because they can produce both electricity and heat. Some hybrid concentrating systems seem simple but are actually complex when compared to the conventional fixed array mount. Seasonal adjustment of the array will require some form of flexible plumbing or ducting of the thermal production.

In hybrid PV, the heat which must be removed from the cells for higher efficiency can be used to heat water or air. Preheating water with the excess heat created by reflecting or concentrating the sun's rays on the cells means a water circulation pump is needed. In addition, a hybrid module with plumbing must be well-constructed to insure trouble-free operation. There has been slow acceptance of both concentrating and hybrid PV for residential use. Whether this indicates doubts PV technicians have for these schemes or the primitive state of the art is hard to say.

Photron, Inc. has developed a low-cost reflector system that enhances the output of a photovoltaic array. The Photron reflector's 0.020-inch thick aluminum bounces additional sunlight onto the array and can increase annual production of small arrays by 30% or more (see Figure 6.13). Though it requires additional mounting hardware and occasional seasonal adjustment, this "fractional sun" (low concentration) trough reflector is a good way to squeeze out a few more watt hours. The manufacturer says that 6 reflectors costing $240 will boost the output of a 4-module array to the equivalent of 5.6 modules.

A word of caution when reflecting. Commercial-grade solar modules are made to be used without significant solar enhancements. Substantial reflecting can cause overheating. Be sure to get verification in writing that your module warranty will remain valid if you use reflectors or concentrators. As with trackers, reflector enhancement performs poorly in predominantly cloudy climates, but since solar input is often less than optimal, properly designed reflectors should not cause overheating.

Pole Mounts

Generally, a pole mount is used when the pole is already being used for some other purpose. For instance, an amateur radio operator could mount his solar modules on his antenna pole. The photograph of the home in Figure 4.1 shows three four-module manually adjustable pole mounts. The units sit on top of the pole and have locking nuts for various tilt angles. Figure 11.6 shows a pole-mounted two-module array being installed by students at Colorado Mountain College.

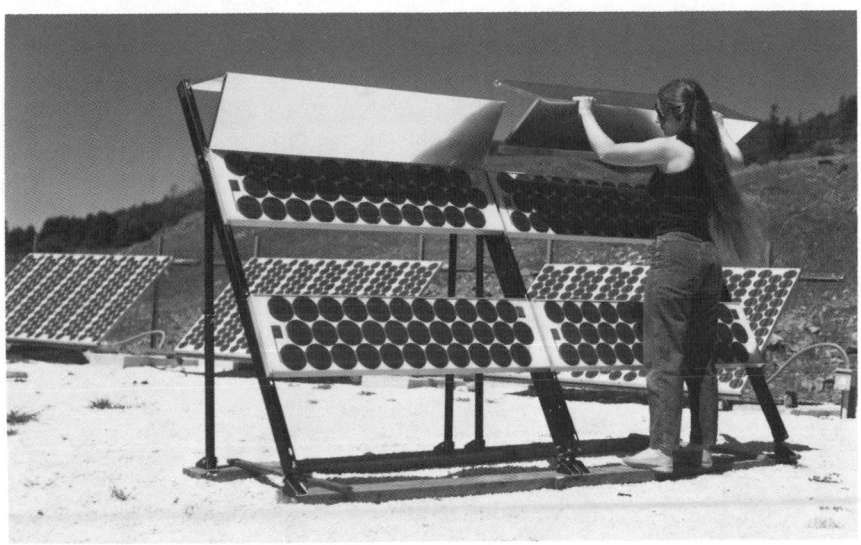

FIGURE 6.13: The installation of reflectors on small solar arrays can increase production. Reflectors should be made of durable, low-cost materials which do not scratch and shed dust and dirt. Fragile glass mirrors are not recommended. The Photron Powerflex combination reflector and adjustable mount provides more power at a lower cost than adding solar modules. (Photo: Photron, Inc.)

Ground Mounts

While ground mounting allows for ease in installation and adjustment, a ground-mounted glass-encapsulated solar module could be damaged by a rock flung from a lawn mower. Theft is another potential problem. (Most people include their solar electric equipment in their homeowner's insurance. Since coverage for fire, thief, vandalism, and accidental damage is reasonable in cost, it is the reasonable thing to do.)

Ground mounts can also be objects of attraction for the curious. At a fair, a ground-mounted array was used to provide power for the musicians. It was an ideal demonstration of PV's portability and power. The solar array also looked like a great play slide to some children. Fortunately, the modules were strong enough to take the load.

FIGURE 6.14: Typical pole mount structure. Some manufacturers use a horizontal framing member at the base of the array, but that eliminates the seasonal adjustment. (Photo: Solavolt International)

Ground mounts do have their place. Many solar arrays used for telecommunications systems on windy mountaintops are ground-mounted. Their low profile and sturdy fastening to concrete footings make them ideal power sources. Also, if you live away from the hustle and bustle, a ground mount is acceptable, since it may be the first and fastest way to get your PV system into action.

Roof Mounts

Roof-mounted solar arrays are the most common. They are easy to fabricate, though store-bought standard roof mounts are available. You can use solar water heater mounting hardware, although it is generally over-engineered for the lighter-weight PV modules.

There are several advantages to using roof-mounted arrays.

1. No additional land or building costs
2. Close to point of energy consumption
3. Less subject to vandalism, theft, accidental damage
4. Usually have an unobstructed sky window
5. Most homes have sufficient south-facing roof

FIGURE 6.15: Typical ground mount structure. Note the adjustable rear legs for seasonal alignment.

FIGURE 6.16: This owner-built ground mount structure shows simplicity in design and ease of construction. Especially with small arrays, it is easier and less expensive to make your own mount. (Photo and design: Bruce Wheeler)

FIGURE 6.17:
Whether ground-or roof-mounted, the solar array frame siting should be set up to minimize snow build-up on modules. This array was raised four feet above ground level to keep snow accumulation from obstructing the modules. (Photo: Robert Sardinsky, Rising Sun Enterprises)

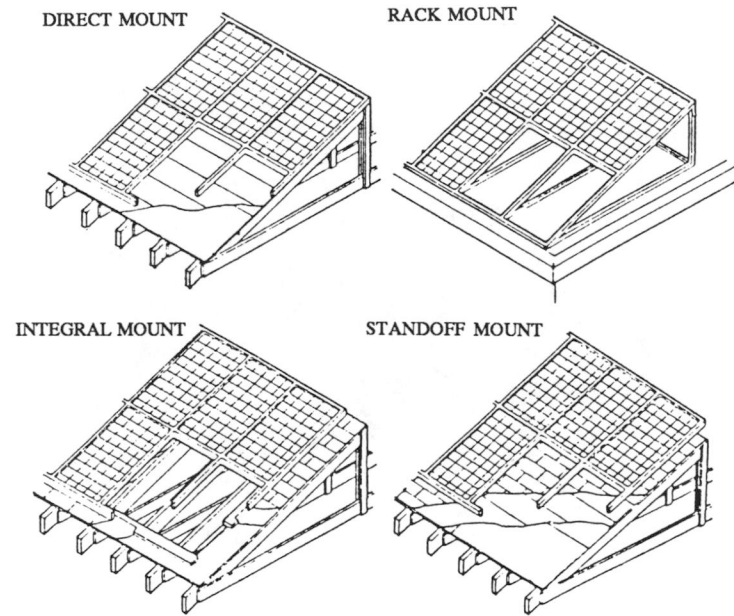

FIGURE 6.18: Four basic array mounting schemes. (T.E.A. Inc. report for Sandia National Laboratories)

FIGURE 6.19: These solar modules form part of the roof covering of this house. To prevent water seepage horizontal seams were first plugged with glazing tape and then caulked. Installer: Peter Talmage. (Photo: Robert Sardinsky, Rising Sun Enterprises)

Rack mounts can be installed over the existing roof surface with a minimum of penetrations. Sometimes called high-profile fixed stand-off mounts, they can be used on flat or pitched roofs. For long life, rack mounts should be metal. These mounts are low in cost, easy to install, able to withstand wind and snow loading if properly designed, permit passive module cooling (since the back of the array is exposed), and can be hinged to provide seasonal tilt. Though foot traffic is rough on all roofing materials, rack mounts do not otherwise interfere with the roof. Properly sealing fastener mounting holes will prevent leaks.

Stand-off mounts are the least expensive. The Northeastern Residential PV Experiment Station developed a very low-cost stand-off to be used for tract housing. Similar to low profile solar water collector mounts, stand-offs have two drawbacks: the roof pitch defines the tilt angle of the array, and future roof maintenance may be difficult. Photron, Inc. has developed a stand-off mount that costs under $40 per square meter.

Integral mounting is often praised because the modules double as roofing material. For this type of array, the modules must be structurally strong. As there is much expansion and contraction in a

FIGURE 6.20: Fastening PV mounts directly to the roof framing members through the roofing material will insure a strong mount. Be sure to use sealants at all roof penetrations to prevent leaks. (Photo: Photron, Inc.)

solar array, it is difficult to make a leakproof integral mount. The concept of using modules for the actual roof to offset some construction costs is appealing during the planning stage, but those who have actually tried to make a leak-free PV roof have found no real savings.

Direct mounting to the roof sheathing makes inside wiring possible and provides a low profile. Back cooling is difficult to provide, except with soffit and ridge venting. Flashing similar to that used with skylights is necessary. The small savings in roofing material and mounting hardware, as compared to rack mounts, has not helped make this scheme popular.

Another form of direct mounting, shingle-style PV modules (not illustrated)—conceived by General Electric, Westinghouse, and others, with a predicted mass production price of $1.50 per watt— never came to be. The actual shingle modules produced for residental testing were very expensive. The many interconnects exposed to the elements (even under the shingles) could collect moisture and cause problems. Large metal plates, such as those manufactured by ARCO Solar for the 7.5-kW 1978 John Long house in Arizona, have also been discontinued.

Tracking has also been tried on a few roof-mounted arrays, and seems to work well (see Figure 6.21). However, the added cost and greater roof area needed to keep the modules from shading each other require a roof large enough to hold the whole configuration. Before considering a roof-mounted tracker, be sure your roof can handle the increased dead load (equipment weight) and live loads (wind, snow, and workers).

FIGURE 6.21: Part of a roof-mounted solar passive tracking solar array. This piston-driven tracker has been installed over the existing roof shingles with stand-offs. Annual production is increased by 30%. Designed and installed by Don Loweburg, Off Line, Independent Energy Systems. (Photo: Off Line, Independent Energy Systems)

It is important to plan ahead when designing your array mount. Leave room for your system to grow. For example, many people using eight modules buy the twelve-module tracker, which only costs about $100 more. Roof-mount racks can also be ganged together to keep the array appearance uniform. Most imporant, make sure your mount is strong. If the mount cannot safely support your weight, do you trust it to support your modules?

FIGURE 6.22: A simple solution to a difficult mounting problem. This 1188-watt array on three separate mounts has a low profile. Located on Maui, Hawaii, and producing over 5 kWh per day, PV powers all of this large home's needs, including satellite dish with motorized tracking, color tv, swimming pool pump, DC refrigerator/freezer, dishwasher, laundry and lights. A back-up generator automatically charges the large 1175-amp/hour battery bank. Designed and installed by James Whitcomb and Rob Hilbun. (Haleakala Resources, Inc., Kahului, HI)

Regulators

The regulator maintains the desired power output from the solar array to the battery bank. It keeps the batteries from being overcharged. This is necessary because properly designed medium- or larger-sized solar electric systems will produce excess power at certain times simply because of differences in solar input (daily and seasonally) or due to weather variations. Some regulators also prevent the batteries from being too deeply discharged by shutting off the load circuits. Low voltage cut-outs make it impossible to use any power until the batteries are recharged.

There are two basic types of regulators: shunt regulators and series regulators. Series regulators utilize a relay or switching transistor in series between the solar array and the battery. Shunt regulators generally are solid-state devices using a transistor to control the flow of array current shunting (turning aside) excess power from the array to a ground or to dissipate it as heat or switch it to another load.

Some regulators also have a float, trickle charge mode of operation. When batteries are nearly fully charged, this two-stage

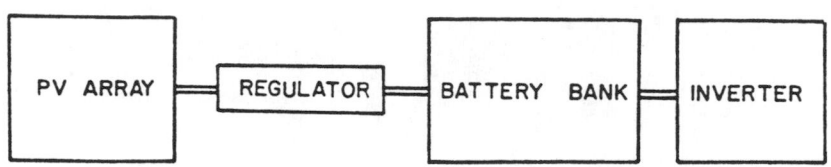

FIGURE 7.1: The wiring of a simple photovoltaic system using a regulator.

regulator will cut off or shunt most of the power being produced by the PV array. However, it will still continue to charge batteries with a float or finishing charge, usually 1% to 5% of the battery ampere-hour capacity. This small charge keeps your batteries "topped off." Some two-stage series regulators have a current limit on the float charge mode.

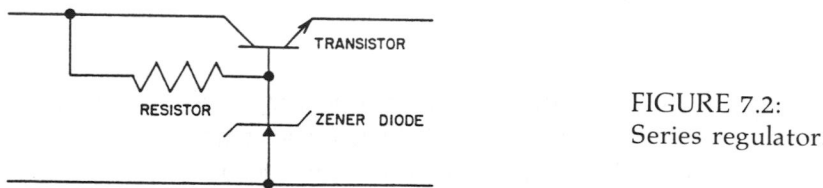

FIGURE 7.2:
Series regulator.

Shunt regulators are like automatic switches which turn the output of the solar array on and off. They are usually simpler and less costly than series regulators. (*Practical Photovoltaics* contains plans for building your own.) If the upper and lower setpoints of a shunt regulator match the battery full charge voltage and charge resumption voltage, a full charge can be obtained.

Do you need voltage regulation? Yes. There's no doubt about it. You will need to regulate your power production to protect your batteries and insure their long and useful life. A solar array sized to handle winter lighting needs will produce significantly more power in summer. This additional power will overcharge batteries. The question is really not if, but what kind of control you need.

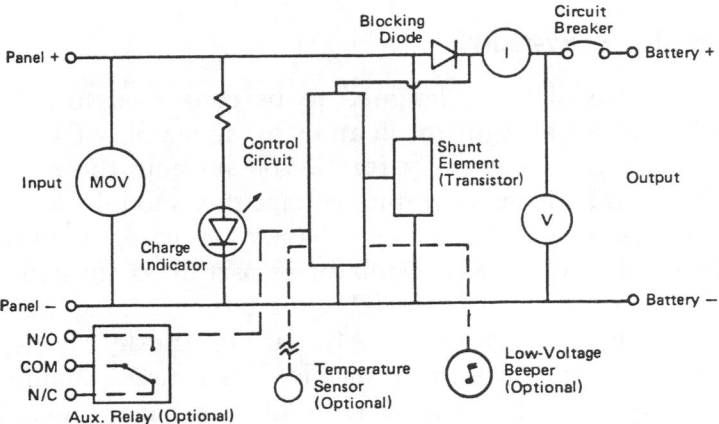

FIGURE 7.3: A shunt regulator. When the battery is fully charged, the shunt transistor conducts across the solar panel positive and negative inputs and no power reaches the battery (Courtesy Specialty Concepts Inc.)

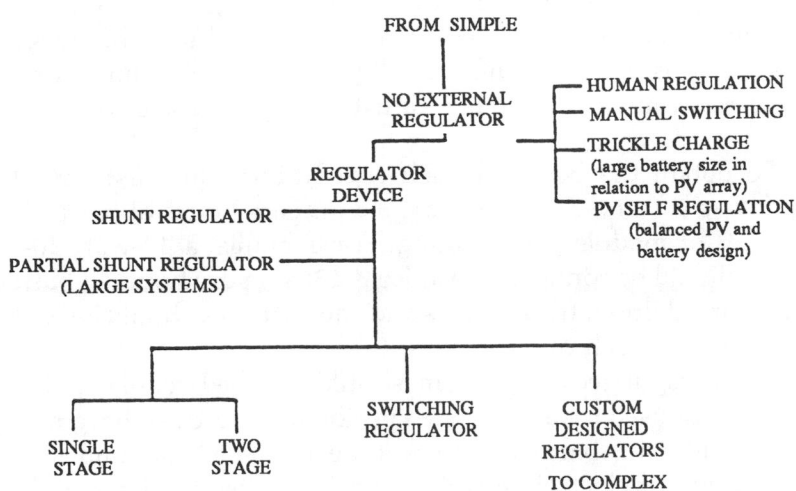

FIGURE 7.4: Regulator decision flowchart.

PV SYSTEM REGULATION

Self-Regulating Systems

Some solar modules are designed to be connected to a battery without an external regulator in an effort to simplify PV systems and to reduce costs. Generally, for 12-volt systems, these modules have 30 cells and a lower operating voltage than modules with more cells in series. The modules do not have a regulator "built-in": no well-designed module has a regulator as part of its construction.

The principle of self-regulated battery charging has been known for many years. Small AC battery trickle chargers for automotive batteries use this principle. These tiny 1/2-amp units keep a car battery up to charge without overcharging because of their low current input and float voltage of 13.5 volts or less.

When charging a discharged battery, self-regulating PV modules produce approximately the same power as other modules. One reason is that a power-consuming controller or regulator, usually with a blocking diode, is not used.

In operation, the charging current provided to a battery by a self-regulating module decreases as the charged battery voltage increases. Thus, a battery at a low state of charge will receive more charging current than one which is at or near a full state of charge. If the system is used during the day, the PV charging current automatically and continuously adjusts to try to replace the energy used.

To obtain the benefit of self-regulation, you must have the proper size battery. Figure at least 3.5 ampere hours of battery for each watt of module power rating. For example, a 42-watt 30-cell module should be connected to at least 130 ampere hours of battery storage. In addition, the modules and the batteries should be within 20°F (11°C) of each other.

A self-regulated PV system should be used regularly. If the system is not used for an extended period of time, overcharging will occur. While the battery may not be damaged, an undesirable consumption of electrolyte will result. If ignored too long, and the battery plates become exposed due to loss of electrolyte, permanent damage will result. Battery plates, once wet, should never be allowed to dry.

An advantage of using a self-regulating PV module is the reduction in module cost (since less material is used in its construction) and a reduction in balance-of-systems costs (due to the elimination of a regulator). There are limitations to the use of these modules. They work well with small systems under 100 watts, but there is a point where matching batteries to modules is not economical. An inexpensive regulator can be a better option than additional batteries which are heavy and take up space.

Blocking Diodes

Now is a good time to look at the varied opinions about blocking diodes and their associated 0.4 to 0.6 voltage drop. Some people say you must have a diode. Others say it is not necessary. Some say you should use zener diodes because they have less resistance and therefore less voltage loss.

The purpose of a blocking diode is to prevent reverse current. That is, the diode will stop any current from leaving the battery bank and going back through the solar module. When can this occur? Whenever the voltage of the array is less than the voltage of the battery bank, current will flow from battery to array. This happens at night.

But is this a problem? Generally, the nighttime losses increase as the size of the array increases. Therefore, a larger array will have more reverse current losses, assuming a voltage difference between the array and battery is present. This is easy to understand. Just as a pond with a large surface area will have more evaporation than a small pond with less area, so too will a larger array have more surface to dissipate power.

However, there are a couple important factors that must be considered. First, what are the losses? Second, are the nighttime losses greater or less than the daytime losses through a blocking diode?

The losses are generally under 1/200th of the daytime production. Factors involved in nighttime losses are the length of the dark period and the circuit resistance and voltage. Since length of day and night vary, a rule of thumb is that if your nighttime periods are longer than daylight periods, use a blocking diode. You can measure your nighttime losses with an accurate digital ammeter. If you measure, for example, 0.12 amps on a 12-volt system and have 12 hours of darkness, then the losses would be 0.12 x 12 x 12 = 17.28 watt hours.

Another rule of thumb is if the daytime loss from a blocking diode is greater than its nighttime savings, don't use it. To estimate your maximum daytime diode loss, multiply your peak charging current times the voltage loss across the diode (usually 0.5 volts) times the locale's peak sun hours.

For example, a 12-volt 10 peak-amp array can lose 25 watt hours production in a 12-hour day through the diode. (10 amps x 0.5 voltage drop x 5 peak hours per day = 25 watt hours daytime diode loss.) With nighttime losses of 1/100th of daily solar production, that same array may lose 72 watt hours of power at night if no diode is used. (10 amps x 12 volts x 5 peak hours = 600 watt hours divided by 100 x 12 night hours = 72 watt hours nighttime loss without a diode.) These figures are only examples. Check with your module supplier for specific recommendations.

Trickle Charge Regulation

If your array's peak output current is less than 5% of the 20-hour ampere hour rating of your batteries, you generally do not need a regulator device. For example, if you are using one 30-cell 30-watt PV module on your sailboat to keep a 105 ampere hour deep-cycle marine battery charged, the module will not overcharge your battery. This is basically the same as self-regulation, so be sure to check the battery electrolyte regularly. Another example of trickle charging that has become very popular is the use of a small module, such as the 5-watt ARCO G100 Genesis amorphous module, hooked up to a car, truck, boat, or tractor starter battery.

Human Regulation

If you are on a limited budget, and are on-site all the time, careful monitoring of your battery with a voltmeter and a good hydrometer can substitute for a regulator device. You can tilt your array to regulate the amount of power going to your batteries, or you can install an on/off switch, checking your batteries frequently. The use of a switch instead of a blocking diode is also possible. Manual switching can even be used to bypass a diode or regulator device so that you can have, on occasion, full current or no current to your batteries. (Full charge current to equalize your batteries is necessary. Read the battery chapter for details.)

Regulator Devices

Most PV manufacturers and some distributors offer proprietary regulators, and most are interchangeable. Shop around for one that suits your specific present and future needs. Remember, it is always wiser to buy a regulator for future needs now rather than having to buy another regulator later because your expanding PV system outgrew the first one. Windchargers also use regulators, but many of the older types are not solid-state. They use energy-robbing solenoids and relays which consume too much current. This is also true of automotive regulators. If you get a schematic for a PV regulator and are handy with a soldering iron, you can build your own, though this generally proves uneconomical.

Specialty Concepts, Inc. Model SCI-1 is a dependable regulator. Other firms make similar devices. The SCI-1 is a two-step series charge controller. That is, it provides full array current to the batteries until they are nearly fully charged. Once the battery bank is almost fully charged, the charge current is reduced to a trickle charge to finish the charge cycle. All wires come into and go from the SCI-1 allowing it to be a junction or DC distribution panel, too.

When using a two-step regulator, power from the PV array, within the current rating of the regulator, goes through the regulator circuitry to charge the batteries. When the batteries are nearly fully charged, an indicator light, usually an LED (light emitting diode), lets you know whether trickle charge is occuring. If you have metering on your system, you will see that charging current has significantly dropped, which is normal. When battery voltage drops below the charge resumption setpoint, the regulator once more goes to the full charge mode.

In the case of the SCI-1, should voltage drop below a user setpoint, low voltage disconnect will occur. Loads through the regulator will be shut down to prevent deep discharging that could possibly ruin the batteries. Life expectancy of batteries is reduced in relation to the depth of discharge. In the case of lead-calcium batteries, total draining can cause irreversible loss of capacity.

When charging brings the batteries back up to the setpoint, the automatically disconnected loads will be turned on again. If loads are larger than the controller's load terminal rating, a secondary external relay can be used to take advantage of the disconnect feature. This is most useful when operating an inverter directly from the battery bank. Simply use a relay with contacts rated for the inverter output. When your battery voltage drops to the preset point, the relay will shut down opening the inverter circuit.

Solar Works! makes the Charge-a-Stat regulator that can switch your array on and off. It can also be used with windchargers and as a diverter to either charge an auxiliary battery bank or another load. The nice thing about the Charge-a-Stat is that it can be repaired by anyone who can read a schematic. It comes in an openable box.

Many manufacturers offer a wide range of options such as temperature compensation, standby generator start-up, low-voltage alarms and audible signals, meters (both analog and digital), circuit breakers, and switches. Regardless of manufacturers' claims, it is essential to regularly test the specific gravity of your batteries with a hydrometer.

TROUBLESHOOTING GUIDE
(Courtesy Specialty Concepts, Inc.)

The following information has been adapted from Specialty Concepts' field troubleshooting guide for SCI chargers. The complete guide is included with every regulator.

General

1. Check system wiring to insure proper polarity and that all connections are sound and with minimum voltage drop due to corrosion or loose connections.

2. Check to see that modules and batteries are in the correct series-parallel configuration for proper system voltage and current. Review instructions and specifications, array output, load ratings, and system sizing to insure that ratings are not exceeded.

3. Check that the array is not partially shaded or dirty.

4. Check all system fuses and circuit breakers.

SCI Charger Model 1™ (SCI 1)

DESCRIPTION

The SCI CHARGER MODEL 1 (SCI 1) is a versatile controller for the efficient use of photovoltaic energy and the protection of expensive batteries. It is available in 12, 24, 36, and 48 volt units for small to medium sized applications with charge currents up to 30 or 50 amps.

The SCI 1 consists of a series-relay battery charge regulator with low-voltage load disconnect, a load fuse, and status lights. The lights indicate "CHARGING" and "LOAD DISCONNECT" conditions, providing system status information at a glance. The SCI 1 is housed in an anodized aluminum chassis, suitable for indoor wall mounting, with a terminal block for up to 10 gauge wire or a spade connector, providing simple installation.

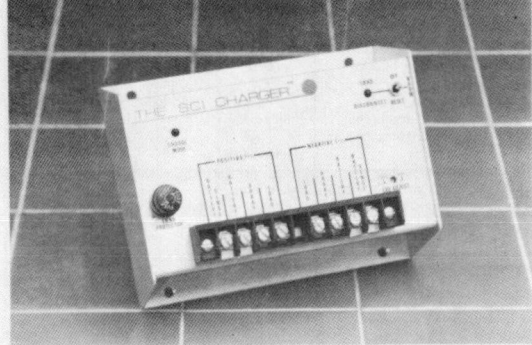

FEATURES —

* Two-step charge regulation (12 and 24 volt)
* Single-step charge regulation (36 and 48 volt)
* Maximum array usage
* Current compensated low-voltage load disconnect
* Adjustable low-voltage load disconnect with reset/disable switch
* Load fuse
* Status lights
* Input noise suppression
* Reverse leakage protection
* Lightning protection
* Reverse polarity protection
* Central wiring
* Many options

OPERATION

(12 and 24 volt units)

CHARGE REGULATION - The two-step control circuit regulates the charging of storage batteries by monitoring battery and array voltage. At sunrise, when array voltage rises, the charging relay enrizges and closes, connecting the solar array directly to the batteries and lighting the "Charging" Light Emitting Diode (LED). The battery will accept as much current as the array will provide, and battery voltage will rise. This is the constant-current (full) charge mode and is the first of the two-step charging sequence. When the bat-

teries reach the full charge termination threshold, the charging relay will open and the "Charging" LED will go out. At this point the float regulator takes over to keep the battery below the float voltage and supply limited current (maximum float current) to the batteries. As the battery approaches the float voltage, the current will taper off, eventually falling to the battery's maintenance current. This is the constant voltage (float) charge mode and is the second charging step.

MAXIMUM ARRAY USAGE - If a load is applied when the charger is in the float mode, the regulator will supply up to its maximum float current to maintain the battery charge. If the load is less than the maximum float current the batteries will still be receiving a net charge from the float regulator. If the load current is more, the battery will supply what the float regulator cannot and the battery voltage will fall. When it falls below the full charge resumption threshold, the charging relay will reclose, reinitiating the full charge mode. This insures that if a large load is applied during the day, maximum use will be made of the power available from the array.

CURRENT COMPENSATED LOW-VOLTAGE DISCONNECT - The low-voltage disconnect (LVD) of the SCI 1

prevents damage from over discharge of the batteries by automatically disconnecting the loads. The disconnect threshold is load current compensated by a factor of 10 mv/amp, and a minimum time delay of 3 seconds is applied to prevent false disconnects. When disconnect occurs, the load relay is energized and opens, and a red L.E.D., visible on the front panel, will light to indicate that the loads have been disconnected. Normal battery charging will continue. At the reconnect threshold the loads will automatically be reconnected and the red L.E.D. will go off. The LVD function has a reset/disable switch and user adjustable set points.

REVERSE CURRENT PROTECTION- About 12 hours after sunrise the charging relay will open and stay open if the battery stays above the reconnect threshold. If the battery goes below the threshold, the relay will reclose and open again every 2 hours to determine if power is still available from the array. This guarantees that the relay will be open every night, preventing reverse leakage current without cycling the relay during the high current part of the day or using an inefficient blocking diode.

The operation of a 36 or 48 volt unit is identical with the exception that no float circuit is included.

FIGURE 7.5: This is a typical regulator specifications sheet listing physical and electrical characteristics and performance. Note the options offered with this unit. (Courtesy Specialty Concepts, Inc.)

SCI 1 SPECIFICATIONS

NOMINAL VOLTAGES

PARAMETERS	UNITS	12v	24v	36v	48v
Charge Current, Max (1)	(Amps)	30	30	30	30
Load Current, Max (2)	(Amps)	30	20	15	15
Array Voltage, Max Voc	(Volts)	22	44	66	88
Operating Temp Range (3)	(°C)	0 to 50	0 to 50	0 to 50	0 to 50
Storage Temp Range	(°C)	−55 to 85	−55 to 85	−55 to 85	−55 to 85
Quiescent Current (4)	(Milliamps)	10	10	10	10
Current Consumption, Charging (5)	(Milliamps)	160	160	80	80
Current Consumption, Load Disconnected (6)	(Milliamps)	140	100	70	70
Full Charge Termination (7)	(Volts)	14.8 ± .2	29.6 ± .4	44.4 ± .6	59.2 ± .8
Full Charge Resumption	(Volts)	12.5 ± .2	25.0 ± .4	37.5 ± .6	50.0 ± .8
Adjustable Load Disconnect Range (8)	(Volts)	11.5 ± .5	23.0 ± 1.0	34.5 ± .5	46.0 ± 2.0
Load Reconnect Span	(Volts)	1.5 ± .3	3.0 ± .6	4.5 ± .9	6.0 ± 1.2
Float Voltage	(Volts)	14.1 ± .2	28.2 ± .4	NA	NA
Float Current, Max	(Amps)	3	1	NA	NA

(1) 50 amp option is available.
(2) Non-inductive.
(3) Operating temperature range may be extended to −20°C to 50°C on special orders.
(4) Both relays unenergized, red L.E.D.s off, typical value.
(5) Charge relay energized, red L.E.D. on, typical value.
(6) LVD relay energized, red L.E.D. on, typical value.
(7) Set points may be specified for nickel cadmium batteries or other special applications.
(8) Decreases by 10mv for every amp of load current.

PART NUMBERING KEY

DIMENSIONS
(In Inches)

EXAMPLE:

Model
Voltage
Options

SCI1 – 12 – AB

MODEL	VOLTAGES	OPTIONS	
SCI1	12	A	Temperature Compensation
	24	B	Load Circuit Breaker
	36	F1	Volt Meter/10 Amp Meter
	48	F2	Volt Meter/20 Amp Meter
		F3	Volt Meter/30 Amp Meter
		H	Generator Start
		I	Array Power Diversion
		L	NEMA 13 Enclosure
		M	NEMA 4X Enclosure (Clear Door)
		N	Load Timer
		O	Low Power Relays
		P	50 Amp Charge Current
		Q	Day/Night Switch

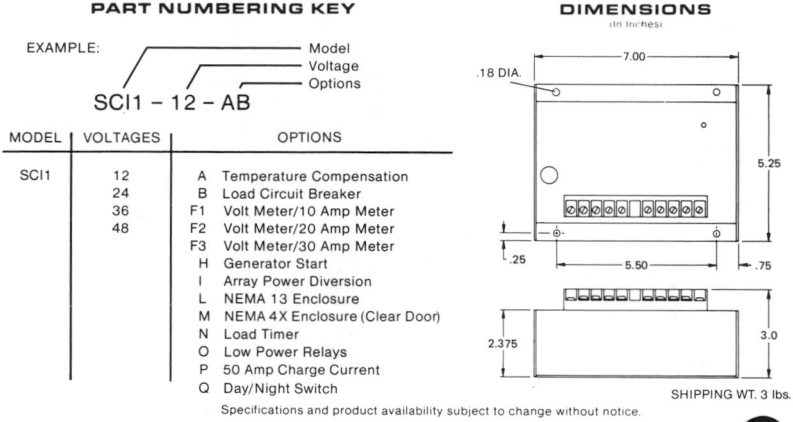

SHIPPING WT. 3 lbs.

Specifications and product availability subject to change without notice.

SCI
SPECIALTY CONCEPTS, INC.

SCI1 7/85

FIGURE 7.6: Solar Works! array and pressure system water pump control panel can regulate and switch 50-ampere charging current for a home power system. (Photo: A. D. Paul Wilkins, Solar Works!)

5. Determine if the regulator is operating at the correct setpoints by measuring the voltage at the battery voltage terminals when relay switching occurs. Allow for specification tolerances and the effect of temperature compensation (-5mv/deg C/batt cell) if your regulator has this option.

Symptom: Battery Undercharged

1. *Possible cause:* Batteries too cold requiring higher voltage to achieve full charge. *Remedy:* Insulate battery bank or replace regulator with unit with temperature compensation.

2. *Possible cause:* Not enough array for load. *Remedy:* Check system sizing and add modules as needed.

3. *Possible cause:* Charge rate too high causing battery voltage to rise too fast before charging is complete. *Remedy:* Add batteries to the system or bypass the regulator with a few modules to increase trickle charge current.

4. *Possible cause:* Battery capacity and ability to accept charge has been reduced by age or abuse. *Remedy:* Replace battery bank.

5. *Possible cause:* Excessive voltage drop to battery caused by high current, small wire, and long wire runs. *Remedy:* Replace wire with properly sized conductors.

Symptom: Battery Overcharging or Excessive Water Loss

1. *Possible cause:* Battery too hot, gassing voltage is lower than normal. *Remedy:* Insulate battery bank or replace regulator with unit with temperature compensation.

Symptom: Regulator Does Not Switch to Nighttime Mode

1. *Possible cause:* Lightning strike or other high voltage source damaged regulator. *Remedy:* Return regulator for repair.

Symptom: Load Disconnects Improperly

1. *Possible cause:* Loads such as inverters can generate electronic noise. *Remedy:* Wire inverters directly to battery bank. Add filtering to load.

2. *Possible cause:* Load has high surge causing battery voltage to drop briefly. *Remedy:* Check load specifications for load disconnect. Add battery storage to compensate for surge draw-down. Load surges should not exceed 25% of battery capacity. Example: 200-ampere surge load requires at least an 800 ampere hour battery bank.

3. *Possible cause:* Lightning damage to controller. *Remedy:* Return for repair.

Symptom: Regulator Relays Buzzing

1. *Possible cause:* Incorrect battery voltage. *Remedy:* Check series-parallel wiring of batteries.

2. *Possible cause:* Improper battery connection. Broken wires from battery. *Remedy:* Check for tight, clean connections. Check wiring.

3. *Possible cause:* Batteries dead. *Remedy:* Check state of charge of batteries with hydrometer. If low, connect array directly to batteries until charged, then reconnect regulator.

Symptom: Array Fuse Blows

1. *Possible cause:* Array short-circuit test performed with battery connected. *Remedy:* Disconnect battery from controller to perform short-circuit test.

2. *Possible cause:* Array exceeds rating of controller. *Remedy:* Add another regulator in parallel.

Symptom: Load Fuse or Circuit Breaker Blows

1. *Possible cause:* Load exceeds rating of controller. *Remedy:* Check surge rating of load. Check for shorts in load circuit.

Chapter 8

Battery Storage

A BRIEF OVERVIEW

Any discussion of batteries becomes complicated because there are so many factors involved. To make things even more difficult, myths and misinformation abound. But the more you know about batteries, the happier you will be. You will no longer see batteries as heavy black boxes filled with dangerous chemicals that somehow "make" electricity. Understanding batteries will help you select the best battery for your PV system and that will save you money.

A battery is an electrical storage device. Batteries come in many shapes and sizes, but they all have one thing in common: they store direct current power for later use. In the Short Course in Electricity (Appendix A), a water analogy is used to compare voltage and amperes to pounds of water pressure and rate of flow, respectively. Electrical resistance (ohms) is compared to the resistance that slows water flow in a pipe or hose. To carry the water analogy a bit further, compare a battery to a bucket. Just as you fill a water bucket for later use, so do you fill a battery with power.

Batteries that are used only once and then discarded are called primary batteries. Batteries that can be used more than once are called reusable or rechargeable batteries. We are concerned only with the rechargeable type known as secondary batteries. PV systems generally use deep-cycle secondary batteries.

Deep-cycle batteries are designed to be repeatedly cycled (or charged and discharged). A cycle is a complete sequence of operation, beginning with a fully charged battery. In use, the battery is "emptied" of its power. Then it is recharged or "refilled." True deep-cycle batteries are designed to handle repeated cycling.

An automobile starter battery is different. It is designed to provide a powerful short spurt of electricity to get the engine turning and running. At that point the alternator or generator takes over and recharges the battery right away. Starter batteries (or any battery) should not be left sitting around partially discharged. Nor should they be deep-cycled. A new auto battery will soon die if deep-cycled.

Batteries consist of cells in series, each nominally rated two volts. Thus, a 12-volt battery will have six each two-volt cells. A battery bank is an assemblage of batteries, sometimes containing many cells connected in series and parallel to obtain the necessary operating voltage.

Just as a bigger bucket means more water capacity, a battery's size is generally an indication of its storage capacity. The larger the battery bank, the more electrical storage. A battery bank weighing 1000 pounds will have more storage capacity than a battery bank weighing 800 pounds. However, weight is not the only consideration.

Allowable depth of discharge (DOD) is also important. A 100-pound battery that can be discharged to 80% of its capacity stores more usable power than a 150-pound battery with a 50% depth of discharge rating.

.The size of your battery bank is dependent on many factors such as the average daily amount of electricity you use, the number of days of autonomy required for your climate, the maximum load at any one time, the temperature the battery bank will be kept at, the type of loads you will be powering, the portion of your load to be used at night.and, of course, your budget. Since most locations have

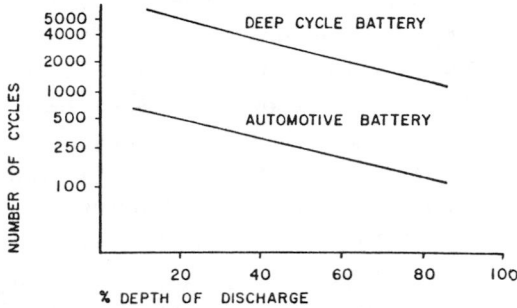

FIGURE 8.1: Deep-cycle batteries will last many years more than automotive starting, lighting, and ignition (SLI) batteries.

less available sunlight in winter, size your battery bank for winter requirements.

After you have determined the size of your battery bank (using the information in the sizing chapter), you are ready to make your purchase. You can buy new or used batteries for your system, though special care must be exercised when buying used batteries. When buying new batteries, be sure to tell the salesperson your intended use and ask for suggestions. If they do not know what you are talking about, shop elsewhere. Few battery dealers are familiar with PV systems but all of them are out to sell batteries. Buy what you need, not what they have on hand. Prices for good batteries do not vary much.

Buying used batteries requires special knowledge. Check battery appearance. Have they been kept clean and out of the weather? Do they show signs of physical abuse? Is the electrolyte clear and clean? Are the plates in good shape and not warped or brownish or white from sulfate? Are the terminals secure and in good shape? Do the cells test properly with a good digital voltmeter and hydrometer or are they completely discharged? How were the batteries used and by whom? Telephone companies usually take good care of their emergency storage batteries and often dispose of them well before the battery's useful life has ended. Is there a maintenance logbook for the batteries showing date of manufacture, service schedule, and so on? Be forewarned that some battery users, especially golf courses and warehouses that use battery-powered forklifts, do not maintain a regular service program. Batteries from these sources may mean trouble and are no bargain, even if free.

A battery's self-discharge rate is typically 10 to 15% per year.

This means that a battery will lose some of its power just sitting around. Life cycle depends on depth of discharge. If you regularly dip deeply into your battery storage, you will not get as many years of service as you would if you just skim a little power off the top and replace it daily. Some high-quality deep-cycle batteries have high (80%) electrical to chemical energy conversion and can be cycled hundreds of times before replacement. Automobile batteries are a poor choice for PV systems. They can only handle a 10% depth of discharge for about 200 cycles. While you may get years of very light cabin use from a starter battery, you are wasting PV power. Starter batteries have a habit of failing during the first real cold spell.

Some so-called deep-cycle batteries for recreational vehicles and boats do not perform particularly well with PV systems. A friend of mine bought some RV/marine deep-cycle batteries that lasted only six months. The batteries were ordered by phone and not tested when received. Purchased right before a cloudy winter, the battery bank was undersized to start. Worse yet, they were not tested regularly to check performance, so they eventually failed. This was not an isolated case. Several other people took advantage of the same sale and also experienced early failures.

Does this mean don't buy RV/marine batteries? No. It simply means be a knowledgeable buyer and test regularly. Batteries can be the weakest element in your PV system. Eastern philosophy teaches that in weakness there is strength. That is, we gain strength through understanding the weakness. And so it is with batteries.

The chemical reactions which occur in batteries can be better understood if you consider your batteries alive. Feed them well, keep them clean, warm and dry, check them regularly, use them properly, and they will prove dependable friends.

Batteries do fail. Eventually the best battery will grow weak and need replacement. Sometimes individual cells, even in new batteries, fail. If you have been testing and maintaining your battery bank properly, you will already be aware of a weak cell or the gradual drop in specific gravity over the years. You will have budgeted for the replacement of the entire battery bank, knowing that it is better to recycle the whole bank and start anew with an unmixed batch of new batteries.

There should be no surprises, except for your battery supplier. If he is still around, he may be surprised at how many years of useful life your batteries have given.

My first good battery bank was manufactured for the telephone company by C&D Batteries in 1968. The telephone company recycled them in 1978. I bought them in 1979 at salvage prices and used them in my Arkansas home. They were still in use when I left in November 1982. I wouldn't be surprised if some of those batteries are still in use somewhere in the Ozarks.

When you size your PV system and battery bank, plan for battery replacement. Your battery bank should last at least six years. If you save one-sixth of the cost of your battery bank per year, with interest you will have the cash when new batteries are needed. If you have oversized your battery bank and use it to only 20% depth of discharge, the same battery bank could last ten or fifteen years. In any case, battery replacement is a factor to consider with your PV system.

Be sure to arrange to recycle your old batteries. While no one would consider dumping them in a ditch and polluting the land, throwing them away is throwing money away. They have value as salvage material because lead is recycled. (As of this writing, lead is worth over 12 cents a pound.)

UNDERSTANDING BATTERIES

While it is believed that the ancient Babylonians used batteries, it is generally accepted that Allesandro Volta discovered the galvanic battery in 1800. Gaston Plante made the first lead-acid battery in 1859. In the last one hundred years batteries have gone from a scientific curiosity to a device found in every corner of the world.

The two main types of batteries are primary and secondary. Primary batteries that are used once—in flashlights, radios, and similar devices—and thrown away can now be replaced with rechargeables. Secondary or rechargeable batteries can be used over and over.

You are already familiar with at least one secondary battery— the automobile starter battery. This shallow-cycle storage battery is designed to deliver high current for a short period of time and then be quickly recharged by the alternator or generator. However, the deep-cycle batteries used in PV systems are very different. They can be more deeply discharged than automotive "SLI" (starting, lighting, and ignition) batteries.

Table 8.1: Lead-Acid Battery Types

Type	General Characteristics	Typical Applications
Automotive (SLI) and Diesel Starting	High discharge rate, relatively low cost, poor cycle life	Automobile starting, lighting and ignition; tractors, snowmobiles and other small engine starting; large diesel engine starting
Motive Power (Traction)	Moderate discharge rate, good cycle life	Fork lifts; mine vehicles; golfcarts, submarines; other electric vehicles
Stationary (Float)	Medium discharge rate, good life (years), some types have low self-discharge rates, poor cycle life	Telephone power supplies; uninterruptible power supplies (UPS); other standby and emergency power supply applications
Sealed	No maintenance, moderate rate, poor cycle life	Lanterns, portable tools, portable electronic equipment, also sealed SLI
Low Rate Photovoltaic (Special Batteries)	Low maintenance, low self-discharge, special designs for high and low ambient temperatures, poor deep-cycle life	Remote, daily shallow discharge, large reserve (stand-alone) photovoltaic power systems
Medium Rate Photovoltaic (Modified Motive Power Cells)	Moderate discharge rate, good cycle life, low maintenance	Photovoltaic power systems with on-site back-up or utility interface, requiring frequent deep-cycle operation

In a PV system the battery bank receives power produced by the solar array and stores it for later use. Batteries do not "have electricity" any more than a bucket has water. Electricity must be put into the battery by charging. You can get no more electricity out of a battery than you put into it. In fact, due to battery efficiency, you'll get a little less out. However, even with that loss, batteries are the best way to store electricity produced by a PV array.

How does a battery work? Basically, a lead-acid battery consists of a series of identical cells made up of plates of two types of lead. These plates are immersed in an electrolyte of sulfuric acid and water. When a battery is charged, there is a change in the chemical composition of the plates and "acid." This change is reversed when the battery is discharged. A battery stores electrical energy in chemical form.

Batteries are either "dry" or "wet." A wet battery has the electrolyte added at the factory. A dry battery has the acid added at the time of purchase or shortly before use. Developed for the armed forces in World War I, dry batteries ship and store easily and are generally fresher. However, to insure long life, dry batteries must be handled and stored properly. If kept in a humid environment, they will age faster and possibly be ruined.

Batteries begin to age as soon as the acid is put in. This aging, called "local action," is a normal process. Wet batteries lose part of their useful life just sitting in a warehouse. Aging of dry batteries kept in a dry environment is delayed until the electrolyte is added. The date of manufacture should be stamped on the battery, wet or dry. Buy only fresh batteries (purchased within six months of manufacture).

As mentioned, batteries consist of cells. Each cell has a nominal rating of 2 volts. Thus, a 3-cell battery is 6 volts, a 6-cell battery is 12 volts and so on. Some large batteries are actually 12-volt shallow- or medium-cycle starter batteries for use on heavy equipment, such as diesel engines. Some small 12-volt batteries, weighing under 50 pounds, are true deep-cycle batteries used to power small trolling boat motors. There are 350 pound or larger "batteries" that are actually large 2-volt cells.

Size is a factor when you transport and install batteries. The 6-volt "golfcart" battery is used in many home PV systems because of its availability and size, as well as its charge/discharge charac-

Table 8.2: Lead-Acid Battery Life

| | | | Life |
| | | Years | Cycles* |
Cell Type	Grid Type	Years	Cycles*
SLI	Lead-antimony	2-5	150-250
SLI	Lead-calcium	2-5	20-50
Motive power	Lead-antimony	5-15	1000-2000
Motive power	Lead-calcium	10-20	750-1500
Stationary	Lead-antimony	15	250-500
Stationary	Lead-calcium	15-24	100-500
Stationary	Plante	24-30	250-500
Photovoltaic	Lead-calcium	10-20	350
Photovoltaic	Pure-lead	5-15	—
Sealed	Lead-calcium	2-5	100-200

*To 80% of discharge at 77°F.

teristics. A friend once got what seemed like a great deal on six large 350-pound 2-volt cells. But by the time he hired two men, a forklift, and a truck to move the batteries, they were no longer a bargain.

Many early PV systems used 12-volt recreational vehicle or marine batteries for energy storage. They are readily available, not too heavy (50 to 70 pounds), and reasonably priced. With the availability of 12-volt appliances and equipment, early PV homes were completely or partially wired for 12 volts. While 12-volt cabin systems helped to popularize PV, they have their limitations and so have not gained general acceptance.

The 12-volt convention is not new. In 1953 American automobile manufacturers began to standardize 12-volt lighting and ignition systems. Earlier cars used 6, 12, 24, 30 and other voltages. Early windcharger systems used different voltages, with 32 volts common in the American midwest.

With all the 12-volt batteries around, it was natural for PV manufacturers to develop "nominal 12-volt modules." The combination of 12-volt PV modules, 12-volt batteries, and 12-volt automotive and marine equipment helped place PV into thousands of applications in remote locations all over the world.

The introduction of efficient inverters capable of powering conventional AC appliances has revealed the limitations of the 12-volt PV system. Twelve-volt equipment requires larger-sized wires placing practical limits on system size. The universal availability of 120-volt AC appliances and tools, and the desire to use them, have opened up a new range of possibilities—and a growing shift away from the 12-volt convention.

OTHER FORMS OF STORAGE

There is a lot of talk in the renewable energy field about new types of batteries and new forms of energy storage. Without discrediting these schemes, I have to emphasize the success of the deep-cycle battery storage system and its near universal acceptance.

While nickel-cadmium batteries (ni-cad) are excellent storage devices for electrical energy, their charge/discharge characteristics and high cost make them impractical for most PV systems. Some experimental batteries may have better features than the standard lead-acid battery, but price and availability prohibit their use. Flywheels, hydrogen, and other gas storage also have limited applicability in residences.

If you want a PV system now, limit your equipment selection to what is available. There is no practical reason to design your system around a concept that hasn't yet reached the marketplace. Such planning only delays the day you begin using solar energy.

This reminds me of the response a successful old farmer would give when I asked why a certain tool or method was not used locally. He'd say, "There must be some reason," meaning that it had been tried and abandoned. Before trying a new storage scheme, be sure it is valid and the equipment available, or your PV power system may be on hold forever. Best to get started right now with deep-cycle lead-acid batteries.

Direct-Drive

Some PV-powered equipment is used without batteries. Driven directly by solar electricity, water pumps, fans, and other devices display amazing simplicity.

A PV-powered water pump using a water tank for storage can perform 24 hours a day. Water pumped during the sunlight hours is stored in a tank or a pond to be drawn as needed. The advantages of direct-driven pumps are that water storage is less expensive than

FIGURE 8.2: This fan and small solar module blow excess heated air from an attached greenhouse into the home to provide supplemental space heating for the home. The fan can also be used to power vent an attic and help keep the house cooler. (Photo: Photron, Inc.)

electrical storage, that a tank costs less than batteries, and that the simple system requires fewer parts and no battery maintenance.

Solar-powered attic, ceiling, and ventilating fans are usually direct-driven. When the sun is shining, the fan turns. The same direct-drive principle is employed on PV-powered, solar thermal water heating systems. A good balance is achieved when the solar heater pump and PV module are matched to the water heater flow rate. The brighter the sunshine, the faster the pump circulates water through the collector. More sun means more hot water.

However, there are limits to direct-drive PV-powered equipment. Since the discovery of fire, man has extended his activities into the night. Over 2 million people work the night shift in New York City. Farmers often start work before daybreak; during harvest they may work all night long. People wake before daylight and go to sleep well after sunset. So a residential PV system must be able to provide power beyond the peak sun hours. It must also provide power on cloudy days and at night. Without battery storage most of the energy striking the solar array would be lost. A properly sized battery bank allows for this by collecting peak sun power and making it available at any time.

This sounds simple but is a difficult concept for people accustomed to utility power. We don't think about the equipment and people necessary to make utility power available. When we

FIGURE 8.3: Residential PV water pumping is cost-effective when the well is 1000 feet from power lines.

produce our own power, the 24 hour-a-day equipment operation becomes our responsibility.

Therefore, sizing the system and using the proper equipment are necessary for reliability. A stretch of cloudy days and an undersized battery bank means low, or no, power.

There's no need for such disappointments if you follow the sizing method in this book. A properly sized battery bank in an energy-conservative home will enable you to enjoy modern conveniences without wasting power or money.

BATTERY CHEMISTRY

In a fully charged battery, the active material in the positive plates is lead peroxide (PbO_2). The negative plates are sponge lead (Pb). Since the active material of both plates is very soft, they consist of a grid that typically contains some antimony or calcium to form a stronger alloy. The electrolyte is a dilute sulfuric acid (H_2SO_4). When the battery is charged, all the acid is in the electrolyte and its specific gravity is high.

As stored electricity is used and the battery discharges, some acid separates from the electrolyte in the pores of the plates. This acid combines with the active plate material becoming lead sulfate ($2PbSO_4$) and leaves behind water (H_2O). When the battery is discharged, the specific gravity is low.

When the battery is being charged, the reverse chemical reaction takes place. Acid is driven out of the plates increasing the specific gravity of the electrolyte. If charging continues beyond the point where the cells can accept energy from the charging current, the electrolyte is broken down into hydrogen and oxygen. This process is called electrolysis and is the principle reason water must be added to your batteries.

The chemical formula is:

$$\text{Discharge} \longrightarrow$$
$$PbO_2 + Pb + 2H_2SO_4 = 2PbSO_4 + 2H_2O$$
$$\longleftarrow \text{Charge}$$

From this formula we might conclude that nothing is gained or lost. Under ideal conditions such would be the case. However, some energy is lost and some sulfate remains in the plates forming a deposit. This gradual deposit or "hardening of the arteries" is part of the normal aging process that occurs when batteries are discharged. When batteries are overcharged or when an equalization charge is applied, outgassing of hydrogen and oxygen occurs. Proper use of batteries simply means minimizing the effects of these two extremes of charge and discharge.

Battery efficiency can be expressed in ampere-hour efficiency and watt-hour efficiency. In terms of ampere hours, the recharge rate should be 110% the discharge rate for 91% efficiency. However, voltage is higher during charging to "push" the current into the battery. Voltage efficiency is about 85%. Combining these two (91% x 85%) gives us a watt hour or total energy efficiency of typically 77 to 78%.

Do not let ampere-hour and watt-hour efficiency confuse you. If you are using ampere hours to size your system, new batteries will deliver 100 ampere hours for every 110 ampere hours of charge. In watt hours, for every 100 watt hours of power put into your batteries, you can expect 77 watt hours out. This applies to new batteries. If your batteries are near the end of their useful life, efficiency can be 50% or less.

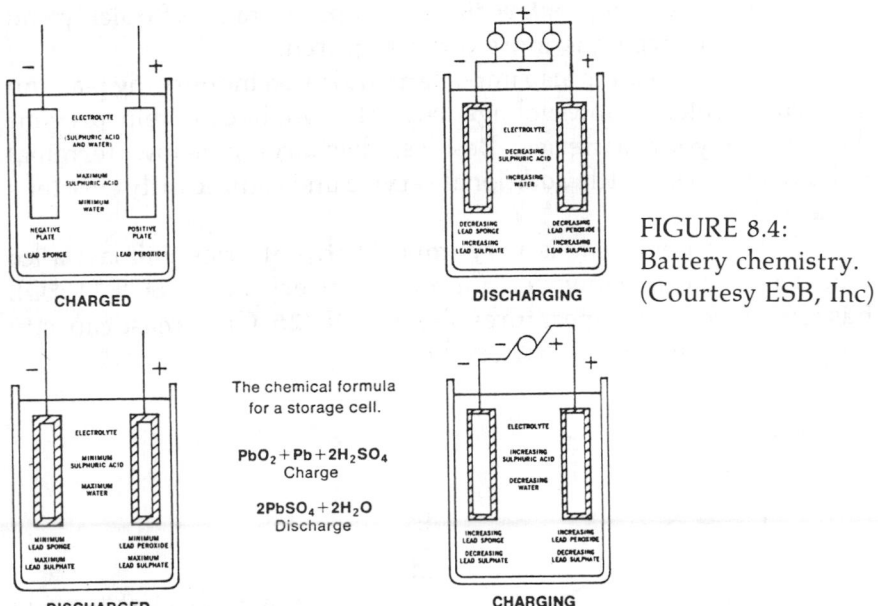

FIGURE 8.4:
Battery chemistry.
(Courtesy ESB, Inc)

The chemical formula
for a storage cell.

$$PbO_2 + Pb + 2H_2SO_4$$
Charge

$$2PbSO_4 + 2H_2O$$
Discharge

All batteries have an actual lifetime of about 20 years no matter how they are used. To the dismay of buyers of used batteries, the capacity of well-maintained used batteries may be so low as to make them a bad deal at any price.

BATTERY CAPACITY

Sizing deep-cycle batteries for minimum 5 days of autonomy at a 40% depth of discharge will give you 6 or more years of service in the southwest. Locations with less solar input require at least 8 to 12 or more days of autonomy to get the same years of usage.

How important is depth of discharge (DOD)? If you have reliable deep-cycle golfcart-type batteries and discharge them to 80% of their capacity down to 12 volts or 1.150 specific gravity, you can expect 400 or more cycles. That can be as low as 12 to 18 months use if you cycle them on a daily basis, which is often the case with golfcarts.

But DOD is not the best way to measure the worth of your battery bank. A better yardstick is available capacity, and that is affected by several factors.

If you have a high *rate of discharge,* capacity falls off quickly and equalization charging is more often required.

Cut-out low voltage is also important and good monitoring plays an important role. A low-voltage cut-off switch can help prevent damage to your batteries. Besides, discharging below terminal voltage gives very little additional service and significantly shortens battery life.

Operating temperature is very important. Batteries will last a lot longer if kept cool. However, a too-cool battery, below 60°F (15°C), has low capacity. Temperatures above 77°F (25°C) increase capacity only slightly but reduce battery life.

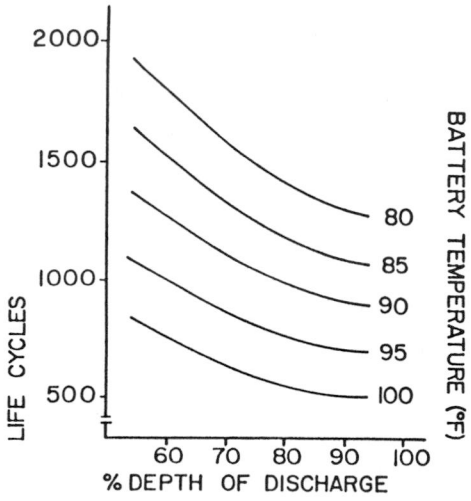

FIGURE 8.5: Effect of temperature on battery life.

The low *self-discharge* of lead-calcium batteries is an advantage this type battery has over less costly lead-antimony batteries. All cells have "local action," the normal chemical reaction which causes discharge when just sitting. Lead-calcium batteries have a lower self-discharge rate, which remains relatively low throughout its life. On the other hand, lead-antimony batteries have more local action which increases about halfway through their useful life.

The last variable is *sulfation.* Sulfation is the natural development of sulfate crystals which form when a battery is less than fully charged. Properly charged batteries do not get a chance to sulfate excessively. Batteries that are partially discharged for long periods

Table 8.3: Effect of Low Temperature on Batteries

Low temperatures reduce capacity but will prolong battery life under floating operation or in storage. Very low temperature may freeze the electrolyte, but only if the battery is discharged (low in specific gravity) at the time. At the temperatures shown in the following table, the electrolyte will not freeze unless the specific gravity is lower than indicated.

Temperature (°F)	Specific Gravity (Approx.)	
	At Same Temperature	Corrected to 77°F
+20	1.100	1.080
+10	1.150	1.130
0	1.185	1.160
−10	1.210	1.180
−20	1.235	1.200
−30	1.250	1.215
−40	1.265	1.225

Courtesy Exide Corp.

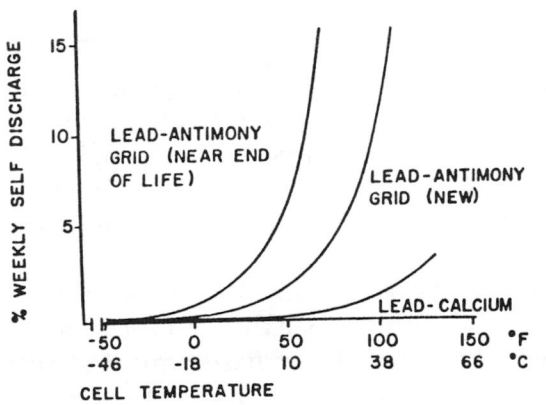

FIGURE 8.6: Lead-acid battery self-discharge rate

of time get permanent sulfate deposits, and it becomes increasingly difficult to return them to their original capacity.

PV users hear stories about "dreaded" battery sulfation. It robs the batteries of power and shortens their lives. It sounds horrible and seems·something that should be avoided at all costs. Well, yes and no.

Sulfation is part of the normal process which occurs when a battery is discharged. Sulfuric acid in the electrolyte acts on both the positive and negative plate active materials to form a new chemical compound called lead sulfate. When the battery is charged, lead sulfate is expelled from the plates and returns to the electrolyte.

Prolonged discharge will cause the formation of sulfate crystals which shorten the life of a battery. Very deep discharging of your batteries can cause the sulfate to expand the negative sponge lead plates separating the lead from the grid to permanently damage the battery.

The only way to avoid sulfation is to keep your batteries fully charged at all times. Since that is impossible with an active PV power system, the next best thing is to have a properly sized battery bank that is not deeply discharged. If your PV production matches your consumption on a daily basis and your battery bank is sized for no more than a 40% depth of discharge, sulfation should be no problem. On rare occasions when you do discharge your batteries deeply, bring them back up to full charge and equalize them.

SELECTION

Sizing your battery bank and selecting your batteries are inter-related. Product endorsements, sometimes the result of limited experience or favors in exchange for discounts and free samples, do not constitute expert opinion. I am amazed by the strong opinions held by people whose experience is limited to a few kilowatt hours of battery storage.

Since 1976 I have used batteries every day. When I bought my first PV modules in 1978, I had already learned the limits of shallow-cycle batteries and discount batteries. I purchased my first real battery bank from a friend who had a 6000-watt 115-volt DC windcharger and lots of batteries. He had six extra used C&D cells bought at salvage prices from a telephone company.

Those batteries served me well, even though they were already ten years old when I got them and their capacity was limited. Since PV modules and the electricity they produce are the most expensive part of a system, you cannot afford to waste power with bad batteries. In retrospect, I probably should have bought new deep-cycle batteries.

FIGURE 8.7: Telephone company type batteries (Courtesy Gould Inc.)

Having used several different batteries and having sold batteries for hundreds of PV systems, I am amazed at how much abuse batteries can take and still perform. Talking with PV users and professional installers, the story is always the same. "I wish I'd gotten good batteries to begin with" or "We installed good batteries and haven't had a bit of trouble."

What should you look for in a battery? (By battery, I mean either battery bank, battery, or cell, unless specified.) First, you must use a true deep-cycle battery. Don't be afraid to ask and get it in writing. Deep-cycle battery manufacturers publish their battery ratings and many have charts with a range of specifications. Compare the specifications of different manufacturers.

Buy from an established, reputable local distributor, if possible. Chances are he sells lots of batteries and rapid stock turnover will insure fresh batteries. If he offers a special deal on some batteries that are "in the back room" and appear dusty and neglected from sitting around, forget it. If they have to "special order" your batteries because they do not stock them, be sure to find out when the batteries were manufactured. Do not buy new batteries that are over six months old. The date of manufacture is stamped on the terminals or on the case.

Tell your supplier how you plan to use the batteries. You may be surprised at the cooperation you'll get. Ask if other alternative

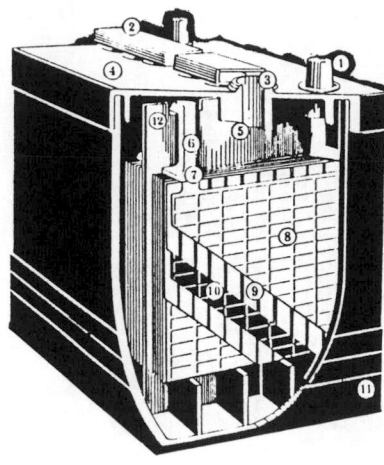

FIGURE 8.8: Cut-away view of a deep-cycle battery.

1. Terminal post
2. Gang vent plug
3. Vent
4. One piece cover
5. Electrolyte level mark
6. Inter-cell connector welded to
7. Plate strap
8. Negative plate
9. Separator
10. Positive plate
11. Container
12. Cell partition
(Courtesy ESB, Inc.)

energy producers have purchased batteries from them. Perhaps they know someone nearby who can give you tips from their experiences.

If you're on your own, look in the Yellow Pages for golf courses or battery distributors. Golf courses use lots of batteries in their golfcarts, and someone there can tell you where they buy them. If there are no golf courses or battery distributors nearby, check with automobile or truck service stations.

A local supplier receiving regular factory deliveries can save you hundreds of dollars in shipping costs. Batteries are heavy and cost a lot to ship. Many PV systems have a ton or more of batteries. Avoid direct shipping costs whenever possible. If you must have your batteries shipped, be sure to have adequate insurance and use a carrier who knows how to handle batteries (which are classified as hazardous materials).

Shopping for price can sometimes result in a bad bargain. The cost of a battery or cell is based on the amount and type of materials used, labor costs, shipping costs, and profit. Similar batteries are usually similar in price. A discount battery may be poorly made, improperly shipped, and badly stored. On the other hand, a high price may just mean a high mark-up.

Most golfcart batteries are reasonably priced. 1986 prices run approximately $80 for 200 ampere hour 6-volt batteries weighing around 65 pounds. These batteries typically have a one-year prorated limited warranty so regular testing with voltmeter and hydrometer in the first few months of use is important to determine if you have any weak cells. Extended warranties are available but may not be worth the extra cost. Some industrial or commercial-grade batteries have 5-year, 10-year, and even longer warranties. Be sure you are aware of all the conditions of the warranty contract and keep an accurate service log on your battery bank.

When shopping, find out why one battery is more expensive than another. Some suppliers sell the same capacity of storage in 6 volts at a lower price than their 12-volt batteries. Often the volume of sales in the 6-volt model is significantly higher than the 12-volt model. You can take advantage of volume discounts by buying a pallet quantity. Ask if delivery is included.

The price tag on your battery bank is only a part of the true cost of your storage system. System configuration, additional storage for greater autonomy, and shallower depth of discharge for longer battery life are all important cost factors. Perhaps you plan to offset a portion of your battery storage with supplemental power from a standby generator or from utility company power. Even if this is the case, scrimping on batteries results in poor performance.

If you have to build battery racks, a battery room, and use special cables, connectors, and venting caps, these additional costs should be compared to a more conventional approach. Do you have to maintain special temperature conditions or some other on-going activity that consumes money, time, or energy?

Maintenance has its costs. Time spent testing and money spent for distilled water, if needed in your area, should be considered. Will auxiliary equipment such as a special battery charger for equalization charges be needed? Is special venting required?

What is the calendar life of the battery? What is its salvage or trade-in value? Is disposal going to be a problem? What are replacement costs during the life of your PV system? It may be wiser to buy twice the number of batteries and go to a shallower depth of discharge or duty cycle. By doing so, you can often more than double battery life.

The answers to these questions will lead you to the selection of the right battery. They will also lead you to someone who can help. Your battery supplier is one such person. Other battery users and experienced professional PV installers can prove information sources as well. The knowledge you gather will insure that your battery bank will last many years.

INSTALLATION AND MAINTENANCE

Long life and reliable power from your battery bank begins with proper installation. Even before your batteries are installed, you can make sure that all will go well by having them shipped by an experienced and insured handler. Inspect your batteries upon arrival. Examine them for damage. Look for leakage or spillage of electrolyte in the cartons, batteries, pallet, or crate. The electrolyte level should be above the plates in every cell.

If spillage has occurred, remove some electrolyte from each cell to put into the spilled cell to cover the plates. This is only an expediency measure. A battery with more than 1/4 inch of plate exposed due to the spillage, or loss of electrolyte, is permanently damaged and a claim should be made against the carrier for replacement. It's a good idea to re-inspect your batteries within 15 days of receipt for possible concealed damage. Again, note the date of manufacture. Be sure your batteries are fresh.

You should never add electrolyte to a cell unless it is proven by prolonged charging (such as the initial charge) that all the electrolyte is out of the plates and the cell voltage on charge is normal. The only thing to put into your batteries is clean water with low mineral content.

If you have dry batteries, check that you have the right electrolyte with the proper specific gravity. Exercise extreme care when adding electrolyte or acid. Wear gloves and safety glasses and protect your clothing from the acid. Do not smoke.

If you must mix acid and water, pour the acid into the water. Never pour water into acid as a violent chemical reaction will occur. Use baking soda and water to neutralize any acid on your skin. A solution of one pound baking soda to one gallon of water will neutralize electrolyte spilt on the batteries, shelves, or floor.

Handle batteries with safety in mind. Remember, they contain acid and they are heavy. Allow only qualified adults near your batteries. Never slide batteries across a rough surface. You might damage their cases. Do not tip or tilt batteries more than 25° from vertical when handling them. Use tools with insulated handles. Lift batteries with straps or carriers. Do not lift batteries by their terminals—you might pull apart internal electrical connections. Be sure to read the safety section in this chapter.

It is best to install your battery bank right away. If you are not ready, have a temporary location prepared for them. This location should be near their permanent location. It should be out of the weather and at room temperature. If batteries are to be kept in an unheated room, be absolutely certain that they are fully charged and kept insulated from the cold. Discharged batteries will freeze and their cases will crack. In a discharged battery, most of the acid is in the plates leaving behind water which freezes at 32°F (0°C).

Fully charged batteries stored at 0°F (-18°C) will hold most of their charge for almost a year. On the other hand, fully charged batteries stored at 130°F (55°C) will become completely discharged in about one month. In general, it is recommended not to store unused batteries for more than three months.

Upon receipt of new batteries, do not add water to cells which are low. It is best to wait until after initial charging and prior to equalization charging so that the water will mix with the electrolyte. What kind of water should you use? Tap water can be used if it is very low in minerals and salts. Check with your local battery supplier to be sure. When in doubt, use only distilled water.

Never use battery additives. Every manufacturer voids its warranty when additives have been used. At best, additives give dying batteries a shot in the arm while shortening any life they may have left. At worst, additives can ruin the plates.

Also, never add electrolyte unless specifically instructed by the manufacturer. If your liquid level is low and you add electrolyte, you will change the specific gravity of the electrolyte. You will also increase the concentration of acid and chance eating up your plates.

If you have to store your batteries, you can give them a freshening charge with a generator or utility power to compensate for self-discharge. You can also temporarily connect your PV modules to the batteries to maintain a trickle charge. (If your PV modules are in a temporary location, be sure they are secure so that they cannot be blown or knocked over). By providing a trickle charge current up to 2% of the total ampere-hour storage capacity, you can maintain batteries in a fully charged state prior to installation.

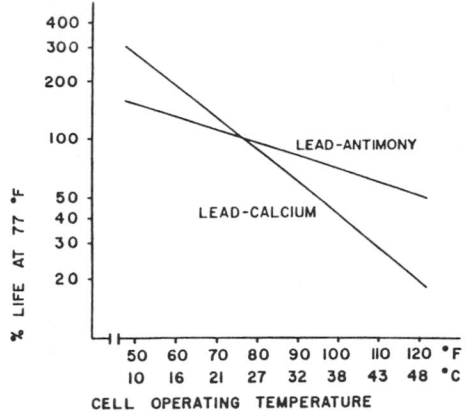

FIGURE 8.9: Effect of temperature on lead-acid battery life.

If you do perform a temporary charge, keep battery temperature under 130°F (55°C). Normal battery operating temperatures are between 60°F (16°C) and 90°F (32°C). Batteries last longer when cool and manufacturers recommend 60 to 70°F as ideal. It is important to keep all of the cells at the same temperature. The chemical reaction which occurs during charging and discharging is temperature-determinant. A 5° or less variation in cell temperatures in a battery bank will reduce the frequency for equalization charging.

You may be warned not to put batteries on concrete floors for the floor will "sap" the battery's power. What actually happens is that a cold floor lowers the capacity of the battery by causing unequal temperatures within the cells. (A concrete floor, for example, is colder than the surrounding air.) Place batteries on a wooden or other temperature non-conducting material.

The same concern regarding temperature should be taken when putting a battery bank against a cold wall such as in an uninsulated garage or a basement. If you do locate your battery bank against a wall, be sure to put some insulation between the wall and the batteries. For fire safety, protect any flammable insulation with a layer of gypsum board (also known as sheet rock or plaster board).

CONNECTING YOUR BATTERY BANK

Before you put your batteries in place on their pallet, shelves, in a box or compartment, or other suitable holder or container, make a sketch showing how they will be situated. Be sure to note which sides face which direction, which terminals connect to what. Note polarity. If you are installing a large bank of batteries in series and parallel, double check your wiring diagram to avoid extra and unnecessary handling. Save your back. Keep your wire runs short and orderly.

With all batteries in place and any spillage cleaned up and neutralized, you can begin making electrical connections. Exercise extreme caution as you will be working around DC voltages capable of welding steel or melting a wrench.

All battery interconnects should be of the proper size using no smaller than 4 AWG stranded and insulated wire (commonly known as welding wire). Check with your battery manufacturer or the wire size chart in chapter 10 for the proper size conductor.

Make your wire runs as short as possible to economize on wire. However, be sure that the wires do not touch each other or anything metal, such as your battery shelves, equipment, or the wall. The use of ring or other connectors securely crimped, soldered, and taped will make the work neater and insure good electrical connections.

Fasten your interconnects tightly but do not deform the battery terminal. Some manufacturers recommend a coating of NO-OX-ID (Dearborn Chemical Company). I personally do not like to put anything on electrical terminals. After a few inspections, you can determine if conditions warrant the use of an anti-corrosive coating. If you do use a greasy coating on your terminals, it is all the more reason to protect your battery bank in a cabinet as dust will collect on the grease. Keep the battery cases free of any coating.

If corrosion has occurred, gently clean lead terminals with 00 grade sandpaper or use a battery terminal cleaning tool sold at most auto parts stores. To clean corroded interconnect wires it may be necessary to first soak them in a solution of baking soda and water. Do not connect your battery bank to any equipment until all cell or battery bank interconnects are securely fastened. Some electronic equipment, such as controllers, inverters, and meters, must be

FIGURE 8.10: Keeping batteries and their terminal connections clean is very important to maintaining optimal system performance. In corrosive environments, such as coastal areas, coating battery terminals with grease retards corrosion. Spray retardants which do not attract dust are also used. (Photo: Robert Sardinsky, Rising Sun Enterprises)

connected in the proper sequence according to manufacturer's instructions.

THE BATTERY ROOM

The battery room or battery area should be properly vented and clearly marked. A lock on the door will prevent children or curious adults from entering. Inside the room, place the batteries on sturdy pallets off the concrete or dirt floor, or on strong shelves. It is a good idea to enclose the battery bank to prevent dirt and dust from collecting on the tops of the batteries. Be sure to give yourself plenty of room to service or replace batteries. If you stack batteries on shelves, leave enough room to get your hydrometer in between the shelves.

Shelves of batteries can be enclosed in a cabinet to keep everything clean, out of reach, and insulated from temperature extremes. The enclosed cabinet also makes it possible to easily vent hydrogen developed during charging out of the building or room.

If you put your batteries on pallets or wooden boards on the floor, be sure the floor can support their weight. An enclosure of plywood with a removable top is a good way to protect batteries located on the floor. Again, venting or insulating from temperature extremes can be made simple by using a dust-free enclosure.

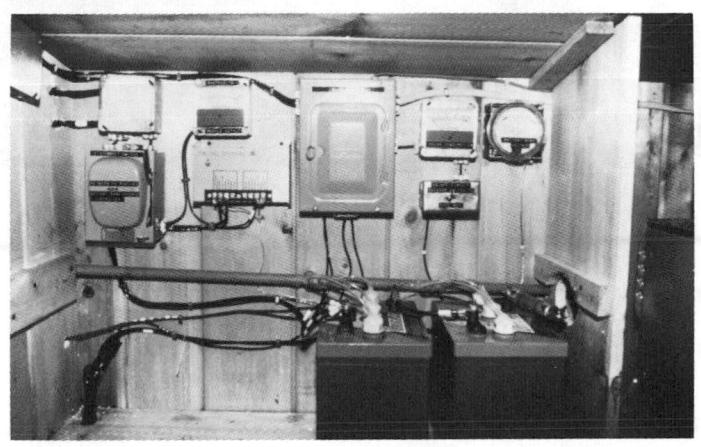

FIGURE 8.11: Batteries, monitoring meters, regulator, disconnect switches and distributor panel are all neatly installed in one place. A clean installation like this makes periodic servicing simple. (Photo: Robert Sardinsky, Rising Sun Enterprises)

Mount all electrical equipment on a nearby wall. If there are switches or relays which may spark, isolate them from hydrogen gas released by the batteries. Wiring should be neatly fastened to the wall in conduit, where required or desired, and run to the regulator, inverter, distribution box, circuit breaker box, monitoring equipment, switches, etc. The wall should be fire-resistant. Gypsum board can be used for this purpose.

Give yourself plenty of room to work in. Don't make your equipment room a cramped, dark, hard-to-access corner. Elbow room of at least three feet clearance is required in front of all equipment. It must be emphasized that no open flames or standing water be in the equipment room. Don't use the equipment room to

FIGURE 8.12: A well-kept equipment room can also serve as, for example, the laundry center, but be sure to provide for adequate battery venting. Batteries and inverter at the Windfarm Museum, Martha's Vineyard, Massachusetts, are side-by-side with home appliances. (Photo: Robert Sardinsky, Rising Sun Enterprises)

store junk or flammables. Use good power plant management and good sense. If some equipment, such as the inverter, requires venting, be sure to mount it where a free flow of cool air is present or place it on a raised open pallet. It is a good idea to mount a fire extinguisher in a handy location.

Although a clean dirt floor is acceptable for the battery room, a sloped concrete floor with a drain will make clean up easier.

Some regulators, inverters, and switching devices have electrical relays and contacts. When the contacts of an electrical circuit open, a spark occurs. Batteries should be isolated from such devices.

Since you are going to take the time to protect your batteries and yourself by locating your batteries in a safe place, it is worth the extra time to be sure that your batteries are kept at 70°F (21°C). If the vented area is unheated, put your batteries in an insulated box. The ideal battery storage area:

1. has good ventilation
2. has no open flames
3. has no electrical sparks

4. is easy to get to for maintenance and inspection
5. is easy to clean
6. is kept at 70°F (21°C), if possible
7. is out of reach of non-authorized personnel
8. has an up-to-date fire extinguisher handy.

BATTERY VENTING

Battery rooms should be properly vented to prevent the build-up of potentially explosive hydrogen gases which are released during charging. Properly sized systems with good charge controllers will limit the amount of outgassing during the charging process. However, topping off and equalization charging—requirements for long battery life—produce some hydrogen gas. Safety dictates that preventive measures are worth taking.

First of all, do not tempt fate. Entering a closed battery room or opening a battery compartment or cabinet, with a cigarette in your mouth or lighted flame (match or candle) is foolish. Some battery technicians open battery compartments and let them air for a few minutes before doing any work. A few minutes is not a long time in the life of a PV system or a person. Better to be overcautious.

Concentrations of 4% hydrogen are explosive. Recommended maximum concentrations for battery storage areas is 2%. Most rooms have an air change every 4 hours. To determine if your battery compartment or area needs special venting, use the following formula:

$$H = 0.00027 \ I \ N$$

where H is the maximum quantity in cubic feet per minute (CFM) of hydrogen evolved while topping off or equalization charging; I is charge rate in amperes; and N is number of series connected cells.

Next calculate the minimum volume of venting air required to maintain a safe hydrogen concentration (2%) level

$$Q = 0.0135 \ I \ N$$

where Q is quantity of air in cubic feet per minute (CFM).

For example, a PV system consists of eight each 33-watt modules and four each 6-volt batteries to produce 24 volts DC. The maximum quantity of hydrogen produced is 0.00027 x 8 amperes x

12 cells = 0.0259 CFM requiring 0.0135 x 8 x 12 = 0.1296 CFM or 77.76 cubic feet per hour. If the average room has an air change every 4 hours, then 77.76 x 4 = 311 cubic feet (total room volume of air). As most rooms are 8 feet in height, a room 6' x 8' (384 cubic feet) would be of adequate size to prevent dangerous concentrations of hydrogen.

It is recommended that you have a vent opening at the bottom of the door and at the highest point in the ceiling. The ceiling should have no pockets such as exposed rafters or frame blocking to trap hydrogen.

If you are unable to provide adequate size and natural venting for the battery area, the simplest way to avoid concentrations of hydrogen is to build a battery compartment of plywood and install a vent pipe similar to those used for gas water heaters or clothes dryers. Place the outlet at the top of the battery box. The pipe should go straight to outside air.

Be sure to put a screen over any vents to prevent mice, bugs, and birds from nesting in your battery box. As a final precaution, put a sign on the door of the battery room or compartment which reads DANGER—BATTERY ROOM—KEEP OUT. A lock should keep out unauthorized personnel and the curious. This may sound like overdoing it, but in the long run you will find that these added preventive measures will give you peace of mind at a negligible cost.

BATTERY TESTING

Keep a complete history of your battery bank and PV system. At first, this may seem like unnecessary paperwork, but over the years this log will provide invaluable information about the performance of your system. Refer to it each time you service or test or modify your PV system.

Note the date purchased, date installed, prices, parts, initial voltage, hydrometer readings, and comments: anything you feel is important at the time should be included. You might find it interesting years from now to note the names of friends who helped you install your system and what the weather was like.

Battery testing begins the day they are received. Three tools are needed: a thermometer, a hydrometer, and a voltmeter. When testing you will look at temperature, specific gravity, and voltage.

STATIONARY BATTERY REPORT

Month of _____ 19 _____

Company _____

Location _____   _____

Cells	Type

Date Installed _____

DAILY READINGS

Day	Bus Volts	Pilot Cell Hydrom.	Eler. Temp.	Day	Bus Volts	Pilot Cell Hydrom.	Eler. Temp.
1				16			
2				17			
3				18			
4				19			
5				20			
6				21			
7				22			
8				23			
9				24			
10				25			
11				26			
12				27			
13				28			
14				29			
15				30			
Pilot Cell No.				31			

MONTHLY READINGS

Cell	Volts	Hydrom	°F			
1						
2						
3						
4						
5						
6						
7						
8						
9						
10						
11						
12						
13						
14						
15						
16						
17						
18						
19						
20						
21						
22						
23						
24						
25						
26						
27						
28						
29						
30						
31						
32						
33						
34						
35						
36						
37						
38						
39						
40						
41						
42						
43						
44						
45						
46						
47						
48						
49						
50						
51						
52						
53						
54						
55						
56						
57						
58						
59						
60						

DAILY READINGS (for lead-antimony)

Record daily, or at other specified interval, the battery floating bus volts, pilot cell hydrometer reading, and adjacent cell temperature.

Float antimony batteries continuously at 2.15 to 2.17 volts per cell (129-130 volts for 60 cells).

Keep panel voltmeter in correct calibration by checking with a known standard every 12 months.

MONTHLY READINGS (for lead-calcium)

Record monthly or at other specified interval the floating voltage and hydrometer reading of each cell and the temperatures of 2 cells in each row.

EQUALIZING CHARGES

Give 1 to 3 months to lead-antimony batteries by raising the bus volts to 2.33 volts per cell (140 volts for 60 cells), for 8 to 24 hours, then resume normal float.

The last equalizing charge was given:

Date _____ at _____ volts for _____ hours.

WATER ADDITIONS

Keep electrolyte level between high and low level indicators by adding approved or distilled water as necessary, preferably before starting an equalizing charge. Note date and total quantity of water added.

Date _____ Added _____ Qts. Total

Date _____ Signed _____

FIGURE 8.13: A sample battery log. Your battery log should be kept weekly for the first two months. Then check your batteries monthly for the rest of the first year. After that, twice a year should be adequate.

By charting the specific gravity and voltage, you will know the battery's state of charge (SOC), depth of discharge (DOD), and capacity. Unlike a gas tank with a fuel gauge, you cannot take a direct reading of the amount of juice in your batteries. However, by knowing the battery's characteristics and its present condition, you can extrapolate how much power you have. That's what records are for.

With an accurate, clean thermometer, measure the temperature of the electrolyte. Specific gravity (SG) of the electrolyte is affected by temperature. Add one point (0.001) SG for every 3°F (1.67°C) above 77°F (25°C). Subtract one point for every three degrees below 77°F.

Specific gravity is the measurement of the electrolyte's density. The more sulfuric acid in the electrolyte, the higher the SG. (When the battery is fully charged, all of the acid is in the electrolyte and the SG is high. When the battery is discharged, the SG is low.) Your fully charged batteries typically will have a SG between 1.265 to 1.280 or higher. Different batteries have different SG, so go by the manufacturer's specifications. If your SG is unknown, use the numbers listed here.

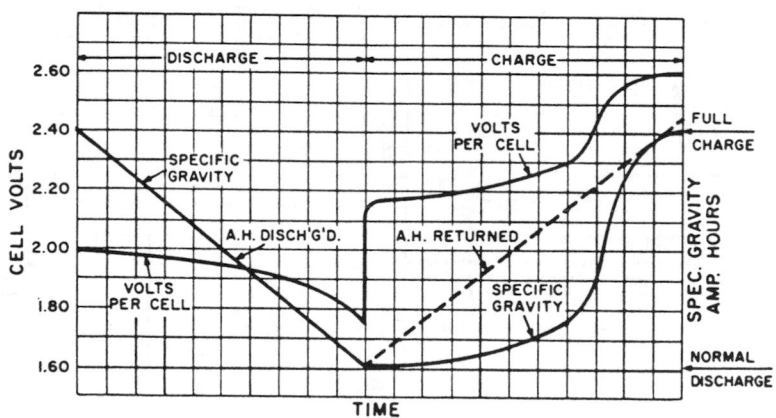

FIGURE 8.14: Typical voltage and gravity characteristics during a constant rate discharge and recharge. (Courtesy Exide)

To read SG you will need a hydrometer. Don't buy a cheap plastic one with floating balls. For under $10 you can buy a good hydrometer which will last for years. The larger the calibrated float and its scale, the easier it is to read.

To take an SG reading, open a battery cell by removing the cap. Stick the clean nozzle of the hydrometer deep into the cell (because the acid is heavy and tends to settle). But be careful. Do not jam the hydrometer into the cell or between the plates because you might damage the cell. At each cell, fill and empty the hydrometer a couple of times. This flushes the hydrometer and will give you a more accurate individual cell reading. Avoid getting dirt into the cell or on the hydrometer nozzle. Don't take electrolyte out of one cell and put it into another. Work clean and avoid contamination.

When should you take SG readings? With an active PV system, it is difficult to make SG tests under ideal conditions since manufacturers recommend letting the battery sit for a number of hours neither being charged nor discharged. There may be a time at night when that condition occurs. However, shutting down the system or waking at 3 am to make tests is not necessary. Allow for less than ideal testing conditions. By keeping good records and testing under similar conditions each time, you can get a fairly accurate indication of your battery bank state of charge.

Here are some guidelines:

1. Plate "surface charge" during and immediately after battery charging will affect readings. Test in the early morning or evening, when loads are light and no charging has occurred for a few hours.

2. If your electrolyte is 1/2-inch low, SG will read 15 points (0.015) above normal and vice versa. If there is no mark to indicate normal electrolyte level, fill to 3/8-inch over the tops of cell separators.

3. SG readings taken during the early stages of recharge will lag as the heavier acid at the bottom of the cell has not mixed with the liquid at the top of the cell.

4. When adding water to a cell, expect several days' lag until the water is diffused to get an accurate reading.

5. The SG of a fully charged cell will drop 0.001 per year due to a slight loss of electrolyte at each reading. Change the pilot (or sample) cell every five readings.

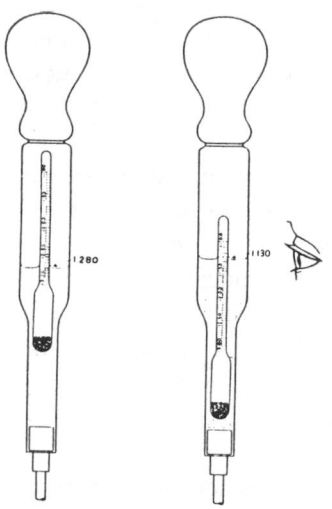

FIGURE 8.15: The correct way to read a hydrometer.

Now you are ready to make your voltage tests. A good digital voltmeter is required since you want to read to 0.001 volts. There are excellent digital multitesters for under $100. I use a Beckman model 77.

First read your total battery bank voltage. Make sure your test probes or leads are clean and that you have a good connection. Wiggle the contacts to be sure. If the reading holds steady, you're doing fine. Record your results.

Take the open-circuit voltage reading for each cell or battery. Note the reading in your log along with temperature and hydrometer reading for that cell. If your batteries are receiving a charge from your PV array, it will be hard to see a relationship. However, there is a direct relationship between voltage and specific gravity of a cell not being charged or discharged by a load. Voltage equals specific gravity plus 0.84 ($V = SG + 0.84$). Final or terminal voltage of a fully discharged battery is commonly 1.85 to 1.75 volts per cell. At this point your batteries are effectively dead.

How often should you test your batteries? Initially, a weekly voltage and hydrometer check of one cell or battery is a good idea. (A different pilot cell should be selected each time.) After two months when everything seems stable, test a pilot cell every month. Four times a year test every cell and record the reading.

Table 8.4: Battery Capacity

Depth of Discharge	Specific Gravity at 80°F (27°C)	Volts per Cell
0%	1.2650	2.100
10%	1.2500	2.090
20%	1.2350	2.075
30%	1.2200	2.060
40%	1.2050	2.045
50%	1.1900	2.030
60%	1.1750	2.015
70%	1.1600	2.000
80%	1.1450	1.985
100%	1.1300	1.750

CHARGING

Proper charging will not produce excessive gassing. It can occasionally raise battery temperature above 110°F (43°C). It can raise cell voltage above 2.4 volts. Your system charge controller or regulator should be designed for your batteries and your use of PV. Here are some guidelines.

If you have a two-stage regulator, that is, one with a different charge rate for the final period of charging or topping off, its finishing rate could be between 4 and 10 amperes per 100 ampere hours of battery capacity (at the eight-hour rate).

Surprisingly, a discharged battery can take a very high initial rate of charge, up to ten times its finishing rate. However, an excessive amount of charge will corrode the grids of the positive plates into lead peroxide which weakens them physically, increases their electrical resistance, and shortens battery life.

"Mossing," the formation of a spongelike deposit on the negative plates and strap, is an indication of overcharging. Sediment falls to the bottom of the battery container when overcharging or charging at a very high rate occurs. Overcharging batteries literally tears them apart.

Your batteries will either be charging or discharging most of the time, so you will not be able to·observe the ideal conditions

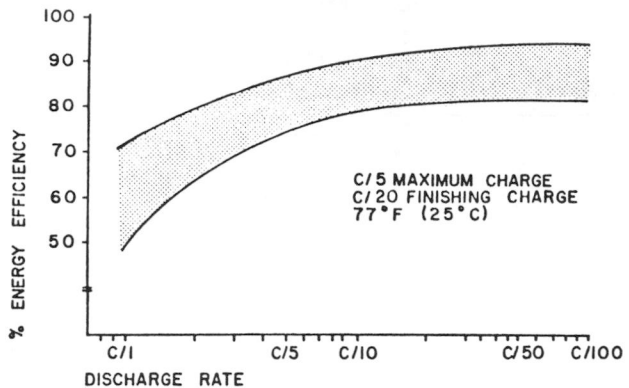

FIGURE 8.16: Typical efficiency range for lead-acid batteries as a function of discharge rate. (Note: C = capacity. Example: for a 200 ampere hour battery, the maximum charge is 200/5 = 40 amps.)

described in the manufacturer's literature. Trying to estimate the state of charge of your battery bank is like guessing the speed of a car by looking at its blur in a photo. However, the more you know about your batteries and how you use them, the closer your approximation of their condition will be.

For instance, float current is a good indication of battery condition and age. Float current for lead-calcium batteries is low, in the 4 to 11 milliampere (0.004 amps to 0.011 amps) range per 100 ampere hours of the 8-hour rating. It rises only slightly as this battery ages. Average voltage for floated lead-calcium cells is 2.17 to 2.20 volts.

Float charge current for lead-antimony batteries is 30 to 50 milliamperes (0.030 amps to 0.050 amps) per 100 ampere hours of the 8-hour rating, increasing to 10 times this rate at the end of the battery life. Lead-antimony float voltage is approximately the same as lead-calcium, 2.15 to 2.20 volts per cell.

An increase of 15°F will double the float current drawn by either type of battery. Also, float current draw of batteries, in general, doubles for each increase of 0.05 volts per cell and decreases by one-half for a similar decrease in voltage. This emphasizes the fact that batteries function best at 70°F and a correct charge rate.

EQUALIZATION

A characteristic unique to PV systems is the rate of charge. Designed to capture and store power throughout the daylight period, there are times when the charge rate is very low, such as early morning and late afternoon. A low charge rate also occurs when some regulators, sensing fully charged batteries, switch to trickle or float charge. At other times, especially mid-day, your charge rate may be relatively high. Thus, a PV system is a variable current charger.

The combination of constant voltage and variable current is a common method of battery charging, but requires a good regulator designed for PV charging. No regulator means close monitoring of your battery bank's condition. An inappropriate regulator, one from a windcharger or an automobile, will waste energy as it usually has the wrong setpoints and is generally less efficient.

Your battery bank will require an equalization charge from time to time. Charging battery banks until regulator cut-out voltage is reached is a standard design parameter of well-balanced PV systems, but battery bank voltage does not indicate voltage differences in individual cells. An equalization charge brings all cells up to a full charge.

Few, if any, regulators provide automatic equalization. To equalize your batteries you must override the regulator and occasionally "boil" your batteries. This "boiling" is actually the production of hydrogen and oxygen from the breakdown of water during electrolysis. It agitates the electrolyte, mixing the acid and any water you might have added. A controlled equalization charge is good for your batteries. It extends battery life and increases capacity. But be sure to follow the manufacturer's instructions for equalization.

There are certain guidelines that let you know when to equalize your battery bank:

- at quarterly inspection when the float voltage of any cell is 0.04 volts below the average for a battery,
- when temperature-corrected specific gravity is more than 10 points (0.010) below its full-charge value,
- after any unusual deep discharge period
- at least once a year.

For lead-antimony batteries the equalization voltage is generally 2.35 to 2.39 volts per cell. For lead-calcium batteries the equalization voltage is generally 2.48 volts per cell.

By monitoring the lowest cell voltage, you can tell when equalization is complete. For lead-antimony batteries there will be no voltage change after 24 hours of equalization. If you use utility or generator power to equalize your batteries, 24-hour charging may be possible. However, an easy way to equalize with PV is to schedule equalization at least four times a year when your batteries are fully charged using the unregulated output of your array.

Table 8.5: Equalization Charging Time

Electrolyte Specific Gravity	Equalizing Charge Time
1.260 to 1.280	None
1.240 to 1.259	4 hours
1.220 to 1.239	8 hours
below 1.220	12 hours

This is when good records help. At least four times a year your battery bank will be fully charged during a sunny period. With the batteries fully charged, bypass the regulator and let the sun equalize your batteries with the array output for 4 to 6 hours. That evening test your battery voltage. If there are still some low cells, equalize the next day.

If, after an equalization charge, the difference between the highest and lowest cell is 50 points (0.050) specific gravity or more, the battery bank may be nearing the end of its useful life.

There will be some water lost during equalization. Be sure to vent your battery area properly because boiling, or electrolysis, is the breakdown of water into dangerous hydrogen and oxygen. Add distilled water, if necessary, a few hours after equalization is complete.

All manufacturers warn against making tap connections. A tap connection is when you use a portion of your battery bank for low-voltage equipment. Some people with 24-volt battery banks use 12-volt DC lights or other equipment by tapping a couple of their 6-volt batteries for power. Unless they regularly move the tap and monitor their batteries closely, they will cause overcharging of the untapped group and undercharging of the tapped group. This can shorten battery life. If you tap your battery bank, monitor the tap closely and equalize regularly.

You can avoid taps entirely by one of three methods. If you must make a tap, use a DC-to-DC converter or voltage dropping resistor on small loads. Better yet, upgrade your appliances. If you have become attached to your low-voltage equipment, or if the cost to upgrade is too high, or if you are getting better efficiency and performance at the lower voltage, set up a separate battery bank specifically for those loads.

Equalization for batteries deeply discharged is greater than for the same batteries not so deeply discharged. When equalization charges cannot be performed regularly, it is better to install a larger battery bank and avoid deep discharge.

If you are not going to use your PV system for an extended period of time, say during vacation, give your battery bank an equalization charge and then float the batteries at the recommended battery voltage. Floating a battery means providing maintenance current once it is fully charged. Be sure your voltage regulator is working properly and that your meters are accurate. Once a year calibrate your meters or compare them to a calibrated meter for accuracy. Most manufacturers recommend a float voltage of 2.15 volts per cell.

While on the subject of float voltage, now is a good time to again emphasize the need for adequate battery bank size. If your batteries are always 0.04 volts per cell or more below the float voltage, your solar array is too small: your consumption is greater than your production. At float voltage, individual cell voltage should be within 0.02 volts. Under such conditions the SG will be at full charge rating.

While all this fussing with the batteries may seem tedious, it actually takes very little time. After the first year, PV homeowners spend about one hour per year maintaining their battery bank. That's less time than spent annually taking out the trash.

Note: Unique to new batteries is the fact that they require a few charge cycles before achieving their full storage capacity.

Over the first few weeks, and usually within 10 complete cycles, your battery bank is getting its chemistry in order. This is good to know so you won't worry about limited capacity or voltage fluctuations beyond your calculations. You can still get an approximate indication of battery state of charge by comparing the specific gravity with the full charge value and the manufacturer's specific gravity drop. Some battery suppliers may not have this information on hand, but be sure to get it even if you have to write the manufacturer.

REPLACING YOUR BATTERIES

The useful life of a battery bank is less than the life of a PV array. Most system designers give a 20-year useful life for PV systems. During that period the array should change very little. The batteries, however, will age and need replacement.

The type of battery, environmental conditions, use, and maintenance all affect longevity. More specifically their self-discharge rate, parasitic power losses, operation partially discharged, non-equalization, and just plain calendar life determine how long your batteries will last.

It is recommended that you buy all your batteries at the same time so that they age equally. When faced with the decision to buy part of your system now and the rest later, it is best to buy all the batteries now and add PV modules as you can. Of course, be sure to have a large enough initial array to keep the batteries charged or you will be defeating this economy measure.

If you must buy additional batteries at a later date, match them to the original set as closely as possible. Mismatched batteries or cells will require equalization more often and may shorten the life of the newer batteries.

The budget-minded have a tendency to undersize their battery banks. This is false economy. Duty cycle is important and will determine the useful life of your battery bank. To put this into perspective, an automobile battery can have a long life even though it may be very briefly discharged at six times its rating. On the other hand, the automotive battery will fail rapidly if cycled deeply. Deep-cycle batteries can provide many years of service but will not perform well at high discharge rates. They are not designed for that type of service.

Even the availability of a back-up generator or utility power to recharge an undersized battery bank does not help in the long run. In fact, in the case of a generator, it is just the opposite. The larger the battery bank, the less often you should have to run the generator. This extends the life of the generator and means fewer oil changes, servicing, and repairs in the same number of years. If you use a generator and battery charger, get the largest commercial-grade battery charger affordable. Running a 4- or 6-kW generator to power a 10-amp battery charger is a waste of fuel (10 amps x 12 volts = 120 watts).

By monitoring your specific gravity and voltage over the years, you will know when your batteries need replacing. Plan for at least one replacement in ten years. If you are using good golfcart batteries at a 40% depth of discharge and taking good care of them, you should get at least six years of service.

Even when your batteries have reached the end of their useful life, they still have salvage value. Do not throw them away. Your battery supplier will credit you for your old set. Lead salvage prices have stablized in the past few years to between 10 and 20 cents per pound. In general, the value of old batteries continues to climb. Include the trade-in value of your old batteries when you calculate the value of your PV system.

BATTERY SAFETY

Low-voltage and low-wattage PV systems are relatively safe. It is hard to get a shock from a low-power system. But in that relative safety is a real danger. Carelessness can be your biggest problem. With a 12-volt system, you can put your finger in a light socket and not get shocked—if you are dry. But the power in that socket can weld the ring on your finger. The current in your battery bank can melt metal. A tool dropped across battery terminals will weld to the terminals. The acid in your batteries will eat your clothes, burn your skin, or blind you. Be careful.

Foremost among the battery dangers is the possibility of hydrogen build-up and an explosion. When batteries are charged, they give off hydrogen and oxygen. This occurs as the batteries become fully charged and continues if they are overcharged or receiving an equalization charge.

The outgassing or "boiling" that occurs can be minimized through the use of a properly designed and calibrated charge controller or regulator. Since outgassing does happen, be sure to vent your battery storage compartment. A well-ventilated storage area can be as simple as leaving your batteries in an open shed, barn, or garage. It can also mean putting our batteries in a special area with ventilation.

Another battery danger is acid. The liquid in the electrolyte contains sulfuric acid. If you've ever spilled battery acid on your clothes, you know that it will eat holes in them. If you get acid on your skin, it can cause a rash or burn. If you get the acid in your eyes, it can blind you. Sounds pretty dangerous. *Handle batteries with care.* Once in place, do not handle them at all.

Under some conditions, such as overcharging in high humidity, "misting," the settling of a fine acid film on the batteries, is possible. It is important to clean this mist from the batteries. Any moisture will cause corrosion and attract dust and dirt. Microscopic bits of metal in dust can accumulate and provide an electrical path from terminal to case.

To see the effect of dirt on battery cases, put a DC voltmeter lead to a battery terminal in your vehicle. Then put the other lead to the dirty case. If you get a voltage reading, it means dirt on the case is conducting stored electricity from within the battery to the case. This discharge shortens battery life. In a PV system, this means wasted power.

A simple cure is to wash your batteries. Dilute the acid with a solution of water and baking soda (sodium bicarbonate). You will see a bubbling chemical reaction as the baking soda reacts with the battery acid. Once the acid is neutralized, flush the battery with fresh water and dry with a paper towel or cloth. Make sure none of the solution gets into your batteries.

Still another battery danger is burns. A battery is a package of stored energy. As long as the energy is untapped, it remains in the battery. Connect something to the terminals and energy is released. Drop a wrench across the terminals and the energy is released rapidly. A golfcart deep-cycle battery has more than a kilowatt hour of stored power. That's enough power to weld heavy steel plate. In fact, DC welding uses the hot arc of electricity created by a short or near-short circuit. It is important not to drop tools or other metal objects onto your battery terminals.

If your batteries are located in a drafty place, they may experience temperature fluctuations. Batteries work best at room temperature (60 to 70°F; 16 to 21°C). If colder, the capacity of the battery bank must be increased to allow for reduced performance.

Storing batteries in a building crawlspace, outbuilding, or attached shed is a good idea. You can even have your batteries in the living space in a closed battery compartment or in a battery closet. Be sure that there is proper air circulation into and through the battery compartment or storage area to prevent the build-up of hydrogen. A louvered door for air-in and a vent for air-out will work well. If you don't have a louvered door, just cut the bottom of the door to provide at least a 1/2-inch gap. Screened, homemade outlet vents twice the size of the air inlet will serve to prevent gas build-up. A screen will prevent bugs, mice, and birds from nesting in your battery room—it will also keep the cat out. A clothes dryer vent, available at most hardware stores, works well too.

Using a power vent to prevent hydrogen build-up is one way to insure against possible hazards. However, a well-designed passively vented battery storage area avoids this use of electricity.

Be careful when moving batteries. They are very heavy: be sure not to strain yourself or drop them on your toes. And be sure to stack them on sturdy, stable shelves or pallets. Wear protective clothing, gloves, and safety glasses.

On the subject of keeping things out, put a lock on the door. No need to have curious children or adults in there. Don't forget to put your cigarette or whatever out before going into the battery room. Candles are not recommended so keep a flashlight (with rechargeable batteries) handy. Avoid open flames. Don't put your batteries next to the gas hot water tank. My repeated precautions are to remind you that safety should be always on your mind.

FIGURE 8.17: This PV-powered manufacturing and research center offers office space for rent, "utilities included." A 2160-watt fixed array and a 240-watt reflector-enhanced tracking array provide 12 and 24 volts DC and AC power from three separate inverters. All lights, fans, refrigerators, security and phone systems are DC. Air conditioners, office equipment, and appliances are AC. (Courtesy Heliotrope General and Simpler Solar Systems)

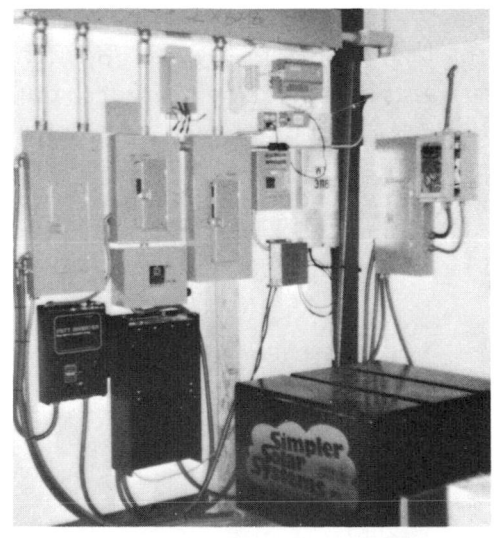

FIGURE 8.18: This power panel consists of (left to right) a 12-volt DC 200-amp distribution panel, AC distribution, AC utility main, Trace inverter and telephone system 24-volt DC distribution, voltage regulator and meters. Below, left to right, are 2.3-and 5-kilowatt Heliotrope inverters and a 2800-ampere-hour battery bank. Utility backup is available for the MIG welder and Dinh air conditioner. (Courtesy Heliotrope General and Simpler Solar Systems)

Inverters and AC

INTRODUCTION

Early on in PV history, the use of inverters was downplayed and most people used direct current. PV systems were costly and small-sized. Inverters consume power, and when PV was two to three times its present cost, that power was too expensive to waste.

Most PV pioneers were willing to put up with DC systems. They either hunted up DC appliances, modified existing AC appliances, made gadgets from scratch, or did without. It was a time of DC lights, fans, and radios salvaged from cars and trucks. Much time was spent scrounging through catalogs and shops looking for military surplus DC motors, and other goodies. Some very creative solutions resulted and a few DC PV businesses flourished for a time.

Even today the DC tradition has been passed on as part of the credo of energy conservation. A guideline for designing small PV systems is, if you can power something directly with DC, do it.

One idea that has grown less popular through changes in technology is record turntable modification. Since better belt-driven turntables use DC motors, it seemed logical to bypass the input transformer and go straight to the motor with 12 volts or whatever the DC requirement was.

For the non-technical: If you trace the power cord from the plug into some devices, the input transformer is the first thing you may find. The purpose of these coils with a magnet is to take the 120-volt AC utility power coming into the transformer primary coil and return reduced voltage from the secondary coil. That reduced voltage is then rectified to DC for use in the low-voltage DC circuitry of the solid-state electronics found in most everything nowadays. For the PV pioneer, much time was spent discussing which devices could be used directly in the DC mode. A voltmeter put across the secondary windings of the input transformer would often tell the story.

Reading the voltage and making modifications meant opening up the device. Electrical hazard warnings and threats of voided warranties notwithstanding, it was strange to open up a case which says "No Serviceable Parts Inside" or "To Be Opened Only By A Qualified Service Technician" and find 12 volts DC just waiting to be powered by PV.

Amateur radio operators and computer tinkerers were already familiar with these exploratory operations. They knew that most solid-state electronics were low voltage, usually 12 volts or less. Thus, ham radios and computers were among the first hi-tech equipment to be PV-powered.

In the early 1980s two trends evolved. Tracking the growth of the recreational vehicle industry, all kinds of DC appliances began to appear. RVers like gadgets, and the manufacturers were accommodating. In one automotive catalog, five models of DC vacuum cleaners were listed. I tried them all. To my disappointment, they were hardly more than toys (though some cost more than small home uprights). They worked, but were designed for small tasks, not cleaning a house in the country where dirt is always getting tracked in.

There are other examples of DC devices that couldn't quite do the job. There were blenders that couldn't crush ice, soldering irons that couldn't solder large wire, drills that broke after a few hours use, bug zappers that missed the big ones. Needless to say, there was room for improvement, and things have improved. Nowadays, we can find better DC appliances—though at a premium price.

Fortunately, another trend was occurring. People were beginning to experiment with inverters. Users were willing to sacrifice the AC appliances they had stored away and to experiment with early square-wave inverters. Some people used old-style motor inverters and got satisfactory service but low efficiency. Some even bought expensive sine-wave inverters costing as much as their entire PV system.

The more technical PV users at that time were building their own. From early designs and testing came a new generation of inverters which was to change the nature of PV use and make the old DC bias obsolete.

WHAT IS AN INVERTER?

An inverter is a device that changes direct current to alternating current. (Converters, sometimes called rectifiers, change AC to DC.) For our purposes, we are speaking of 12-, 24- or 48-volt DC power inverted to 120 or 240 volts AC. (See Figure 7.1)

Don't let other approximate AC voltages confuse you. You will hear 110 volts or 115 volts or even 117 volts. The range is 110 to 120. However, actual readings may be greater or less. There are two conventions 110 and 120. Old timers use 110, but the catalogs typically spec electrical equipment at 120 and 240 for utility voltage. Unless your equipment specifically requires something other than utility power, 220, 230, and 240 volts are also interchangeable. The simple test is if you can plug it into grid power, call it what you will—it's 120 or 240 volts AC.

Europe has different voltages, generally 230 volts, 50 Hertz (named in honor of an early experimenter in electricity). The difference is the frequency at which the current alternates (thus, alternating current or AC). In the U.S. we use AC at 60 cycles per second or 60 Hz.

While the purist may want to operate his PV-powered home entirely DC, there are some limitations. Besides limited availability of appliances, larger wire is required to carry the same power load. For example, a 200-watt load at 12 volts needs wire large enough to carry 16.6 amperes, whereas, the same load at 120 volts is only 1.66 amperes.

Too often in PV discussions we hear a lot about inverter efficiency losses and very little about the cost of large wire used for comparable DC loads. We won't mention wire loss inefficiencies because of undersized conductors. That kind of penny-pinching is, in fact, just throwing good money away. Sometimes those losses and costs can be as much as the difference in the cost of using an inverter. Factored into cost somehow should be the time spent making do with DC, but that is not often the case.

Inverters come in all sizes, shapes, and price ranges offering a vast array of options. Some inverters produce the simple square wave suitable for most loads. While there is reason for concern about the quality of square-wave AC, it will do the job. Surprisingly, most computers will operate on square wave. Some computers have power conditioning equipment built into their power supplies which allows for almost any quality of electricity. This is done because grid power is so variable. I powered my Apple II Plus and two disk drives and Epson printer with a very simple 550-watt square-wave inverter satisfactorily.

Recently, my office was moved to the front of the building which is serviced by the utility company, while my PV system remained connected to the old office. One morning I turned on my computer and a surge or spike zapped the computer power supply resulting in costly repairs. When I got the computer back from the shop, I ran a 120-volt AC line from my old PV system to the front office, specifically dedicated for my computer. Needless to say, I am relieved to be back on smooth and reliable PV power again. (When I'm on the road, I also avoid unregulated grid power from ruining my computer. A second battery in my car and a 300-watt inverter are my portable office power supply and Uninterruptable Power System. If I have to stay on location a few days, a portable PV array provides all the power my "office on the road" needs.)

More costly inverters put out a modified, or stepped, square wave which more closely matches grid power. Some test equipment needs the exact type of power produced by the utility company (60-cycle reference sine wave) to operate. In that case, a costly sine-wave inverter is necessary. The early sine-wave inverters were actually square-wave devices with ferro-resonant transformers to smooth out the flip-flop square wave. Thus, efficiency was sacrificed in the transformer to produce pure sine-wave inversion.

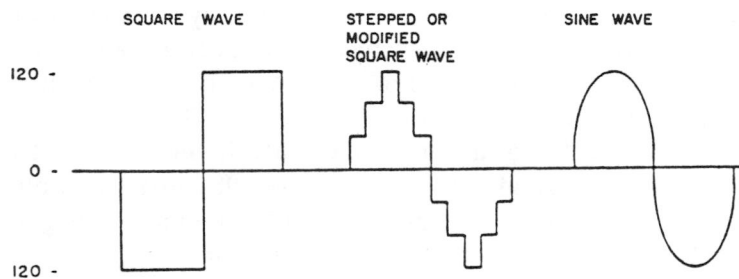

FIGURE 9.1: The wave forms of various inverters from the most primitive simple square wave to pure sine wave. For most applications, the stepped wave will work well and reduce inverter costs.

Now there is a new generation of inverters: relatively lower cost, very high-efficiency digital sine-wave inverters. They can power anything—just like grid power—and they are changing the way PV will be used.

So far we have been discussing solid-state, or electronic, inverters. An old standby is the rotary or motor inverter. This device is a motor which drives an alternator. The DC power runs the motor. A common shaft ties the motor and alternator together, and the output of the alternator is sine-wave AC.

If rotary inverters put out a nice sine wave, why aren't they used more often in PV systems? The simple fact is that they are not very efficient. At less than 65% efficiency, and as low as 30% efficiency with loads 20% of the rated output, rotary inverters are not particularly suited to PV. This is primarily due to the amount of energy required to move the motor and alternator. PV electricity is too precious to waste on spinning parts. But for the occasional, short-duration sine-wave load, a rotary inverter may be just the thing.

Also, rotary inverter output voltage varies in direct proportion to the DC input voltage. This means that as the battery discharges, the output voltage falls with it. Even though the output is a sine wave, you can see voltages varying from as low as 90 volts to upward of 140 volts AC.

One nice thing about rotary inverters is that they can really take a beating. They handle motor-starting surges well. I know of cases where rotary inverters were used until they began to heat and smoke. After cooling off a while, they were called back into service and performed without any problem. You can't beat that for durability.

On the other hand, a good solid-state inverter should have enough protective circuitry to do the same. Of course, it shouldn't have smoke coming out of it, but if it can't take motor surges and occasional overloading, then it is under-designed.

I mention under-design because inevitably an inverter is called on to handle loads beyond its rated capacity. People have a tendency to pinch pennies when buying an inverter, opting for the smallest size they can get by with. Then they push it to the maximum. Inverter manufacturers know this. For that reason, wise inverter builders factor in extra capacity to insure long life. But don't rely on this "fudge factor." Keep the use of your inverter within its factory ratings.

Solid-state inverters may also have an automatic demand on/off. This means that you can turn the inverter on and leave it on, using relatively little power, until a load comes on. When it senses a load, the needed power is applied. Without a load, it waits. In the on and no-load state your inverter should use little power. If your inverter does not have a low no-load mode, be sure to turn it off when not in use. Do not waste your PV power.

What can you power with an inverter? Basically anything that operates from the grid can be operated with the modern solid-state inverter. Resistive loads like coffee makers, toasters, and hair dryers will operate with no problem, as well as incandescent lights. Fluorescents may not work well if your inverter does not interface properly with the fluorescent ballasts.

Some induction or brushless AC motors have very high starting surge requirements. Well pumps, garbage disposals, dishwashers, refrigerators, air conditioners, and washing machines all have high starting surges. These surge requirements can be as much as five times the normal operating power requirement. Thus, the 300-watt motor load of a refrigerator may not even start if your inverter is rated less than 1500 watts.

Some motors do not work well with some inverters. Why is this? Is it the fault of the motor or the inverter? The answer is both. Motors are made as cheaply as possible nowadays. Most manufacturers leave out what is commonly called the motor-run capacitor. The job of this capacitor is to smooth out the interaction between the load (motor) and the power source. Grid power with its pure sine wave does not create as much bad feedback (inductive reactance) with such loads, but some inverters do.

If you find that motors are not running up to normal speed with your inverter, try putting a 3- or 4-microfarad 400-volt electrolytic capacitor across the motor windings. Such a capacitor may be purchased for under $10 at most motor stores. Be sure to note polarity when installing the capacitor. If you don't know what you are doing, ask questions. Also, be sure to put the capacitor on the motor side of the switch so that it does not load your inverter when the motor is off. If you use an inverter designed for reactive loads, this should not be a problem.

SELECTING AN INVERTER

There are a few basic guidelines to use in selecting an inverter for your PV system. First, you need to know your power requirements. We always come back to this. The power requirements identify the loads and how long they will be used. Your power requirements list will give you an idea of what will be operating on AC and what AC loads will be operating at the same time. The inverter must be able to handle the combined AC loads which will be operated simultaneously. In fact, the inverter must be able to handle the surge of these loads, too.

Note the type of loads to be powered by the inverter. Are they motor loads with high starting surges? Is complex electronic equipment to be powered? How about kitchen loads? Will you be running the garbage disposal and dishwasher at the same time? If so, don't forget to include the well pump as you will be using water, too. And finally, don't forget the refrigerator. It can come on while all these other loads are running.

Total your automatic loads and surges. Then add up demand loads, such as dishwasher and washing machine, and your convenience loads, such as blenders and toasters, separately and in possible combinations of simultaneous operation. After you have done that, you may find that you will have to monitor your combined loads to keep the size and cost of your inverter to within reason. (Bill and

Jackie Perleberg, you'll remember, do just that with very little trouble.)

There are a couple of other reasons to look at your power requirements when considering an inverter. When you size your system and allocate your budget, the inverter should be considered in your first purchase. Although inverters come in a variety of sizes, it is false economy to buy a small inverter you will soon outgrow. Buy the inverter that will suit your future needs now.

There's a good reason for up-sizing. When starting out, you may have lots of construction or remodelling work to do. A bigger inverter will help you through this period with ease. It sure is nice to have a quiet inverter instead of a noisy generator when you are doing carpentry. And besides, what if your generator breaks down? Then you are faced with a repair job before you can do the construction work.

Once you are settled in, the over-capacity of a larger inverter bought for construction purposes still has merit. Operating inverters at or near capacity may give high efficiency, but also may lead to shortened inverter life. Some over-capacity is good insurance.

On the other hand, a small inverter may be all you can afford, especially if you are installing PV in stages. Be sure to plan each additional part of your wiring though, or you will have imbalanced circuits. If you use a couple of smaller inverters for your loads, you have insurance in the form of redundancy. If one inverter should fail, you'll still have the other one. However, if the bigger one fails, will the smaller one adequately operate important loads?

This brings us to cascading inverters. While no inverter can be paralleled on the same circuit without one burning up the other, some inverters can be tandemed for increased capacity.

In 1985, Heart Interface began marketing inverters which could be tandemed for bigger loads. They did this to improve efficiency and to eliminate the costly 5000-watt inverter. Unless your loads are always big, a 5000-watt inverter is not such a good idea. It just doesn't make sense to use a 5000-watt inverter to power a 100-watt load. On the other hand, if you need an inverter to handle a 5000-watt job, you must use one. Cascading or tandem inverters get around this mix. By stacking up to four 2500-watt inverters, you can handle big loads. The stack will be called on only if needed.

Should you ever have a problem with one of your cascading inverters, you will still have the others. Not a bad idea. For planning purposes, cascading inverters make sense, too. Let's say you are just getting started and want to power your home with PV. Your eventual full complement of loads will be large. However, in the beginning you may be operating on a limited budget or requiring a limited amount of power. By using the first of a cascading inverter set, you can squeeze by. Later you can add on as needed.

A word of warning: Never put two AC inputs on the same circuit or something will burn. That means never feed your generator or grid power into a circuit being powered by an inverter at the same time. Wire your home so that this can never happen. Never feed generator or inverter power into the utility grid without the power company's permission. In the first place, it is illegal. More importantly, it endangers power line workers.

Buying an Inverter

When you buy an inverter, get one that can be repaired. And hope that the company that manufactured it and the person who sold it to you will stay in business. Buying close-out inverters is unwise. Where will you get technical support when you need it?

It is always a good idea to ask your seller if you can call for technical information after the inverter is in place. If you have hired someone to install your system, get a service contract that includes replacement. This may cost extra but it may be worth it because should the inverter fail you won't be without power.

All inverters must be tested when installed. Some may have to be adjusted for the specific installation. Be sure to ask "What if . . . ?" If you don't like the answer, shop elsewhere. A reputable inverter sales outlet should satisfy your every need.

Disadvantages

With all this talk about the wonders of inverters and their benefits, what are the disadvantages? We will not even consider poor quality, low-efficiency inverters. Of the good inverters, there will inevitably be problems, usually in the first few months, as they "burn-in" or get adjusted to regular use. Other problems occur because of bad installation, undersized DC input wires, rough handling, and overloading.

A disadvantage already mentioned is your reliance on the inverter to power all your loads. Should your inverter fail, even if fully warranted and the repairs or replacement are cost-free, you will be without all or some of your load-carrying capacity. If everything electrical is AC, then you are out of luck—and power. Let's hope you at least planned in DC emergency lights and water to carry you through any waiting period for parts and repairs.

A less important disadvantage of inverters is efficiency. This has become a secondary consideration since the advent of 90% plus efficient field effect transistor (FET) inverter technology. But it still should be considered. If your system is small and your budget limited, perhaps an inverter is not for you. A 10% loss can be costly if you are operating on a shoestring.

Conversely, an inverter makes wiring a lot cheaper. You can use readily available low-cost standard wiring and hardware. Switches and breakers and all the other goodies used in AC homes are relatively inexpensive. The savings on the wiring along with the convenience and savings in using standard appliances make shoe-string DC almost a thing of the past.

THE PV/GRID CONNECTION

It is possible to install a medium- to large-sized residential PV array and remain hooked up to the utility power grid. The advantage of this type of installation is that the battery storage system is eliminated. When the sun is not producing enough power to run the home's electrical appliances, or the peak load is greater than production, the grid-connect PV home gets its power from the utility company. When PV production exceeds consumption, the home is credited or actually paid for the power it produces and puts into the utility grid.

Although there are a few thousand grid-connect wind power homes, there are far fewer grid-connect PV homes. The two main reasons for this are economics and a lack of marketing. It has been difficult to present an economic argument for PV/grid homes because of their cost compared to paying a monthly electric bill. But like so many things that are not "cost-effective," PV/grid homes could have been marketed to the affluent for other than economic reasons. Unfortunately, they were not. On the other hand, Wind-

FIGURE 9.2: Jon and Eileen Giltner live in Boulder, Colorado, and use both utility and PV power. Their PV system powers the 12-volt solar water heater pump, fan, and lighting. Four 35-watt modules charge two 6-volt 240 ampere-hour batteries. Installer: Dr. Mark McCray. (Courtesy Rocky Mountain Solar Electric)

works, maker of the Gemini Synchronous Inverter, did an excellent job building a reliable grid-connect inverter for wind systems and did a good job selling it. In fact, they looked so good that a Wisconsin utility company bought the company.

In addition, Windworks was able to get parity pricing for their customers. They did all the system engineering with the approval of the utilities and, in exchange, customers usually got one meter installed, which ran backwards when the system produced power. Nowadays we generally see two meters on these systems—one for buying power and one for selling. Parity pricing for alternative power systems is pretty much a thing of the past.

So the development of grid-interconnect windchargers proceeded smoothly with intertie equipment costing about as much as a battery bank. Parity pricing and the elimination of battery maintenance were important selling points.

A few Gemini inverters were also used on PV systems. Today we see wind and hydro grid-interconnect systems using a variety of inverters similar to the Gemini.

Synchronous inverters are line-commutated, line-feeding inverters which change DC power to AC at standard line voltages and frequency. In operation, all the available DC power is converted to AC. If more power is available from the DC source than is required by the home, the excess flows into the AC grid where it is used by others. If less power is produced than is being used, the difference is provided by the AC grid.

FIGURE 9.3: The Department of Water and Power (DWP) of Los Angeles, California, Optimum Energy House is open to the public. Featuring energy conservation features and solar water and space heating, it recently was retrofitted with 64 modules. (Photo: F. B. Orner)

These inverters have circuitry capable of handling unregulated DC power input. For PV arrays, where the maximum power output is not a function of a single variable, automatic tracking circuitry seeks the highest output by incrementally varying the loading of the array while monitoring the power output. Therefore, the array's highest wattage regardless of a standardized voltage is what the electric meter sees.

Do not confuse power tracking with physically tracking the sun. Power tracking is electronic peak power-seeking circuitry built into the power-conditioning equipment. In some DC PV water

FIGURE 9.4: The Los Angeles DWP demonstration house has two utility meters. During the day, the PV array's output is sold to the utility. At night, the surplus electricity is bought back. No battery storage is used in this system. (Photo: F. B. Orner)

pumping systems, peak power tracking is used to match array output to optimum pump motor operation.

The installation of synchronous inverters is generally beyond the scope of the average homeowner. An electrician is needed to do the job. In addition, utility company engineers will want to be involved to insure that the system is safe and producing power equal in quality to grid power.

The belief has been that when PV prices drop, more grid-interconnect systems would be installed. However, when PV modules were selling at half price (as they effectively were when the tax credits were available), there was no rush to install these systems.

Another factor that has limited the installation of PV grid-interconnect systems is that utility companies considering grid interconnect of power fields have the advantage of economy of scale. Why would someone interested in PV and able to afford a grid interconnect system do so if their utility company was going to do the same thing on a larger scale at a lower price per watt? Logically, by paying your monthly electric bill, you could buy utility solar electricity and support a much grander step toward a PV society.

As it turns out, few utilities are willing to put in PV power fields. Those utilities that have, publicize their efforts out of proportion and give the impression that they are "going solar" for their customers. Thus, rate payers are discouraged from the

FIGURE 9.5: An American Power Conversion 2-kW utility interconnect synchronous inverter (upper right) is used to link the Optimum Energy House (Figure 9.3) to the grid. Additional monitoring equipment provides data for utility company engineers. Cooperative utilities help promote PV and improve their public image. (Photo: F. B. Orner)

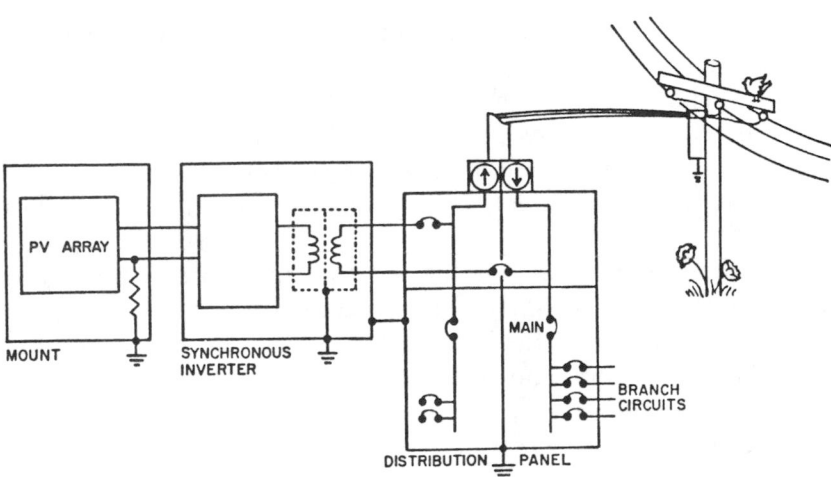

FIGURE 9.6: The synchronous inverter interconnects the solar array to the home and the utility company. At the main distribution panel, power can flow in two directions. During the day when an excess of PV power is being produced, it is sold to the utility. At night or when loads exceed PV production, power is bought by the utility.

PV/grid connection because they mistakenly think their utility company has done it for them.

But even with short-term economics against them, some true pioneers have put in PV/grid-interconnect systems. Not only did they have to convince themselves of the merit of PV, they also had to convince their utility company. Now that a few people have made the grid interconnection, we are seeing more cooperation from the utilities. In fact, some of the most publicized grid interconnections have been those done by utilities.

Steve Strong, architect and designer, has done some excellent work on grid-interconnect homes. All have been costly, but they have paved the way for general acceptance of the concept. Steve's first grid-interconnect PV home (built in Massachusetts) cost about the same as non-PV homes in the same neighborhood. His work has been featured on the PBS series "The All New This Old House." Perhaps one of the most important impacts of his "Impact 2000" PV-powered home has been on Boston Edison, the utility that commissioned the job, for they are now able to see that PV can work in their locale. It is also providing valuable design guidelines for other architects willing to follow in Steve's steps.

In conjunction with the groundbreaking grid-interconnect homes Steve has done, he has completed a number of stand-alone homes with all the conveniences and appliances a family could want. Whether working on a small cabin-type system or on a larger scale, his main goal is to bring PV into the mainstream. The work he has done is helping make this happen.

FIGURE 9.7: The Boston Edison IMPACT 2000 demonstration home was featured on the 26-week PBS series The All New This Old House. Designed by Solar Design Associates, Boston, Massachusetts, this home generates 4.3 kW peak. With no battery storage, the system is a utility buy/sell, providing 100% of this large and luxurious home's spring and fall power needs. (Photo: Robert Sardinsky, Rising Sun Enterprises)

Chapter 10

Wiring

SAFETY FIRST!

Wiring a PV-powered home is no different than any other wiring job when it comes to safety. It is very important to follow proper wiring practices to avoid shocks, fires, and other hazards. If you live in an area where building inspections are required, consult with the local authorities. Living in a wilderness area, however, does not free you from safety concerns. In fact, remote home dwellers' freedom from the interference of building inspectors must be tempered with the responsibility of providing for their own safety. Sloppy, inadequate, or unsafe wiring is an invitation to tragedy.

Safety is based on knowledge and knowledge rests on a foundation of fundamentals. Once you know what is required for the safe wiring of a conventional utility-powered home, you can modify available equipment to serve your special needs. There are many easy-to-understand instruction books on conventional home wiring. Sears has good, inexpensive how-to booklets for the do-it-yourselfer, and all public libraries have home wiring books. *How to Be Your Own Home Electrician* by George Daniels is a favorite reference.

Table 10.1: Wire Types

Type Letter	Trade Name	Max. Operating Temp.	Application Provisions	Insulation	Outer Covering
T	Thermoplastic	140°F 60°C	Dry Locations	Flame-Retardant Thermoplastic Compound	None
THHN	Heat-Resistant Thermoplastic	194°F 90°C	Dry Locations	Flame-Retardant, Heat-Resistant Thermoplastic	Nylon Jacket or Equivalent
THW	Moisture and Heat-Resistant Thermoplastic	167°F 75°C 194°F 90°C	Dry and Wet Locations	Flame-Retardant, Moisture- and Heat-Resistant Thermoplastic	None
THWN	Moisture- and Heat-Resistant Thermoplastic	167°F 75°C	Dry and Wet Locations	Flame-Retardant, Moisture- and Heat-Resistant Thermoplastic	Nylon Jacket or Equivalent
TW	Moisture-Resistant Thermoplastic	140°F 60°C	Dry and Wet Locations	Flame-Retardant, Moisture-Resistant Thermoplastic	None
SE			Service Entrance		Flame-Retardant, Moisture-Resistant
UF	UG Feeder & Branch-Circuit Cable-Single Conductor	140°F 60°C 167°F 75°C		Moisture-Resistant Moisture- and Heat-Resistant	Integral with Insulation
USE	UG Service-Entrance Cable Single Conductor	167°F 75°C		Heat- and Moisture-Resistant	Moisture-Resistant Non-Metallic Covering

Type Letter	Trade Name	Max. Operating Temp.	Application Provisions	Outer Cover
NM	Romex	194°F 90°C	Dry Locations (i.e., indoors)	Plastic Sheath Flame-Retardant Moisture-Resistant
NMC	Romex	194°F 90°C	Dry or Moist Locations* (i.e., laundry rooms, basements) Not outdoors or buried	Solid Plastic Sheath Flame-Retardant Moisture-, Fungus-, Corrosion-Resistant
UF	Romex	194°F 90°C	Wet Locations* Outdoors Direct Burial Corrosive Locations	Plastic
Armored Cable	BX	140°F 60°C	Permanently Dry Locations Locations needing protection from penetration	Flexible Steel or Aluminum
SO			Dry or Damp Locations Portable Tools and Appliances	Oil-Resistant Moisture-Resistant Ultraviolet-Resistant
SJ & SJO			Hard Usage	

*Not permitted: in storage battery rooms; as service entrance; embedded in concrete; where exposed to sun.

Wiring ties the components of a PV system together. It is the conductor through which your electricity flows. Proper wiring means proper performance, so don't think of wiring as a minor element. Since wiring usually costs far less than other components in the PV system, skimping is foolish. It makes no sense to throw away power because of undersized wires.

The size of a wire determines its capacity to carry current. The larger the wire, the more current it can carry; the larger the wire, the smaller the number ascribed to it. A 6 AWG wire is larger than a 14 AWG wire. (AWG stands for American Wire Gauge and is a standard for wire sizing based on the cross-sectional area.) Like a hose, the bigger the cross-sectional area or diameter, the bigger the wire conductor.

Wire comes in various metals. While you may be able to get a good price on aluminum wire, use it cautiously. Check with your local building inspection office since many areas forbid or restrict its use. Mixing different wire materials can cause problems. Use special fasteners to tie dissimilar wire together to avoid electrolysis.

I consider only copper wire acceptable. I also consider not more than 2% wire losses acceptable. Although 5% wire loss has been the standard for residential PV systems since its early days, the *National Electrical Code* calls for 2% loss or less. The DC voltages from solar modules are higher than 12 volts, and this has been used as an excuse to allow for 5% wire resistance losses. However, that resistance means loss of power. The reason the wire in your 12-volt system seems big is that it must carry ten times the current needed to perform the same task at 120 volts.

Now is a good time to dispel a myth about solid versus stranded wire. All things being equal, there is no difference between the two. Electricity travels through the wire conductor, not on its surface. While current does skim the surface of wire carrying thousands of volts, that is not the case in home use.

The main difference between solid and stranded wire is ease of working. Stranded wire is more flexible and easier to bend and run through conduit or wrap around a screw terminal. Solid wire, especially large diameter sizes, can be difficult to work with (though needle-nosed pliers will help).

Another difference between solid and stranded wire is that solid wire must be securely supported at connections to avoid flexing and eventual breakage. If you run solid wire to your PV array, be sure to fasten it so that wind does not cause movement. Of course, all wires, solid or stranded, should be fastened and not hanging loose.

Solid wire is almost always used in home wiring. Romex (a brand name like Kleenex) has become so standard that electricians will call almost any home wiring romex. It is typically 12/2 or 10/2 with ground, which means the white outer insulator holds three wires. One is bare and is used as a ground wire, the others are separately insulated. In AC circuits the white insulated wire is the neutral wire, and the black is the hot wire.

Wire insulation can have one or more of several characteristics. Some wire insulation is acceptable if used in conduit, metal, or plastic piping. Other wire can be used outdoors overhead. Some wire designated UG can be used underground or for direct burial. Some insulation can resist ultraviolet (UV) rays. Still other wire is oil-resistant.

While this book will touch on AC house wiring, you are encouraged to read at least one of the recommended books or booklets (see Appendix D) which covers specifics like three-way lights, running conduit, pulling wire through walls, placing outlets, etc. The DC wiring in this book is very basic and provided to get you started with a small PV system.

However, if you have any doubts about your ability to wire your system or your home, *do not* attempt to do the job. Electricity can kill and bad wiring can cause fires. If you have any doubts, hire a professional electrician to either do the job or check your work.

The wires used in a PV system are:

Module interconnects
Array wiring
Battery interconnects
Inverter wiring
Interface wiring
House wiring

Starting from the sunlight converted to current flow within the solar cell, the first actual wiring encountered in a PV system is the cell interconnect. This wire is built into the solar module and carries current from one cell to another, in series, to increase voltage. Cell interconnects are internally connected to the modules' plus and minus terminals.

The *module interconnects* are the first wires in your system to bring solar electricity into your home. Just as cell interconnects tie the cells together for higher voltage or current, module interconnects tie the modules together. If you are installing a 12-volt system and using 12-volt modules, connect your array in parallel. For 24 volts, you must connect the same modules in groups of two in series.

To change my eight-module office system from 12 to 24 volts, I re-wired the array. First, I removed the parallel module interconnects. Then I made four two-module groups by interconnecting them in series. Next, I interconnected the four series groups in parallel. Should a module fail, the array will still work, less the failed module and its partner.

An alternate method of 24-volt wiring is to interconnect each half of the array in parallel and then tie the two halves together in series. This alternate method has the advantages of simplicity and using less wire. Its main disadvantage is that should one interconnect in the middle of the array fail, part of the array will be cut out of the circuit. This usually is no problem as wires have high reliability. Just make sure your connections are as reliable. Be sure you have sized your parallel wires to carry all the current.

Systems of 48 volts, or more, are wired in the same manner. First, develop the necessary voltage with series groups called strings. Next, parallel the strings to develop the operating current. In utility interconnect systems, there can be as many as 15 or more modules in series for over 200 volts DC.

Module interconnects must be "weather wire," since they are exposed to the elements. Spade or ring connectors can be soldered to the wire ends to give a secure connection at the module terminals. Be sure to fasten your module interconnects so that they do not flop around in the breeze.

The next element in the wiring of a solar electric system is the *array wire*. This generally means the wire that comes from the array and goes to the battery bank. This wire, therefore, has the regulator or charge controller somewhere in its path. The array wire can be very large if you are running low voltage and high current a great distance. The array wire may be the most expensive wire run in your entire system, including the home wiring.

Most modules do not have adequate terminals for large wires. Their terminals are generally designed for 10 AWG. If you have to use larger wire, there are two ways to make the connection. The standard method is to bring the large wire to a separate junction or terminal box which should be located as close to the array as possible. Then one or more smaller conductors are used to tie the modules and array wire together.

An alternate method is to pigtail a small wire onto the end of the large array conductor. This short pigtail is soldered to insure a good electrical connection and then taped for insulation. The pigtail can then be fastened to the small terminal screw.

Remember, when you use smaller wires because of undersized regulator or module terminals, you are restricting the flow of electricity. It's like putting the squeeze on a garden hose. Pinch it too much and you have trouble. Too much resistance can cause loss of power or can heat up the insulation and possibly cause a fire. Always consult the wire size charts.

The wire from the array to the regulator does not need to be weather wire its entire length. You can use another suitable type once it enters the house or conduit. If you bury the wire from the array to the regulator or house, use direct-burial UV-stabilized wire the entire length.

Since the regulator is usually located indoors near the battery bank, indoor wire can be used. The wire from the regulator to the battery bank carries the entire output of the solar array. Don't put in Romex 12/2 just because you are doing inside wiring. Again, consult the wire chart.

An important consideration when planning your wiring is voltage and the building codes. Code calls for wire in conduit if the voltage is 32 volts or greater. There is a reason for this. Foolish disregard of safety codes is not a sign of defiance against authority. Even if you live miles beyond the reach of inspectors, the laws of physics still apply.

With power to your batteries, we are ready to look at *battery interconnects*. Commercial stationary and forklift batteries use standard interconnects. These are sometimes wire, but more often strips or bars of metal specifically designed for the battery bank.

Most golfcart battery banks use 4 AWG "welding wire" for battery interconnects. Kept as short as possible, this wire has insulation that can resist battery chemicals, but it is not acid-proof. (This is the same wire used for welding cables, thus its name.) With proper ring or spade connectors, this 4 AWG stranded wire is easy to work with and will serve most needs.

Should you require long wire runs to tie your batteries together, larger wire might be needed. However, proper planning and layout of the battery storage area and battery bank should prevent unusually long-battery interconnects.

You will need to bring power from the batteries to your inverter *(inverter wiring)*, if you are using one. Otherwise, with a 12-volt DC system you will either go from your regulator load terminals or batteries to a DC load panel *(interface wiring)*. This panel may be a simple terminal strip, an automotive fuse block, or DC circuit breaker box. The wire tying your batteries to the DC load panel must be large enough to carry the combined loads in the home. Be sure to fuse DC output from your battery bank.

When wiring batteries to the inverter, follow the inverter manufacturer's recommendations for wire size and length. In general, inverter wiring should be as large and as short as possible. In my office I use 2/0 (00 called double aught) wire, 2.5 feet in length.

Standard house wiring methods are used from the inverter to your AC load panel circuit breaker or fuse box *(interface wiring)*. The PV DC side of the system is covered in the *NEC* Article 690 (See Appendix E).

WIRING SPECIFICS

A standard service entrance located on the exterior wall, inside near the battery bank, or preferably, near the utility or generator service entrance is recommended. Grounding and lightning arrestors should be located at the same place. A manual disconnect which can also serve as a manual transfer switch (if needed) should be located at the service entrance on the exterior wall (or be otherwise accessible). The array ground and negative bus from the array and neutral AC must remain connected to ground even when the transfer switch is opened.

The positive leg of the DC system should be properly fused at 110% of full current. This fusing can be a part of the disconnect switch or in a separate unlocked enclosure. Although conduit is

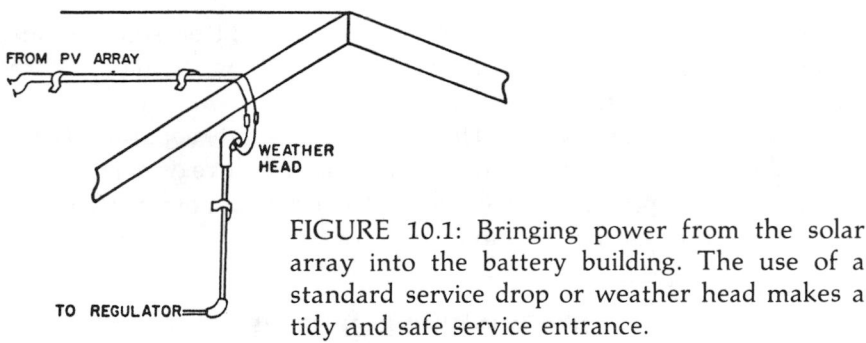

FROM PV ARRAY

WEATHER
HEAD

TO REGULATOR

FIGURE 10.1: Bringing power from the solar array into the battery building. The use of a standard service drop or weather head makes a tidy and safe service entrance.

generally not required for circuits under 32 volts, it is recommended on the battery to inverter circuit. Safety is worth the additional cost. Most 5000-watt inverters require 48-volt DC input; conduit must be used on the inputs to these inverters.

The regulator or charge controller should be protected with a circuit breaker or fuse. It is recommended that a quick disconnect plug in the positive leg of the DC circuit at the battery location be included. A similar disconnect should be included in the DC positive leg to the inverter. A good inverter will be both thermally and electrically protected from overheating and shorts.

The AC circuits from the inverter to the generator or utility should be protected with a circuit breaker. AC circuits to the house are installed as recommended in the *National Electrical Code*. The use of conduit, a main distribution box, a main circuit breaker and branch circuit breakers with adequate grounding is required.

All circuits, switches, breakers, and boxes should be clearly labeled. Identification tags are important. You can't remember every detail of your wiring. The day may come when someone else has to work on the system and you are not available. It is a good idea to keep an equipment log with a record of receipts, warranties, dates put in service, maintenance checks, repairs, modifications, notes, and so on. This log should be kept near the main breaker box or control panel.

While stranded wire is preferred, solid copper wire can be used, though it is a little harder to work with. DC conductors are, by convention, red for positive and black for negative. AC conductors are black for phase A, red for phase B, white for neutral, and green for ground. All splices and terminal ends should be soldered and wrapped with electrical tape or heat-shrink tubing so no bare wires are exposed. Wire nuts can also be used.

Bypass diodes can be used in the array to isolate separate strings of modules. In the event of a module or string failure or partial array shading, the bypass diode will allow the array to perform without damage to the rest of the modules.

COPPER WIRE PROPERTIES

The *National Electrical Code* contains useful information regarding the properties of electrical conductors. Some wire types and sizes may not be available locally. To meet current-carrying capacity and to avoid excessive resistance voltage losses in DC circuits, it is possible to double-up or triple-up conductors and "make" a larger conductor. For example, you have a wire run using Romex AWG 10/2 with ground. Each #10 conductor has an area in circular mils of 10,380. By twisting all three conductors together you get a total of 31,140 circular mils which is greater in current-carrying capacity than one #6 conductor.

This method works well when adding modules to an array. By coupling the existing plus and minus leads and running another new wire of equal size (to keep the circuit balanced), you can avoid the cost of replacing the whole length of array wire. Be sure to check the insulation on each leg of the array wire run for bare spots that will cause a short circuit. Also, take special care at terminals to insure that all conductors are securely fastened.

There is a practical limit to the number of conductors that can be doubled-up, as well as other restrictions. For instance, conductors in a specific circuit must be run in the same conduit. That is, you cannot split circuits. The danger in doing so is that unbalanced, mismatched, or short circuits may occur. Good wiring means keeping conductors paired or grouped neatly together. Be sure your conduit is large enough to add wires later.

Another limit to doubling conductors is cost. It may be less expensive to add a single lead to a doubled-up lead when making an addition. However, proper planning and sizing of conductors and conduit should make doubling-up unnecessary. When you design your system and make your parts list, allow for system expansion

Table 10.2: Copper Conductor Properties

AWG Wiring Size	Area Circular Mils	Ohms 1000 ft
14	4,110	3.07
12	6,530	1.93
10	10,380	1.21
8	16,510	0.764
6	26,240	0.491
4	41,740	0.308
3	52,360	0.245
2	66,360	0.194
1	83,690	0.154
1/0	105,600	0.122
2/0	133,100	0.0967
3/0	167,800	0.0766
4/0	211,600	0.0608

and additions. It is always easier and less costly to put in the right size wire initially.

WIRING THE 12-VOLT PV-POWERED HOME

The following information has been provided by Steve Willey, who lives in Idaho in a unique home powered by two windchargers and photovoltaic panels. He has extensive electronics knowledge and practical experience, and teaches classes in energy conservation and solar energy. Professionally known as Backwoods Cabin Electric Systems, Steve provides information, design consultation, and equipment for the home alternative energy user.

How To Get It Together

The easiest way to connect your PV panels to your batteries is to use a meter and fuse box like the one made by Backwoods Cabin Electric Systems (Figure 10.3). Hang it on the wall and connect the wires from your solar panels, batteries, and five house wiring circuits. The interconnects described here are already done within the box, and the parts mentioned are mounted on the front. One meter shows how much current is coming in from the solar panels; another

Table 10.3: Copper Wire Size Chart

Amperes	14	12	10	8	Gauge 6	4	2	1/0	2/0	3/0
120 Volts										
2	225	350	575	900	1500	2400	3825	4800	5800	7300
4	100	175	275	450	725	1200	1912	2400	2900	3650
6	75	120	175	275	450	725	1275	1600	1934	2434
8	55	85	145	225	355	570	957	1200	1450	1825
10	45	70	120	190	300	480	765	960	1160	1460
15	30	45	70	120	190	300	480	765	960	1220
20	X	35	55	90	145	225	360	575	725	915
25	X	X	45	70	115	180	290	460	580	730
30	X	X	X	60	95	150	240	385	485	610
40	X	X	X	45	70	115	180	290	360	455
50	X	X	X	X	55	90	145	230	290	365
30 Volts										
2	56	87	144	225	364	572	900	1440	1820	2290
4	25	43	68	113	182	286	450	726	900	1145
6	18	30	43	75	119	188	300	482	610	765
8	13	21	36	56	88	143	225	363	450	570
10	11	17	28	45	71	114	182	288	364	458
15	7	11	17	30	47	75	120	192	240	306
20	X	8	13	22	36	56	90	144	182	229
25	X	X	11	17	28	45	72	115	145	183
30	X	X	8	15	23	37	60	96	122	153
40	X	X	X	11	17	28	45	72	90	114
50	X	X	X	X	13	22	36	57	72	91
12 Volts										
2	22	35	57	90	145	228	360	580	720	912
4	10	17	27	45	72	114	180	290	360	456
6	7	12	17	30	47	75	120	193	243	305
8	5	8	14	22	35	57	90	145	180	228
10	4	7	11	18	28	45	72	115	145	183
15	3	4	7	12	19	30	48	76	96	122
20	X	3	5	9	14	22	36	57	72	91
25	X	X	4	7	11	18	29	46	58	73
30	X	X	3	6	9	15	24	38	48	61
40	X	X	X	4	7	11	18	29	36	45
50	X	X	X	X	5	9	14	23	29	36

Note: All distances in feet from point to point and include roundtrip 2% resistance.

FIGURE 10.2: Steve and Elizabeth Willey's solar- and wind-powered home, Sandpoint, Idaho. This unique home with two windchargers and a solar array has all the comforts of the average American home—but uses less than one-tenth the power. (Courtesy Backwoods Cabin Electric Systems)

shows how much current is being used in the house. There are switches to shut off either current. The diode necessary for PV panels is already hooked up to the meter box. A diode that will accommodate more panels is available.There is an indicator to test the diode and another to let you know if anything electrical has been left on in the house when you leave. Another option is a connection to recharge flashlight or radio rechargeable ni-cad or gel-cell batteries from your house batteries. On the front panel are six snap-in fuses—one for the battery and five for power to your household lights and outlets.

The circuits involved are really pretty simple. The following sections will explain how to set up your own panel-to-battery connection, and how the parts work.

Equipment Availability and Specifications

Wire

For your rooftop PV array, consult the wire size chart in Table 10.3. Secure the wire to the mount to prevent its weight from pulling against the terminals on the panels. There should be some slack at

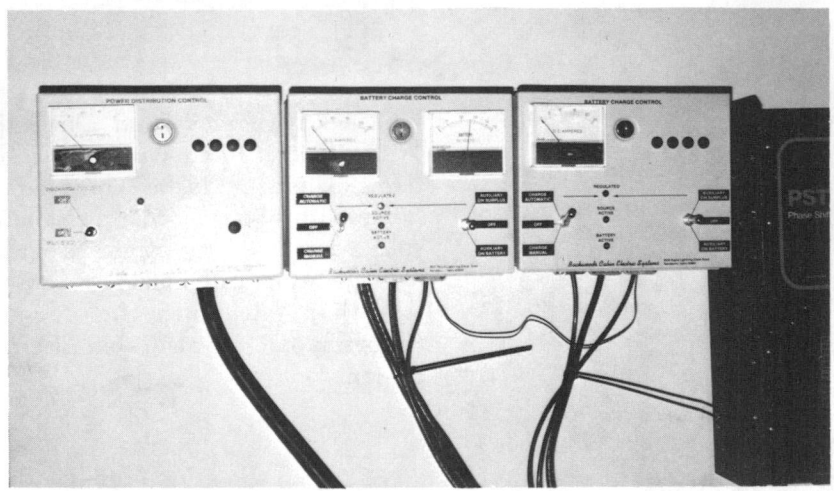

FIGURE 10.3: Backwoods Solar Electric Systems equipment from left to right for distribution, regulation, and monitoring. Meters and indicator lights show battery, array, and load status. (Courtesy Backwoods Solar Electric Systems)

these connections to avoid strain. Use single separated pieces of this wire to connect individual panels, making note of the color used for positive and negative so that you will connect it correctly at the other end. Be very careful not to reverse the battery connection during installation. That would result in the battery and panel pushing in the same direction in the charging loop, and too much current would flow for the survival of the panel. Observe all + and - markings on panels, batteries, and diagrams.

Diode

Each diode in the Backwoods Cabin panel has a rating of the maximum current it can carry in the forward direction and maximum voltage it can withstand in the opposite direction where it will not pass current. Most will withstand 50 volts or more. I use 200-volt rated diodes for a safety margin since the cost is only a little more. Diodes come in current ratings of less than 1 amp to 500 or more. Select one rated for more current than you ever plan to run through it. I use 12-amp diodes for up to 8- to 10-amp charge circuits, and 55-amp diodes for above that. The 12-amp diodes must be on a heat sink surface about 12 inches square. Try the power transistor heat sink and heat sink compound available from Radio Shack.

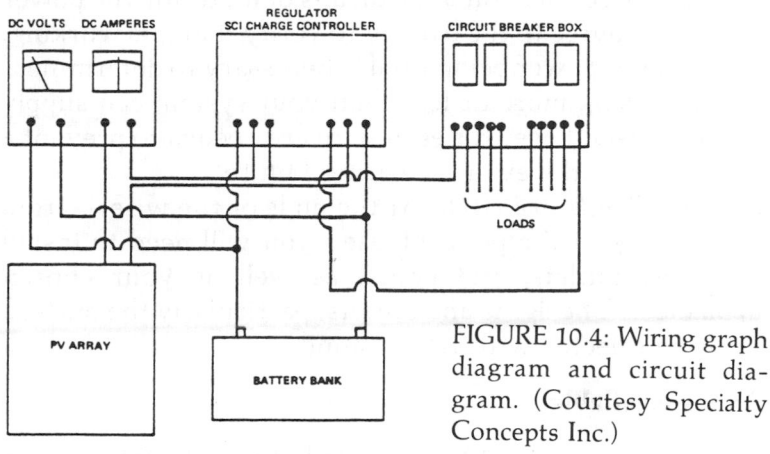

FIGURE 10.4: Wiring graph diagram and circuit diagram. (Courtesy Specialty Concepts Inc.)

Meters

Meters cost from $10 up to $50. I use the less expensive ones which are accurate enough for battery charger needs. Automotive ammeters with 0 center are impossible to read at 2-, 4-, or 6-amp charge rates. Their scale ranges from 0 to 30 or, more often, 60 amps in half the scale width.

Fuses and Holders

Auto parts stores have fuse blocks which hold one to six fuses each and have screwdriver connections for the Romex house wiring. Use the appropriate AGC-type fuse for each house circuit.

12-Volt Home Wiring Specifics

After establishing a working power source (e.g., wind, solar, hydro, gasoline) and obtaining the proper type and quantity of batteries, there are three more things you must do.

1. *Wiring.* To complete the circuit, wire inside the house to bring power to the point of use, and wire from the battery to the source of power.
2. *Connections and Fuses.* In an orderly manner, connect all wires to the battery terminals. This provides not only a safe mechanical connection but a means of metering the power from the source to the battery to verify that it is working. Metering of power being used is necessary to determine if you are using more or less than your system can supply continuously. Fuses for each wire are needed to prevent a house fire in the event of wiring failure.
3. *Switches, Outlets, and Lights.* At the ends of the wires—from the battery to the point of use—you will need to install switches, outlets, and lights, as well as your chosen appliances. The lights and outlets, particularly the outlets, must be selected with DC in mind.

Wiring Types Available

To carry the current flow at 12 volts, wire has to be larger than the wire used in 120-volt house. If you conserve energy, you may be able to get by on the wiring used in a 120-volt house because your power requirements are less.

If the wire used is too small, or too long, the current flow in it will be restricted. Using just a small amount of current, the flow may be fine but when the current increases, such wiring will restrict the larger flow—lights dim when other lights are turned on.

To avoid this, use the proper size wire. #12 Romex is standard house wiring and is the best bargain. #10 is slightly larger, #14 smaller. Any will work, but the larger the better for long runs or for higher current uses (motors or inverters). Romex is usually three-conductor, having a black, white, and extra safety ground wire. Lamp cord (#18) is considerably smaller and easier to work with, and costs as much per foot as Romex. It is *not* recommended.

Meters, Fuses, and Interconnects

It is best to use a separate run of Romex to each area of the house and wire to code standards. Otherwise you can supply several outlets and lights from one Romex run from the battery, *if the* outlets are fairly close together, and *if* the total distance is not over 25 feet, and *if* no use is planned for over 5 amps. If you use the smaller #14 wire, make that 15 feet.

Start at the battery area and run wires as directly as possible to each location rather than looping a long route around the house. At minimum, run wire half way around the room in one direction, and another wire half way around in the other direction. In multistory buildings, it is easy to run wire straight up a wall to an outlet and light switch on each floor.

Where all the wires come together, you need to connect them to the battery and provide a separate fuse for each and a current meter (ammeter) for the total.

Figure 10.5 shows how to connect a set of branch circuits to a 12-volt DC storage battery. The following parts are used.

1. Automotive-type fuse block ECHLIN FB6260 from NAPA auto parts.
2. 5-amp fuses (circuit breakers for 120 volts won't be accurate at 12 volts and cost more to mount—not recommended).
3. Ammeter 0–10 or 0–15. Auto-type reads -30–0–+30 and will not be sensitive enough—not recommended.
4. Standardized outlets. Polarity is important.
5. Switch. You may want a main switch to turn all power on.
6. #10 or larger wire. For connecting the common + and - to the battery. Be especially careful when wiring from the battery to the main fuse, as there is *no* safety fuse.
7. The negative wires must be fastened together to a wire to the negative battery post. Use a copper strip with a row of screw connectors or a 1/4-inch bolt with washers between each wire. First mount it to the plywood with one nut, then the wires and washers and then the second nut. Do not bunch all the individual wires on the battery terminal.

Make all connections neatly, stripping no more insulation than is required to make the connection. Be sure no wires can move and touch. All these parts can be mounted on a one-square-foot piece of plywood. A larger piece of plywood can accommodate the meters for your charging system, too.

Connect the positive wire of each Romex run to the bottom

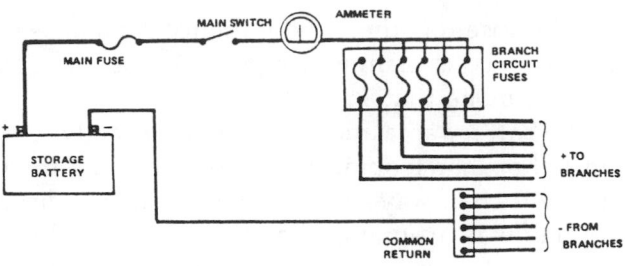

FIGURE 10.5: Simplified wiring diagram for batteries to fuse/control box.

screw of each fuse; that is your positive power source to that Romex run. If there is an accidental short circuit (+ and - wires touching), power will try to flow through the wire from the battery in almost unlimited quantity (amperes). Only the slight resistance in the wire limits it. The wire gets hot, sometimes red hot, and can cause all sorts of problems. However, since you just wired in the fuse, any flow of current in excess of the fuse rating will melt the fuse immediately, and the power is thereby disconnected. Your peace is preserved! Always use fuses in the line, and keep spares around so you aren't tempted to replace a blown fuse with something that doesn't limit the current potential, like nails and bolts.

The negative wires from the branches are all connected to each other and to the negative post of the battery. An electrical supply store may be able to sell you a copper strip with a row of screws or clamps on it for connecting a number of ground wires. If so, great. If they say they don't know what you are talking about or they will get it in a few weeks, there is another way. Get a 1/4-inch bolt, several inches long, threaded all the way. Then, with a washer between each or by soldering on wiring tips, you can clamp all the Romex wires together and to another wire that goes to the battery's negative post. Make the single wire to the battery a larger size and keep it under three feet in length, since it has to carry *all* the power that is in *all* the other lines combined.

Recharging Ni-Cad or Gel-Cell Batteries

Portable battery-powered devices can be recharged from 12 volts DC if:

1. the voltage of the battery pack is less than or equal to 12 volts. (If higher, it will not charge but rather run down into the house battery, since voltage [pressure] flows to lower.)
2. the amount of current flowing from the 12-volt house outlet into the portable batteries can be limited to a very small amount. (Most penlight, C, and D size cells can only be recharged at 0.020 to 0.200 amps, usually measured in milliamps [1/1000 amp]. So, 20 to 200 milliamps [mA] is usual.)

Determine the voltage of the device:

1. It may be listed on the label.
2. It may be listed on a charger that came with it for AC charging.
3. Open the device and see the battery type: each penlight cell or C or D cell is about 1.5 volts.
4. Measure it with a meter (volts).

Determine the current (rate at which the battery can be recharged without damage):

1. It may be on the label or instructions if the device originally came with built-in batteries.
2. The charger that came with it, if any, may list its maximum current. Or using the charger, cut one wire and insert a meter to measure current, 500 mA scale.
3. Penlights charge at about 89 mA; C or D cells about 125 mA; lantern-type 6-volt gel-cells about 200 mA.

The easiest way to limit the current to your determined value is to make a cord and connect this device to your 12-volt outlet. (This may require a special plug, which you can make by cutting the cord from the AC charger that may have come with the unit, as with cordless shavers and electric drills.)

Make your recharging cord by connecting two flashlight bulbs in series with the plus and minus wires going to the special plug. In other words, each wire coming from your 12-volt DC source will have a small flashlight bulb connected to it and between the 12-volt source and the item you want to recharge. The purpose of these bulbs is to drop the voltage and give you a 0.100 ampere (100 milliamps) charge rate. The nice thing about using light bulbs to drop voltage is that they let you know if you are charging and help to determine proper polarity. For safety's sake, be sure to make your bulbs, wires, and clips short-proof. Use insulating tape and cover all bare wires.

The correct bulbs are Radio Shack 272–1112 or 272–1118 which will fit into a 272–324 holder. You can also use 272–1487 in a 272–324 holder. Mount the holders in a plastic pill bottle for durability.

Operation is simple. When correctly connected, the bulbs will glow very dimly as the battery charges. You may have to leave the colored cap off the bulb holder to see the glow. If you are uncertain as to the polarity of the connection, reverse the wires and compare the bulb brightness. If the bulbs glow both ways and one is a lot brighter, the dim glow is correct. If it only glows in one direction and that glow is very dim, that is the right connection.

Since ni-cad batteries can be recharged over 500 times, these initially expensive batteries prove inexpensive in the long run if properly maintained. It is best to run them all the way down before recharging, and to recharge them slowly over a 24-hour period. They tend to form a "memory." If continually recharged or run down partially, they begin to limit their reserve power to the between-charge amount. If you must charge them when they are not yet empty (say for a trip), it's ok once in a while—just not all the time. Ni-cad may not be the wisest choice for flashlights in winter, since they quit working temporarily if their temperature goes below freezing.

Gel-cells are different, they are lead-acid batteries with gelled electrolyte instead of liquid. Sears has a 6-volt lantern cell, the kind with two spring contacts on top, that is excellent in a flashlight. This type can be recharged about 250 times. But, like car batteries and unlike ni-cad, don't run them all the way down: keep gel-cells fully charged. Always recharge after use or at least once a month. These batteries charge best slowly, four hours for each hour of use.

Entertainment Equipment

Adapters are available to run tape decks and radios from a 12-volt battery, when the unit operated needs less than 12 volts. Some radios are designed for a lesser voltage—usually 6, 7.5, or 9. Do not run these radios directly from 12 volts if they require a lower voltage. You may get away with it for a while with a 9-volt unit, or for a few minutes with a 6-volt unit, but when they fail from over-voltage, the repairs will be major.

Adapters usually come with cigarette lighter socket plugs. You can install a socket in the house for this purpose, but it's better to solder a wire onto the voltage converter and be able to plug it into the regular house outlets.

Another possibility is to make the converter. It will cost less and will likely be a better unit, made to suit your specifications.

Do not leave converters plugged in when you turn off the radio. Some continue to consume power (you can tell if they are warm). Converters sometimes fail placing full voltage on the radio, causing great damage. Also never leave a converter connected to your 12-volt power source unless it has a load connected to it. Converters can short out and cause a fire.

It may be more worthwhile to put rechargeable (ni-cad or gel-cell) batteries in the radio and operate it from the batteries, recharging them overnight for free by connecting them to your house 12-volt outlet. By using a voltage dropping resistor, you can tap lower voltages from your battery bank to power radios, ni-cad chargers, etc. The formulas for determining the value of a voltage dropping resistor are:

Voltage Difference/Load Current Rating = Resistance
Voltage Difference x Load Current Rating = Wattage

Example: To operate a small radio, you may need 9 volts and 300 milliamperes (0.300 amperes). You are using a 12-volt battery bank. The difference in voltage is 12 - 9 = 3 volts.

3/0.3 = 10 ohms   3 x 0.3 = 0.9 watts

Thus, a 10-ohm resistor rated for 1 watt of power will do the job. Most technicians double the wattage for a safety margin in heat dissipation capacity.

Stereo Equipment

The easiest way to operate a stereo with 12 volts is to use an automotive-type radio and tape player and good-quality speakers, 8-inch diameter minimum, preferably 12-inch. You might as well get the best you can afford because speaker quality is compromised in auto stereos. Some are junk, 4 watts power with 8% distortion admitted. An economical way to good sound are 12-inch coaxial speakers (available, unmounted, through Radio Shack). They must be mounted in a box, cabinet, or closet door to sound as they should. The air movement caused by motion of the cone will simply go around back to fill the void if not enclosed.

Another way to operate a stereo is a little more trouble but results in far better quality. Transistor equipment uses DC. A regular plug-in AC stereo system converts the AC to lower voltages of DC. If the stereo you already have uses only a few reasonably low DC voltages in its circuits, it is possible to supply these to it directly by injecting the right voltage into the circuit where needed. A lot of sets work off one or two basic voltages and from these derive several others. Thus, supplying it with the basic voltages needed runs it the same as does AC. And very small amounts of power are needed so that the voltages over 12 volts DC can be supplied by additional rechargeable batteries (motorcycle batteries or lantern-type gel-cells). These will last from 1 to 4 months and then can be recharged overnight from your deep-cycle main batteries. The only hitch is if the stereo needs many or high voltages, like +80 or -80 volts DC, thus requiring additional batteries. Mine takes +24 and -24.

Television is more of an energy consumer. If you have a small PV system, only 12-volt sets should be considered. It has a little heater that glows red which takes as much power as the rest of the set. A black and white set uses 1 to 2 amps, which is not bad—just about the same as a good 12-volt reading light. (Zenith and Sony make good 12-volt sets.) Color tv is available for 12 volts but takes four times the power. That means you can watch black and white tv five hours for every one hour you run a color set.

Turntables that use DC motor-belt drive frequently use a 12-volt motor and thus a wiring change allows operation directly from the batteries. Some cassette stereo recording decks are also equipped with 12-volt motors. Wiring modifications need to be done by someone very familiar with such procedures, and a circuit diagram is helpful. I can't recommend a specific model number because they keep changing.

Some of the larger portable cassette/am/fm units can be powered from 12 volts, or 9 or 6 volts derived from your 12-volt system through an adapter. CB and amateur radio equipment is available for use directly from 12-volt DC.

Conclusions

The main difference between wiring a 12-volt PV-powered home and a grid-powered home is that PV homes require larger wire to carry the DC current. This is due to the increased resistance at lower voltages.

When wiring a PV-powered home, you must take into consideration the voltages and currents used in each circuit. It is wise to plan for maximum loads so that your wiring will be up to future usage. The initial cost of the wiring will be offset by the flexibility of the system as your needs change and as you expand your production and consumption of solar electricity.

The wiring from the PV array must be sized to carry the peak current from the array to the battery bank or regulator/control board. The wires from the battery bank to the individual appliances must be sized to carry the appliance load and must be properly fused for protection against short circuits in the wiring or the appliances.

Special outlets, switches, and other electrical wiring equipment are available from recreational vehicle suppliers. This special equipment works well but is not absolutely necessary. You can use standard, inexpensive outlets, plugs and other readily available equipment if you adapt them to your special needs.

Polarity is critical with a direct current system. If you decide to use standard outlets, be sure to wire *all* the outlets in the house in the same way. That is, fasten the white wire onto the same terminal at each outlet, likewise with the black, and then place a red dot to indicate the positive side of the outlet. Do this with plugs, too. Thus you will have both polarity and uniformity.

Many people use cigarette lighter-type outlets and plugs to insure proper polarity and avoid accidents. Some plugs and outlets have large and small slots, and others have configurations that make

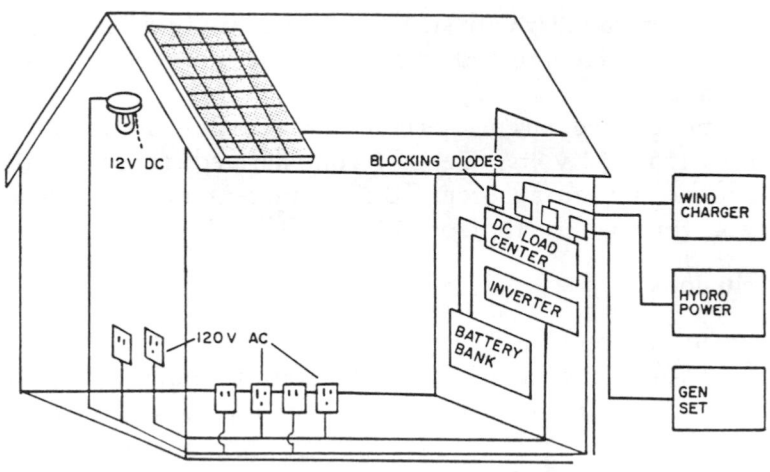

FIGURE 10.6: Simplified schematic of house wiring showing both 12-volt DC and 120-volt AC outlets, as well as several sources of electric power.

it impossible to reverse polarity. These can be used, but they cost more than the conventional type. The non-interchangeable type must be used to comply with code.

An important consideration in the selection of outlets is the possible confusion to visitors in your home. If there is a possibility of an accidental reverse polarity when plugging in an appliance, then it is best to prevent such an occurrence by using commercially available DC outlets and plugs. In addition, if you have an inverter and 120-volt AC outlets, be sure to make it impossible for someone to plug a DC appliance into your AC outlets and vice versa. Some outlets have covers (outdoor-type or childproof) which help avoid mis-pluggings.

A simple way to insure proper polarity and prevent mismatches between DC and AC plugs is to wire all AC three-prong plugs and outlets black to the small prong, white to the large prong and ground to the round prong. Match the AC outlets throughout the house for this polarity. For the DC circuits, wire the negative to the round prong and the positive to the large prong. Match the DC outlets to this polarity. Mark all outlets either AC or DC. Do not mix AC and DC on the same outlet box: use separate boxes. Using different color cover plates for AC and DC outlets will also help to avoid confusion. (This method is not code-approved.)

Switches are no problem. Some people use automotive switches, which are specially designed to carry DC current. Others use standard house switches. However, when a switch in a DC circuit is opened and closed, the contacts will arc, causing their eventual deterioration. To avoid this, you can put a capacitor across the switch-wire connections. PV equipment suppliers and electronics supply houses can help you select the appropriate capacitor. To avoid or lessen the switch contact arcing problem, use older-style switches—the loud clicking (and less-expensive) ones. Do not use "silent" mercury switches.

Some people use circuit breakers for their PV wiring. Circuit breakers are automatic switches, and even the best have the contact arcing problem. For this reason, most 12-volt DC homes use automotive fuses. They are inexpensive, readily available, and offer more flexibility in system sizing. Fuse blocks can often be found for next to nothing in auto salvage yards. If you wire a room for the eventuality of larger loads, you can start with small value fuses and change them as the load increases to insure a safe circuit at all times.

GROUNDING AND LIGHTNING PROTECTION

Grounding prevents electrical shock. In the case of a fault, short circuit, or other problem in wiring, equipment, or appliances, grounding will reduce the chance of electrocution, fire, or explosion. Grounding also helps eliminate static interference or electromagnetic noise created in some equipment.

Ground the house as you normally would with a 3-wire 240-volt AC service neutral grounded on the line side of the main contactor. The house should have three-wire circuitry for proper appliance grounding. Ground the negative side of the 2-wire DC system. This is common grounded to the same metallic system as the AC neutral ground. Also common ground all equipment in the system with chassis grounds.

The house wiring neutral should be grounded at the house panel with at least 6 AWG wire at the panel neutral bus. This neutral bus should not be tied into the panel enclosure. All equipment should be chassis-grounded with at least 6 AWG wire

FIGURE 10.7: A lightning protected solar array located in White Sands, New Mexico. In locations with a high probability of lightning strikes, rods and grounding are required. (Bill Cirrito, Electrasun Energy Systems, Tucson, AZ)

and this ground wire connected to the house earth ground. A 6 AWG or larger ground wire should be used from the array negative to the house ground bus below the service panel.

The metal frame of the solar array and mount should also be grounded. All four grounds—the negative DC leg, the AC neutral, the equipment chassis, and the array frame—must be attached to a common ground, *not* separately grounded. If separate ground rods are used at two or more locations (i.e., the array and the house), a ground conductor must be used to tie the separate grounds together.

There remains some controversy about grounding the negative leg of a DC circuit. Old windcharger builders did not earth-ground the negative conductor. Instead, they grounded the windcharger tower. This was protection against damage by lightning. The belief still holds that a grounded negative in a DC circuit will provide a path for lightning, causing possible damage to DC equipment and batteries. There is evidence to prove this to be true.

However, the *National Electrical Code* requires grounding on the negative leg of the circuit. An alternative to the hard-wired earth ground is an isolated ground created by the addition of a 0.1-megohm resistor between the DC circuit and earth ground.

The isolated ground resistor creates a high resistance path to ground protecting anyone who happens to be touching the DC circuit positive and a grounded chassis or frame. In the case of higher DC voltages which are becoming common in PV utility interconnect systems, the isolated ground will prevent a fatal shock from occurring. It will also prevent an arc from occurring between the array circuit and the frame, a situation which might be a fire hazard in a non-isolated ground system.

The isolated ground circuit can also be used as a test point to determine whether there is leakage between the frame and the array circuit. By indicating a ground fault and the magnitude of any current flowing from array to ground, you can more easily locate which array string is the problem.

The use of copper-clad steel ground rods is standard. These rods are 8 to 10 feet long by 3/4 inch in diameter. They should be driven into the ground at the point where power enters the building, and the ground wire fastened securely to the rod with a clamp. It is a good idea to tag the ground wire at the ground rod.

If you live in an area where it is impossible to drive a 10-foot rod into the ground (sometimes the whole world seems solid rock), then an alternative method must be used. Either clamp the ground wire to a metal water supply pipe that has a good earth ground or seek local advice. In some dry rocky regions it is necessary to dig a trench and bury a measured length of bare wire to provide an adequate earth ground. The local utility company or an electrician should be able to help.

Although most of us live in locales where the chance of a direct lightning strike is very low, it is important to protect your system from the induced effects of nearby lightning. If you live in a lightning area, be sure to check with local authorities and neighbors to see how they protect tv antennas and other equipment which rise above average height.

In any case, nominal surge protection in the form of varistor

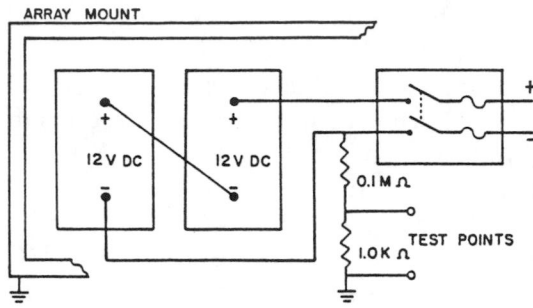

FIGURE 10.8: The 0.1 meg-ohm isolated ground resistor will help prevent shock should you touch the positive DC circuit and the solar array frame. The 1.0 kilo-ohm resistor is a convenient reference test point. (Adapted from SAND 85-7020)

surge arrestors across both AC and DC entrance cables is recommended. This protection can be tied to system earth ground providing a shunt for any transients or side effects of nearby lightning.

Protection against nearby lightning strikes can be accomplished by a GE V275 LA40B Varistor from the positive array terminal and GE TLP175 Spark Gap Type Lightning Protectors or a similar device on each phase line of the AC circuits. Again, it is important to point out that local codes should be followed, not the general instructions in this book.

Chapter 11

Installation, Testing, and Maintenance

INSTALLATION GUIDELINES

Now that you are familiar with the components, we can look at putting the PV system in place. Each home is different, of course, so it is impossible to cover all the possible arrangements. However, a few basic guidelines will help you think through your installation and enable you to avoid pitfalls that might require later modification. As in all building projects, it is a good idea to go over the steps before starting work. Write down each step and make notes of ideas and suggestions. At the same time, develop a materials list so that when you do start, all hardware, fasteners, wires, and parts are on hand. If you live far from stores, this is particularly important. An unfinished installation is an invitation to trouble. You can be sure that rain or snow will catch you at the most inopportune time.

The key rule to remember is: Keep It Simple. Your system will change with your needs, and a simple straightforward, easily accessible installation facilitates modification. A straightforward installation will be a safe installation. Make all electrical connections in accordance with proper practices. Make sure all connections are mechanically tight and that you use the correct fasteners. Soldering requires some special skill, but don't be afraid to try. Wire nuts and fasteners will also work if done properly. Visit an electrical supply house and ask to see their wire connectors, stand-offs, and terminal blocks. Become familiar with what is available in your area. Don't be afraid to ask questions. Most suppliers will be curious about your PV set-up and will want to help.

Equipment Location

The first step in the installation procedure is selecting the location for your equipment (review the section on site selection in chapter 6). Be sure to read the manufacturer's recommendations carefully. It is in everyone's interest that the equipment you buy be used to its best advantage. Most manufacturers provide free installation guidelines and instructions. Some offer assistance by telephone should you run into problems or have specific questions. Distributors and dealers will also help. It is wise to buy your equipment from someone who actually uses the hardware.

Site constraints require additional consideration. Since the electricity from the PV array is low-voltage DC, you must take into account the loss of power due to long wire runs from the array to the home. In general, a pole mount with heavy wire is a better choice than cutting down a shade tree. However, there are limits to this sort of thinking. Unless you can get a good deal on cable, or your wiring budget is unlimited, there is a point at which you have to compromise in site selection. Check the wire size chart and local wire prices to determine the most practical distance between your array and home. And while you're at it, think about underground cable as compared to unsightly and unsafe overhead wire runs.

Accessibility to the installed equipment is important in siting the array, the battery bank, and the regulator/control board or box. A short wire run to an unhandy niche in a dark, crowded corner is false economy. If you can't move around freely in the battery and controls area, everything from accident and injury to plain aggravation will make you wish you had spent the extra dollars for longer and heavier wire to reach that open access spot. I mounted my first regulator on a wall in the corner behind a table. Each time I changed the system to test new methods and products, I had to move furniture, grab a flashlight and a mirror, and work upside down in a tight place. Eventually, I wised up and made the controls and junction box more accessible.

As always, safety first! Be sure your array and batteries are disconnected while working on your system. See chapter 8 for a complete discussion of battery location.

The location of the regulator and controls is equally important. Some regulators and inverters require proper ventilation to keep the internal parts cool. Dirt and dust on controls and contacts can be a problem. If your regulator or inverter must "breathe," isolate it in a dust-free area. Hardware stores sell electrical boxes for equipment mounting and protection. Don't put your controls in or near a closet where someone might use them for a clothes hanger. Never locate your control box and regulator directly above a heat source.

A note on mounting. Simple temporary ground mounts—legs attached to the array frame—are suitable for testing purposes. However, you are taking a chance if you leave this set-up for too long. I have used both fixed and movable roof mounts. The fixed mounts were attached to wooden or metal stand-offs of 2 x 4 pressure-treated lumber blocks or to channel steel or aluminum bolted to the roof rafters. Then the array was fastened with bolts to the stand-offs. The movable array was mounted in the same way except that metal legs from the sides of the array were added. These legs made seasonal adjustments possible. Look at a few billboards and signs to see what kinds of metal frames, angle iron, tubing, and hardware are used in your area.

Obvious, but of utmost importance for roof mounts, is to be sure to fasten the array securely. The best way to check the array mount is to grab it and pull. Hang on it if you must to test the strength. A periodic inspection of the fastenings is a good idea.

Don't forget to protect your batteries, open wires, and fasteners from falling objects. Children and the curious can be all hands. Keep things out of reach, safe, and secure.

PV INSTALLATION EXAMPLE

Each installation is different. This example sugests methods which will help you think through your own PV system. In chapter 5 we sized a PV system for a home in Colorado (see Figure 5.4). The initial system consisted of:

    18 each 55-watt solar modules
    40 each 6-volt batteries
    1 each 30-ampere regulator
    1 each 2500-watt inverter

Frame, Modules, and Mounts

For this installation a roof mount is selected. The primary voltage to power the inverter is 24 volts DC. The solar array is to be located on the roof of the home. The batteries and equipment are in a separate room in an attached shed on the north side of the house, a structure shared with the washing machine and water pump. There are no local restrictions or zoning regulations regarding height, size, or appearance of arrays.

An alternative downsized system could be a passive solar mounted 12-module array, a back-up generator, the same size battery bank, an inverter with a built-in battery charger (or the same size inverter and a separate battery charger). The regulator remains the same to allow for expansion. The battery bank is kept the same size to insure long life. The auxiliary charger must be adequate to keep the generator running time to a minimum. The tracking array is situated in a sunny location 50 feet from the house.

Given the choice of wood or metal array framing, always choose metal. Wood mounts do not provide a grounding or lightning path and are subject to weathering which weakens them.

Since these modules have aluminum frames, aluminum is selected for the array frame. Although in this case there is little likelihood that the array will be seasonally adjusted, the mounting hardware is made to be adjustable. The dimensions of the modules are 1 x 4 feet. The modules have weatherproof interconnect boxes on the back which make mounting easy.

This house is in a snowy region, with a roof pitch of 55° angle to the south with direct access to the sun.

There are two easy ways to check your roof angle. If it is possible to stand to the side of the house (east or west) approximately level with the roof, you can sight the roof angle with a protractor in your hand at arm's length. A more accurate method is to use a 24-inch bubble level and a sliding T bevel. On a ladder or on the roof, hold the bubble level parallel to the ground butting its end to the roof. Mark the angle between the roof and the top edge of the level with the sliding T bevel. Now use a protractor to measure the angle of the sliding T bevel.

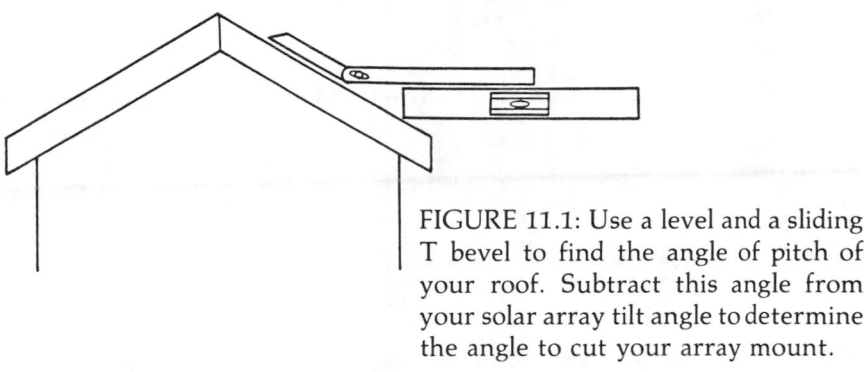

FIGURE 11.1: Use a level and a sliding T bevel to find the angle of pitch of your roof. Subtract this angle from your solar array tilt angle to determine the angle to cut your array mount.

A stand-off mount, the simplest and least costly configuration for a roof of the proper tilt, is used. To allow for a summer tilt of 25° (latitude -15°), the top of the array is hinged with lower legs bolted and stowed away under the array frame. The array is situated on the roof so that two people on separate ladders can reach it from the eaves to raise the array to 25°. In doing so, the lower legs drop into position to be fastened to the angle aluminum stand-offs.

All of the stand-offs are securely bolted into the roof framing members on the corners, at no more than 4-foot centers. The stand-offs are 2 inches high, enough to provide underside module cooling in warm weather without creating a large snow dam in winter.

The frame can be made in a shop and all hardware (hinges, stand-offs, and fasteners) are assembled on the ground to insure fit. The modules are also tested while on the ground for proper fit. Re-drilling module mounting holes while on the roof can be dangerous.

With all hardware and tools handy and at least two helpers, mount the array on the roof. It is a good idea to have a few extra people on hand when the array is being installed, but wait until the work is done before having a party.

It should be noted that the modules could have been wired after being fastened to the roof-mounted frame as the upper edge of the frame is hinged, providing access to the underside of the modules. If the array mount was not hinged, the modules would have to be fastened to the mount on the ground and pre-wired.

FIGURE 11.2: Assembly and pre-wiring of PV modules for a residential solar array. (Courtesy California Energy Commission)

Equipment Room

With the frame and modules safely stowed, awaiting the day they will be mounted on the roof, it is now time to work in the equipment room. First, lay out the location. The inside wall adjoining the house is selected for the battery rack. Make a sturdy framework of angle steel. (Wood may be used, but it must be able to support the weight.) The first shelf of batteries should be a few inches off the ground to make cleaning underneath easy. Each shelf is spaced so that a hydrometer can be used to check the batteries. Always make it easy to check your batteries.

Batteries are placed carefully on the shelves with proper polarity so that wire runs are as simple and as short as possible. These 40 batteries weigh 2800 pounds (10 parallel strings of 4 in series). To leave room for servicing, the homeowner has put 16 batteries on the two lower shelves and 8 batteries on the top shelf.

FIGURE 11.3: A pre-assembled solar array being moved into place. (Courtesy California Energy Commission)

That means that each lower shelf holds over 1100 pounds. Batteries are heavy.

The battery bank can now be wired. Using 4 AWG welding wire with the proper crimped and soldered connectors. The batteries are first wired in series groups. This means each four battery string will have three short battery interconnects connecting positive to negative. Next, the strings are wired in parallel. Starting from the lower left, make a short parallel connection from the front group to the back group. This is done across the bank so that all four strings are paralleled. The same is done on the other shelves.

What we have done is to first turn 6-volt batteries into 24-volt batteries, and then turn the entire bank into one large interconnected 24-volt battery bank.

Caution must be taken when working with batteries. Keep tools clear of the terminals and wires. This battery bank has enough power to run an all-electric home for one or two days. The power in it can weld thick steel plate or severely burn you. It is a good idea to wrap wrenches with insulating electrical tape. A low-cost battery-operated smoke detector is a good investment. (See chapter 8 for information on venting and other precautions.)

Next, the regulator is mounted on the wall between the battery bank and where the wires from the array enter the room. The

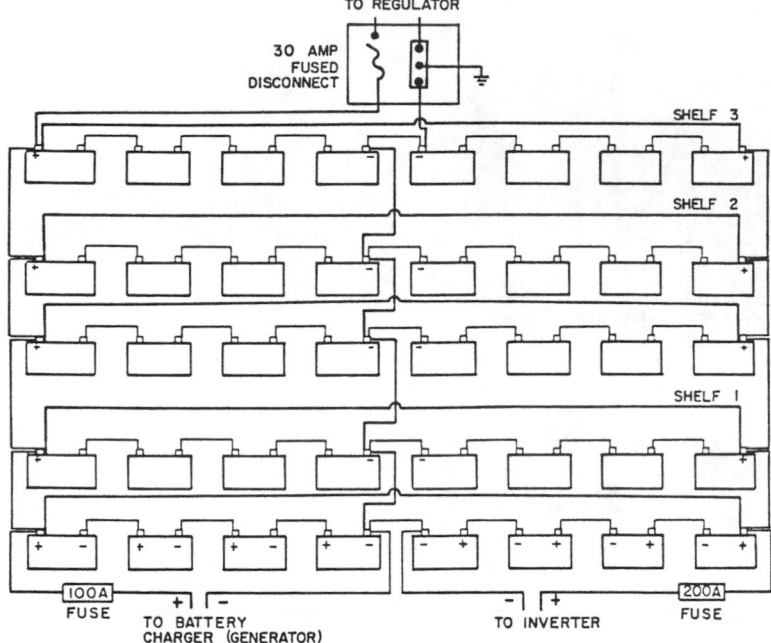

FIGURE 11.4: Battery bank layout. Space is allowed between shelves to permit testing specific gravity with a hydrometer.

inverter is mounted on a shelf or table as close to the batteries as possible. If AC power is needed to complete the job, the 1/0 AWG wire from the battery bank to the inverter can be connected at this time. Be sure the inverter is off before making the connection. Follow inverter installation instructions to the letter and test the unit before using it. A safety cartridge fuse between the inverter positive and the battery bank is important should the inverter on/off switch fail.

Roof Mounting

With the battery bank and regulator installed, we are ready to work on the roof and put the system into operation. Prior to installation, check the roof framing to insure that it is adequately braced. Although this array weighs under 300 pounds, additional rafter braces were nailed in the attic to support the weight of the workers.

On a nice windless day, when the roof is dry, assemble the mounting hardware, frame, modules, and necessary tools. Use wooden ladders and scaffolding if other live wires are in the area. Take especial care when working on the roof. Use safety ropes. Don't toss things from the roof heedlessly, for the groundworkers' sake. Try not to damage the roofing material.

Locate the roof framing members and mark the placement for the stand-offs. If you're not sure how to do this, hire a local tradesman skilled in carpentry and wiring. When the stand-offs are secured, make sure all roof penetrations are sealed and water-proofed.

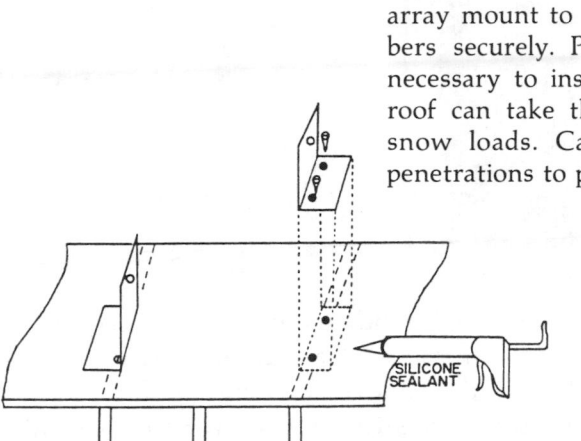

FIGURE 11.5: Be sure to fasten your solar array mount to the roof framing members securely. Put in extra framing if necessary to insure that the array and roof can take the additional wind and snow loads. Caulk and seal all roof penetrations to prevent leaks.

SILICONE SEALANT

Pass the modules carefully, one by one, to the roof and bolt them into place noting layout and polarity. Because modules produce power as soon as the sun strikes them, cover with blankets or, better yet, tape heavy paper over them until connected.

With 10 AWG interconnect wires in place, the next step is to hook up the array wire. As the terminals of the modules cannot accommodate wire larger than 10 AWG, a junction box is needed on the roof. This box is located above or to the side of the array, fastened to the roof and sealed to prevent leaks. The modules wired in either two module series strings or combined strings (making sure that the current-carrying capacity of the wire combinations is

FIGURE 11.6: The two-module array used in the Colorado Mountain College design example is installed by students on a remote mountain cabin. (Photo: D. Stewart)

correct) are brought to the junction box where all series strings are paralleled. All interconnect wires and wires running to the junction box are securely fastened to the roof or the array mount.

The array mount is grounded with a 6 AWG wire and this wire is brought into the junction box. Thus, the junction box now has array + and - and ground separate terminals. Wire or conduit can be run either through the inside or outside of the house in accordance with the *National Electrical Code*, and local codes, to the equipment room.

In the equipment room, install a junction box and a 30-ampere fused disconnect switch and wire the array to it. The ground wire is taken from the junction box and run with 6 AWG to a ground rod located at the side of the house. Chassis grounds from the regulator and the inverter are also run to the junction box. An isolated ground resistor is used to ground the array negative leg.

A second 30-ampere fused disconnect switch is wall-mounted and + and - wires from the battery bank are brought to this switch.

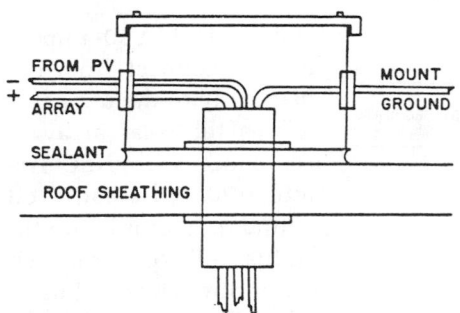

FIGURE 11.7: Roof-mounted junction box.

With both switches off you are ready to wire the array through the regulator to the battery bank. All along you should note and double-check voltage and polarity. The voltage at the array disconnect switch should be the array open-circuit voltage of (2 x approximately 21 volts) 42 volts. Battery voltage is between 24 and 26 volts.

Follow the regulator instructions. Generally, the connection sequence is:

1. Connect battery negative to regulator
2. Connect battery positive to regulator
3. Connect array negative to regulator
4. Connect array positive to regulator

Now you can remove the covering from the array. You have installed a safe, well-grounded solar electric system and are charging the batteries. If the system has monitoring meters, you should read between 25 and 30 amperes on a sunny day.

With the inverter off, the next step is to tie the house into the solar electric system. AC wiring from the inverter to the home distribution circuit breaker box should be in conduit and wired in accordance to standard house wiring practices.

Tracker and Generator Installation

The alternative tracker and generator system is installed in almost the same way. The main difference is in siting and mounting the tracker and generator housing.

At the selected site, dig a hole at least three feet deep. Place a flat rock or slab at the bottom and position the pole for the tracker mount. Pour concrete into the hole around the pole, bracing it

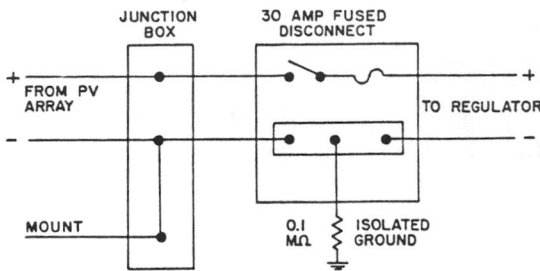

FIGURE 11.8: A 30-ampere fused disconnect from the solar array to the regulator isolates the solar array in accordance with *NEC* Article 690. The isolated ground resistor is located in the fused-disconnect switch box on the negative bus.

straight and plumb. It is a good idea to put a metal cross brace (below ground) through the pole to keep it from turning in high winds. Paint the entire pole with tar or some other rust inhibitor.

Fasten conduit to carry the array positive and negative wires and a ground wire to the side of the pole and down into the ground. A weatherproof junction box at the pole will simplify fastening the modules to the array wire. The conduit to the equipment room should be buried. After a few days, when the cement has set up, lift the tracker onto the pole. Cover and mount the modules, and fasten the interconnects. Connect the array to the junction box as we did the roof-mounted array.

In the equipment room, connections must be made from the standby generator located in a shed safely away from the house. Generator location and housing must be in compliance with safety and fire prevention regulations. In the absence of local regulations, use good judgment and check with your generator supplier, volunteer fire department, or forest service. Do not take chances with flammables.

Conduit is run from the generator shed to the equipment room. A transfer switch is required to select either generator or solar. (This system will also work in the city with utility power.) The simplest transfer is a manual safety switch rated for the generator and inverter. These switches come in rainproof boxes and carry 30, 50, 100 or more ampere ratings on the two or more sets of contacts. AC hot, neutral, and ground are brought into the box. A single-contact switch can be used if grounds are common bused.

To make the transfer manually, bring the generator up to speed for the proper voltage. The switch is then moved from Inverter-to-Distribution panel, across Center-Off, to Generator-to-Distribu-

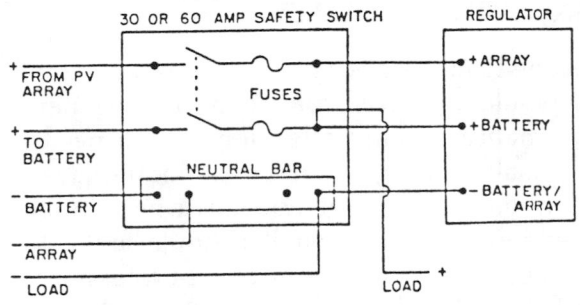

FIGURE 11.9: A fused safety switch used as an array disconnect.

tion panel. Motor loads and sensitive equipment should be off when making the transfer.

If you have an inverter with a built-in standby battery charger, you have the option of automatic transfer. By wiring the inverter battery charger input to the generator through a time delay, you can start the generator and the inverter will automatically become a battery charger. There are cautions to observe with this set-up. First, the power from the generator, when running, will be routed through the inverter. Be sure the inverter is not forced to carry loads larger than its capacity. A dedicated 20-ampere circuit on the 2500-watt inverter output will prevent this from happening. Second, be sure you have installed a time delay on the battery charger to prevent the standby transfer relay from chattering and eventually burning out. One way to get around limited inverter vs generator capacity is to dedicate large loads to the generator only.

A simple transfer switch, used in recreational vehicles and PV systems, is available for under $150. There are also more expensive transfer switches, such as the Photron Tranzac which is adjustable and has built-in monitoring and control circuitry. The use of such a transfer switch and a generator with remote starter makes it possible for all switching to occur at the house. Better still, setting the cut-in and cut-out voltages makes transfer automatic.

For systems with a generator- or grid-powered separate battery charger, the charger is hard-wired into the house circuits just like any major appliance. In all cases with inverters, generators, transfer switches, and battery chargers, the manufacturer's instructions must be followed. If you are in doubt as to whether you can do this work yourself, check with a PV installer or electrician.

HIRING A PV INSTALLER

If you decide to hire someone to install all or part of your PV system, treat that person as you would any hired contractor or workman. Expect professional performance, prompt attention to your needs, timely service, and reasonable charges. In some locations an electrician's license is required, especially if house wiring is to be done. If permits are needed for the job, the installer should be able to get them and meet any inspection requirements. Some professional contractors are also bonded.

There is nothing in a PV system that should baffle an experienced electrician. However, the newness of the technology and working with DC will tend to slow down those unfamiliar with PV systems.

If you want to get your hand in on the job too, be sure the installer knows this. Some technicians will not work with the homeowner because it slows them down; others won't because their lack of skill will be revealed.

To insure that you get a first-class job, ask for references. Talk to previous customers and look at one or two installations. If you like what you see and hear, you will probably be happy with the work done for you.

The quality of work done shown in this book ranges from very good to excellent. These systems were selected not just because they are typical of what is being done, but also because of the amount of thought and attention these installers invested.

The do-it-yourselfer should understand that well-planned and organized work done safely will result in a quality installation. Take your time and think. If in doubt, call in a professional.

MONITORING AND TESTING

In the normal course of events you will hardly notice your PV power system. After a while you will take it for granted. Should a problem occur, however, it won't be much of a surprise because you've kept a relaxed eye on things.

There are basic tests which can be done to check system performance. Although performing these tests is the responsibility of the installer and is not generally done more than a couple times during the life of a system, they are described here so that you will have a better idea of what to expect from your system, and how the tests are performed.

Solar array performance is tested by disconnecting the positive lead from your solar array to your battery bank (with the regulator out of the circuit) and connecting a suitable DC ammeter in series, observing the correct polarity. Reconnect the battery positive lead. The ammeter should now be reading the solar array output under load. In full sun at noon or 1 pm, the current reading should be equal to or greater than 70% of the array current rating.

FIGURE 11.10: Colorado Mountain College student tests the output of a passive tracker-mounted solar array. (Photo: Bob McGill)

If you are reading less than 70% of the loaded array current rating and it is possible to adjust your array tilt, disconnect the array positive and negative lead. Connect the ammeter in series to the two disconnected array leads. Read the short-circuit amperage for no less than 70% of the rated output. Adjust the array for maximum reading. While reading short-circuit current, with your voltmeter across the array leads, take an open-circuit voltage reading. This reading should be at least 80% of the array open-circuit voltage. For multiple strings, perform this test on each string.

If the readings are less than specified, isolate the problem. An open circuit or short circuit in the wiring will be easy to identify as readings will differ dramatically from the expected. If readings are low, isolate the problem string and then the problem module or modules. If the modules are under warranty, call or write your supplier for instructions as to what tests to perform before returning the equipment.

If your array and regulator are performing properly and your power production matches your consumption, your battery bank specific gravity should remain at or near a full state of charge. If this is not the case, it may be necessary to test your battery bank self-discharge rate. This requires taking the batteries out of service.

The Solar Power Corporation (Exxon) Installation and Maintainence Manual states: "Properly installed solar electric generator systems should only require regular maintenance visits once a year." You'll probably want to check out your installation more often, especially when it's new and you are familiarizing yourself with the set-up. PV systems actually work better if left alone.

Test your modules when you first receive them. Prior to testing, inspect all equipment for manufacturing defects and shipping damage. File all papers, receipts, and warranties in a safe place.

Monitoring equipment is essential to assess the status of your PV system. Even with a simple system, you'll need to know the system voltage, current, and state of charge of your battery bank.

The following monitoring devices will give you the whole picture:

1. array ammeter (charging current)
2. DC voltmeter (DC system voltage)
3. production and consumption ampere-hour or watt-hour meters
4. AC voltmeter
5. hydrometer (to measure battery state of charge)
6. multitester (analog or digital) to test your system
7. a load resistor (to test module output) (optional)

A multitester is an essential tool. Available for under $100, this meter will pay for itself many times over as you test resistance, voltage, current, and so on. You can use it to check lighting circuits and wires for shorts or opens, and even do tests on your vehicle. The multitester should have several ranges so you can check AC volts, DC volts, DC current resistance and continuity. Be sure to get one that reads at least 10 amperes DC.

The array ammeter gives you an indication of how your system is performing. The array output ammeter on some regulators not only shows array charging current, but also lets you know when the regulator is on, off, or in the trickle charge mode.

By combining the use of the DC voltmeter and hydrometer, you will know the condition of your batteries. Once you have the feel for the charge–discharge characteristics of your system, an occasional glance at the voltmeter will give you a good indication of the average specific gravity of the battery bank. This, in turn, lets you know your battery's depth of discharge.

A number to remember is 0.84. Take your battery voltage and divide it by the number of series cells in a battery bank string. Then subtract 0.84 and you will have the average specific gravity. For example, a 24-volt (nominal) battery bank is 24.8 volts. There are 12 two-volt cells used to make up the 24-volt bank. $24.8 \div 12 = 2.066 - 0.84 = 1.226$. The battery bank is at 70% capacity. (See the chart on page 199.)

To learn how much energy you are consuming, install a kilowatt-hour meter in the AC circuit between the inverter and the AC distribution breaker box. You are already familiar with this meter: It is the glass-encased counter that the utility company uses to bill you. These meters sell for under $35.

Another tool for monitoring a primarily DC system is a DC ampere-hour meter. For around $300 the conservation-minded person with a limited budget can watch and match production and consumption.

While it is possible to have a PV system without metering and monitoring equipment hard-wired directly into the circuit, it is important to be able to occasionally "take your system's temperature." That's why a good multitester is essential. If you do decide to have lots of meters in your system, be sure they can be switched off. Such parasitic loads will gobble up precious PV power.

The load resistor can be used to test each PV module for

output under simulated in-service conditions. I use a variable wire-wound resistor capable of handling the capacity of most modules. Use the Ohm's law chart in Appendix A to determine the size resistor you will need for testing. Resistors can be purchased at electronic supply houses or you can make a resistor with a length of wire of the proper resistance (as determined by the multitester). Some older cars have adjustable resistors in their starting circuits or variable-power resistors in their heater fan controls.

Now for testing. First test your modules' open-circuit output. Measure the voltage and current at noon on a sunny day with no haze. Record your readings for each module. Next, put your load resistor across each separate PV module's positive and negative lead or terminal. Voltage is measured parallel to the load; current is measured in series with the load. Record your results. These initial tests are important if your modules have a wattage degradation warranty. Readings should be at least 70% of manufacturer's rating for current and 80% for voltage. Date each test result sheet and keep them with your records.

Test your batteries with the hydrometer following manufacturer's instructions. (Be careful not to drip any battery acid on yourself.) Measure the specific gravity of each cell once a month during the first year. Your batteries should be fully charged initially.

Next, test the entire PV system. Measure the current and voltage output. Test and record voltage and current with your regulator and batteries in the circuit as a load. You should probably test the entire system several times the first year; after that once or twice a year is sufficient.

You may want to test your batteries for self-discharge. This test takes a week, but gives a good indication of the remaining useful life of used batteries. To self-discharge test, disconnect the batteries and charge them with a standard automotive charger at a current rate not exceeding the battery's capacity in ampere hours divided by 20 hours. For example, a 200 ampere-hour battery should be charged at 10 amps or less. When all cell voltages are 2.3 volts, the battery is fully charged. Record specific gravity readings for each cell. Keep the batteries out of use for a week at room temperature and test for specific gravity again. Good cells will have a self-discharge rate of less than 0.015. When buying used batteries, it is advisable to perform a self-discharge test.

MAINTENANCE

Maintaining your PV system is very simple. Check your battery electrolyte level and replenish with distilled water as necessary. Don't overfill. Clean and tighten battery posts. Keep your batteries clean since dust and grime can conduct current between the positive and negative terminals and the case. To retard corrosion, use the treated felt rings sold in auto parts stores or spray-on protectants on the posts and connectors. Once a year check all appliances, wires, cords, connectors, plugs, and outlets. Repair or replace as necessary.

Check your PV array for dirt build-up. If you live in town or in a pollution zone or on a well-traveled road, you will probably have to wash your array one or more times per year. Arrays with 15° or more tilt usually stay clean with rain. When you do clean your array, use a mild soap or plain water and a soft cloth. Do not use solvents or strong detergents.

How does dirt on solar arrays affect electrical production? Dr. John Shaefer, director of the photovoltaic division of the New Mexico Solar Energy Institute, Las Cruces, New Mexico, reported a 6% increase in performance after detergent washing a portion of a 17-kW array near El Paso, Texas. A few days later a gentle rain washed the other portion of the array and the difference in output between the two was undetectable. He concluded that detergent was a waste of money. He also cautions that dust, dirt, and pollutants in other locations may require detergent washing.

I found the same to hold true on an eight-module array in a smoggy part of Los Angeles across the street from a dirty and dusty railroad right-of-way. At first I cleaned the array with paper towels and water. Now I simply wipe the glass face of the modules twice a year with a dry paper towel, knocking off any accumulated dust. My conclusion is that the danger of falling off the roof outweighs the need for module cleaning.

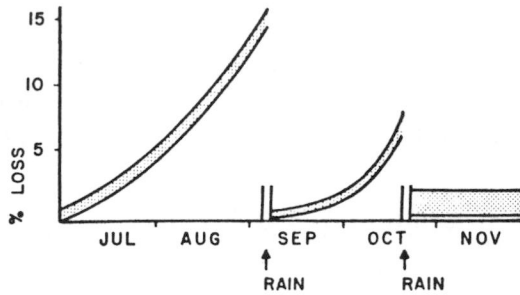

FIGURE 11.11: A 2.2-kW array installed in 1978 at Southern California Edison offices at Rosemead (urban Los Angeles) showed a gradual degradation of power in summer. After the first rains, power output went back to normal. Power degradation between rain storms for this system was low, and the system is self-maintaining.

FIRE SAFETY

Fires can happen anywhere. If there is a fire, be sure the electricity is off. PV/utility interconnect homes using high DC voltage have required disconnect switches from the array to the regulator or utility intertie inverter. Access to wires and quick disconnect switches or plugs are a good idea for any system, as is a fire extinguisher.

Dangerous currents are present on the AC side of the PV system—from the utility company or generator or inverter. Always assume that a wire is hot unless you have personally disconnected and tested it. Cut-out switches, circuit breakers, and safety switches should be plainly marked, painted, and tagged.

Remember, you have an electric power plant and you are the power plant safety engineer. You are also the maintenance and maybe even the rescue crew. Make your jobs easier through proper planning.

FINE TUNING YOUR PV SYSTEM

This section first appeared in *Solar Age* (July 1985; now *Progressive Builder*). It consolidates system maintenance information for those who have already installed PV, and provides the PV newcomer a useful summary that will make system planning and designing simpler.

With over 15,000 photovoltaic-powered homes in the United States, it's obvious PV is here to stay. This troubleshooting guide should help you get the most out of your home power systems.

Let's start with the sun. When your PV system was first installed, did you point your array toward the sun? Sounds basic but there may be room for improvement. Most PV modules used on homes receive diffused sunlight. These modules have significant output even 20° off true south. However, if your compass was a little off, your mounting structure a little mis-aimed, and if you failed to compensate for the declination (difference between magnetic north and true north), the accumulated error can be robbing you of power.

While checking your array orientation, take a look at trees and buildings that may have sprouted since installation. Does the sun have a clear path to your array all day long—even in winter? Is it time to trim that tree—or remove it? Perhaps it would be easier to move the array. Some people have switched from roof mounts to passive tracking pole mounts and gained better solar access, increasing electrical production by over 30% annually.

Is the array clean? Wipe your hand across the superstrate. There may be an accumulation of dirt that rain and snow can't remove. Clean the array face with a mild soap. You may be losing over 6% of your production to air pollution stuck on your modules.

Now check the module interconnects. Are they secure? A loose wire is a bad connection causing power loss. Are the terminals tight, clean, and bright? Corrosion on wires and terminals may be robbing you of power. Are the wires the right size? Some systems were installed when the rule of thumb was 5% wire loss, less stringent than the current *National Electrical Code* standard of 2% resistance loss.

Next: the regulator or charge controller. If it's an electronic solid-state "black box," you probably have assumed that everything is ok. If it has meters for monitoring production, it's a good idea to compare them with a digital multitester. Look at terminals and wires and be sure they are clean and secure. If your regulator has vent ports, are they free of dust and dirt? People have had problems with wasps (locally known as mud-daubers) nesting in their controls. Take a flashlight and look into the vent ports to see if any uninvited guests have taken up residence. If salt air or humidity are problems in your area, perhaps it's time to open your controls and spray them with LPS.

Now you're ready to tackle the battery bank. If you've been performing regular battery maintenance, then you know your battery's condition. If not, pull out your hydrometer and digital multitester and check the state of charge. Are all of the cells within 0.175 volts of each other? Are hydrometer readings for cells approximately the same? Is the electrolyte in all of your cells at the "full" mark? If you have been adding distilled water often, perhaps your regulator is overcharging your batteries. It's worth checking out and perhaps recalibrating your "black box."

If you have a battery box, does it protect your batteries from extremes in temperature? Perhaps you should add insulation. Have you installed the vent for your battery storage area? Have any creatures made the battery box home? Now look at your battery interconnects and terminals. Is there corrosion? Look closely because some corrosion can form at the junction between terminal and connector. Perhaps it is time to upgrade your wire-wrap battery connections with crimped and soldered ring connectors or some other, more secure, fastener. In fact, crimping and soldering ring, spade, and other connectors throughout your system will cut resistance losses. Again, be sure wires are properly sized.

Test for voltage from battery terminals to battery case. Dirt and dust on the case can contain current conducting material. Misting of electrolyte from overcharging will cause dust to adhere to your batteries, and you might be losing power because of that. A thorough cleansing with baking soda and water (one pound baking soda per gallon of water) will solve that problem and greatly extend the life of your batteries. Be sure to keep the baking soda out of the batteries and wipe everything dry.

Have your batteries been cycled to death? Frequent and very deep discharge of your battery bank will shorten its life. As batteries age, their self-discharge rate increases putting a parasitic loss on your system. New batteries are about 90% efficient. That is, for every 110 ampere hours put in, you can expect 100 ampere hours of stored energy. Batteries near the end of their useful life can be as low as 50% efficient. That's too much power to lose.

The battery bank capacity may need to be upgraded. If you are thinking about getting new batteries, consider doubling the size of your battery bank. All things being equal, twice the battery storage in one big bank should last more than twice as long as two battery banks half the size. At 80% depth of discharge, most deep-cycle batteries have about 400 useful cycles. At 40% depth of discharge the same batteries will have over 800 cycles. These same batteries cycled only 20% can have between 1600 to 2000 cycles.

In a properly sized PV home power system, battery drain is replaced on an almost daily basis. In other words, daily power production should match daily consumption based on climate and long-term weather conditions. (Since weather is variable, occasionally your battery storage is used more deeply.) Most battery banks sized to 40% depth of discharge have a useful life of 6 years or more. Doubling the battery bank can extend the useful life to perhaps 15 years. The verdict isn't in yet, but battery banks with more than adequate capacity for the load and the climate seem to perform well beyond expectations.

Have you been using your battery and controls area as a storage room? For safety's sake, keep it clean and dry and don't store flammables near the electrical system. Solar-powered homes share the same outstanding safety record as the solar industry. Let's keep it that way.

If you are using an inverter, are the wires properly sized? Is the inverter adequately ventilated? Have you been following manufacturer's recommendations to insure its long and trouble-free life? The same goes for any auxiliary battery charger you may be using. Perhaps it's time to upgrade your inverter or battery charger. Some heavy-duty industrial battery chargers have been replaced with improved models that are more efficient. Greater inverter and battery efficiency means more useful power from your PV system and less back-up generator running time.

When was the generator's last tune-up and oil change? When was the last engine rebuild? Perhaps it's time to phase the generator out completely. If it's time for another rebuild or replacement, consider going 100% solar. Now that you've learned to live with a PV home power system, conservation habits have become natural. If you're not using the generator very much, there's no sense hanging on to it.

In dual voltage systems, it's a good idea to look at your house

DC wiring. As you've added appliances and loads, your wires may now be undersized. Low-voltage DC wire losses are a major factor in inadequate DC lighting and can shorten the life of some DC appliances. Check all wiring, cords, plugs, and terminals. Remember, safety first means safe wires. Look for spots where mice have nibbled on the wire. One house suffered major problems because the wiring crew ate fried chicken for lunch.

Finally, look at your fuse or breaker box. Have you and your PV system outgrown an undersized distribution service? Has your expanding system turned into a spaghetti bowl of wires? Is everything properly tagged and labeled?

As you can see, being your own power plant manager is an important job. If your PV home power plant installation was done properly, managing it should take less than one hour a year. If it's taking longer, then it's time to fine tune. The reward for fine tuning is improved efficiency, more useful power, and longer system life.

Combining PV and Other
Electric Systems

Many people now using solar electricity previously used other alternative energies. Homesteaders, when first moving to a new locale with no conventional electricity, have used their automobile or truck battery and alternator to furnish small amounts of power for the necessary lights and music.

A few years ago, there was a growing "cottage industry" supplying dual battery kits for this purpose. Daily driving charged a deep-cycle battery in the same manner recreational vehicle coach batteries are charged. The output of the alternator is split by a battery isolator charging both the starter battery and the RV coach deep-cycle batteries. The kit provided a junction box and plug so the home could be plugged into the second battery. Do-it-yourselfers also assembled their own dual battery systems.

From such modest beginnings solar modules were added so that the time between vehicle trips for driving and charging could be extended. Eventually, the batteries became stationary as the addition of more PV modules eliminated the need for the dual battery system.

Some people moved away from the grid and took along a generator. This gasoline, diesel, or propane generator was used to power tools during construction and large appliances and water pumps. The addition of a battery charger, battery bank, and inverter greatly expanded the generator's efficiency.

FIGURE 12.1: Temporary living quarters while building your home can be PV-powered. This array provides all the electrical needs for the trailer and power tools. (Photo: ARCO Solar, Inc.)

Now the generator could be used to charge batteries while it powered other things. Dual use better utilizes lightly loaded generators and means increased fuel efficiency. At night or when loads are light, the generator is not needed because the battery bank and inverter will do the job, which means less generator running time and less frequent generator servicing or repairs. As the generator's use lessens, its useful life is extended. Solar modules can be added as the budget permits.

Eventually, solar modules replace the generator. If the timing is good, the generator should last as long as needed. Such wise planning, however, is the exception. People usually keep a generator in service until it becomes a financial burden.

A few people have dabbled with thermoelectric generators as a power source. These devices, based on a semiconductor version of the thermocouple, have begun to catch on with people who have high heating requirements. Some thermoelectric units operate on propane and can be modified to run on biogas. Some attach to the firebox of a wood, coal, oil, or gas heater.

In water-abundant mountain locations, hydropower is used in conjunction with solar. The hydroelectric power plant can be fairly large with the familiar dam and spillover. It can also be a small generator that attaches to a pipe from a spring or even a kitchen faucet. For those who have gravity-fed spring water, a faucet-attached generator makes sense. Imagine turning on the tap to produce electricity. Of course, such a system will work only when the home is sited in the path of a controlled flow of water.

Small human-powered generators resembling exercise bicycles have also been linked to a PV-charged battery bank. For the physically active, or those who wish to be, this is a good idea. You can build a pedal-powered generator from spare bicycle parts and a small generator. But be forewarned: if you use a car alternator or generator, you've got a steep hill to pedal. The average human at work is equal to one-third horsepower. At 100% efficiency, this means you could pedal over 200 watts into your battery bank. But for how long? A half-hour on an electric exercise cycle may be all you can do in a day.

THE PV STANDBY SYSTEM

Photovoltaic power systems can be categorized into four basic types: Autonomous Direct, Autonomous Battery, Utility Interconnect, and Standby.

Autonomous direct systems are used to power water pumps, fans, and other equipment directly from the solar array when the sun is shining. The advantage in using PV power directly is eliminating battery servicing and maintenance. The wiring is also very simple. In water pumping systems, it is less expensive to store water in a holding tank than to store electricity in batteries. This system uses DC equipment.

Autonomous battery systems are the most popular and are used in remote locations requiring power either 24 hours a day or when the

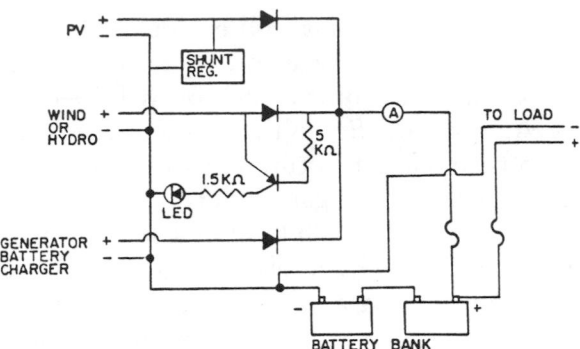

FIGURE 12.2: This circuit diagram shows several sources of alternative low-voltage electricity feeding into a single battery bank The blocking diodes are the heart of the system, isolating the power source until its output voltage is higher than the common system voltage. The optional indicator light comes on whenever the source is delivering power. Notice that the shunt voltage regulator is connected to the source side of the blocking diode.

sun is not shining. By using storage batteries, power is available whenever needed. Equipment can either be DC or standard AC with the use of an inverter.

Utility interconnect systems use a synchronous inverter to couple the PV array to the utility company. The PV array provides power during sun periods and any excess is sold to the utility company. During non-sun periods, needed power is purchased from the utility company.

The *Standby* system is an autonomous system used in conjunction with utility power or a generator. It is sometimes called an Uninterruptable Power System (UPS). PV charges batteries which are also charged by the utility or a fossil fuel generator. Automatic or manual switching can be used to give priority to PV. This means that when the batteries are fully charged and the sun is shining, any loads through the system will get their power from the sun. Should the loads be greater than the PV production, the utility or generator will carry the load. Thus, PV can be used to offset a portion of the utility bill. An obvious advantage is that the PV system can be expanded to eventually replace the·dependency on utility or generator power. During power outages, the standby system carries either all or a selected portion of the normal loads. The block diagram, Figure 12.3, shows a standby system with a manual transfer safety switch.

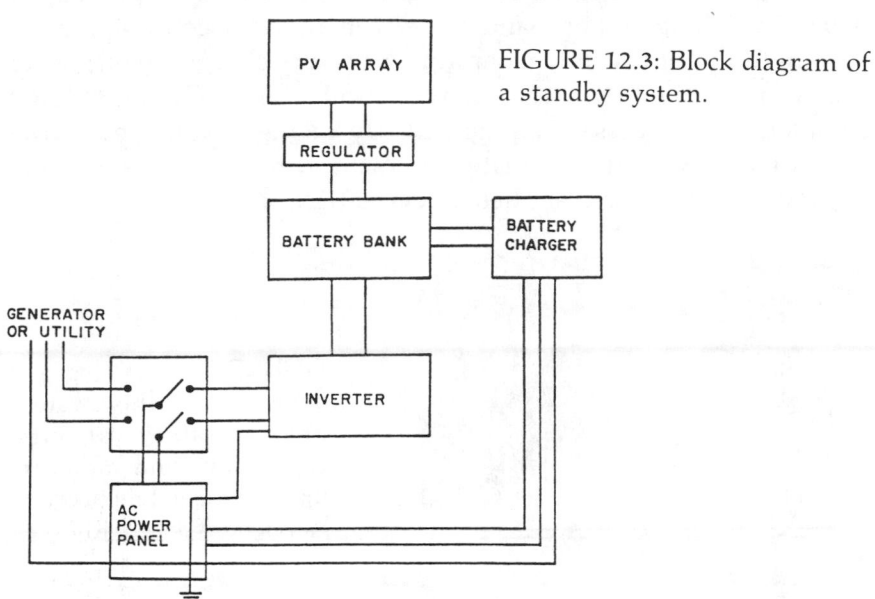

FIGURE 12.3: Block diagram of a standby system.

If a generator is used with the standby system, while the generator is running, the PV battery bank is charged by a conventional battery charger and the generator is more fully used to its capacity. Should the generator fail, the PV system is there to take its place. As more modules and batteries are added, generator running time is reduced and eventually eliminated.

Operation of the standby system is straightforward. Power is obtained from two sources—utility (or generator) and PV. Under normal conditions two events can occur. When the storage batteries, charged from PV and utility/gen, are fully charged, power will come from PV if the sun is present. Thus, the system acts like a backward UPS in this mode. It "simulates" a utility/gen outage or power disconnect, and PV takes over. In the event of a real utility/gen power failure, the system automatically switches over to

PV for the duration or the capacity of the battery storage. A properly designed system should have adequate battery storage.

Installation of the standby system is straightforward. Figures 12.4 and 12.5 show the equipment room for a standby system which provides most of the power for a large home in Maine. PV is set up as usual. Switching, either manual or automatic, is used to tie the PV system to the utility or a generator. A battery charger (built-in like the Heart unit) or a separate charger (I.B.E., Sears, Graingers, etc.) completes the necessary equipment. By taking advantage of low-rate utility power or an on-site generator, users can incrementally expand the PV system as their budgets permit.

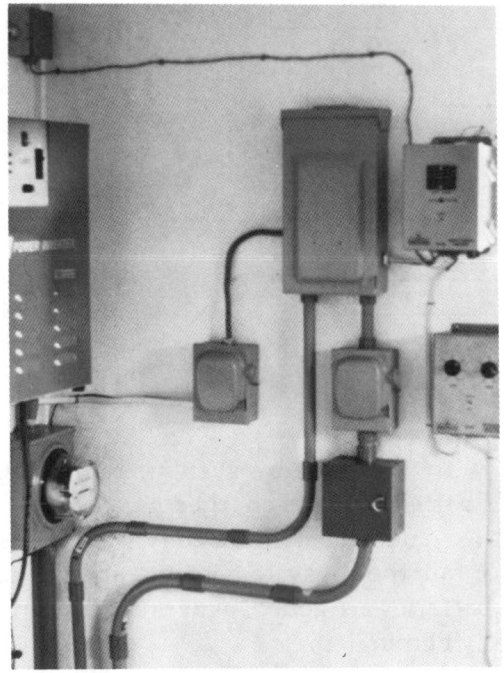

FIGURE 12.4: Wiring and controls for this system provide safety cut-outs, regulation, and monitoring. (Talmage Engineering, Kennebunkport, ME)

Photron, Inc., of Willits, California, offers a pre-assembled power board which simplifies the installation of PV systems tied to a utility or a generator (see Figures 12.6, 12.7, and 12.8). Photron's owner, Laurence Jennings, believes that a safe installation requires a skilled installer. However, finding someone with the proper skills is not always possible. Therefore, pre-wired, well-documented power

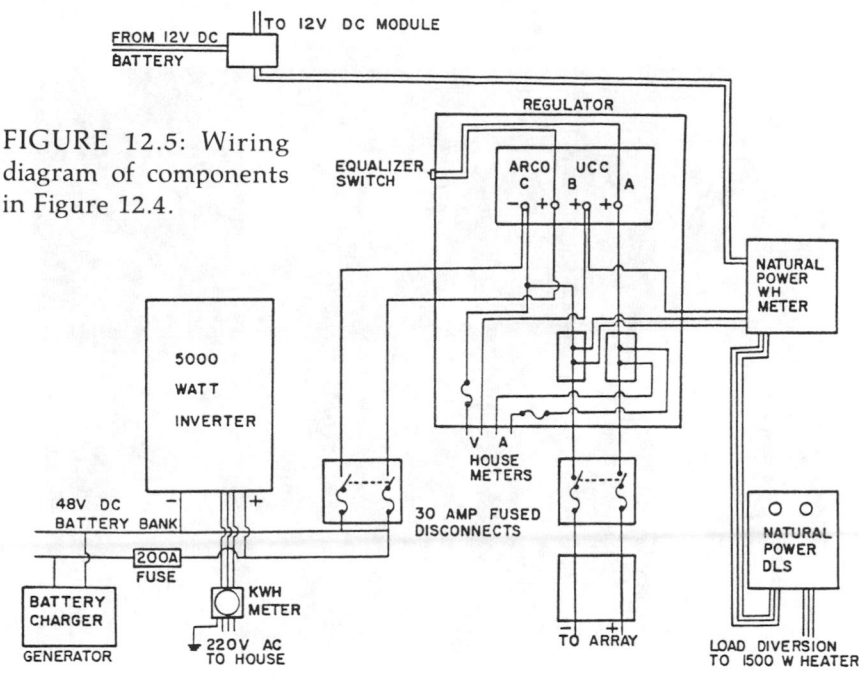

FIGURE 12.5: Wiring diagram of components in Figure 12.4.

or control boards make it possible for the handyman with some electrical skills to install a PV and utility (or generator) system.

I have coached hundreds of absolute novices by mail and telephone. If the do-it-yourselfer buys quality equipment from a knowledgeable supplier and is willing to take the time to do a clean, safe, and thorough job, installing pre-assembled power boards or individual components is possible. In fact, doing it yourself will give you a better understanding of the equipment and its limitations. This knowledge helps to insure proper use which means long and trouble-free system life.

It doesn't matter what this set-up is called—a standby system, photo/gen set, UPS, gen-set plus—it's all essentially the same.

While many system designers use a fossil fuel generator with PV, stating cost-effectiveness and greater generator efficiency, this practice is only an expediency. The justification for using a generator stems not from the efficiency and cost-effectiveness of the generator, but from the additional equipment used with the generator to reduce its running time. If we follow this line of

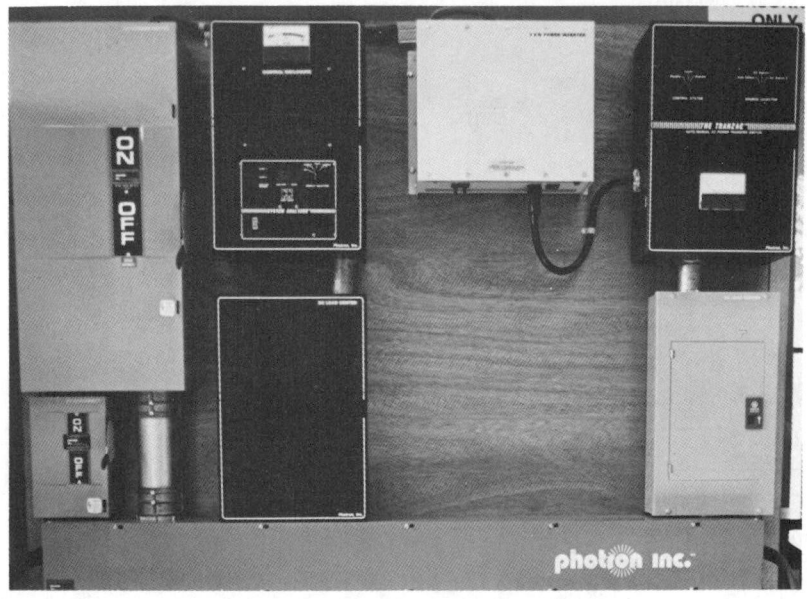

FIGURE 12.6: The Photron Powerboard is a complete assemblage of balance of system equipment and can start and stop a standby generator, transfer loads, and provide digital monitoring. This unit also has a 3-kW inverter. Pre-assembled, pre-tested, and pre-wired components can make installation simple. (Photo: Photron, Inc.)

reasoning, the ultimate in cost savings and efficiency would be to totally eliminate the fossil fuel generator.

Fossil Fuel Generators

When people first started using PV for home power systems, there were two strong social trends emerging. The first was a desire, regardless of cost, to use this gentle technology and not polluting, wasteful, and costly generators. The second trend grew out of a growing realization of planetary ecological limits. Fossil fuels may have seemed cheap before OPEC, but the price to be paid for non-renewable energy technology must be measured in more than dollars.

The first PV-powered homes, often occupied by people seeking a simpler life, seemed energy impoverished to many. But as these

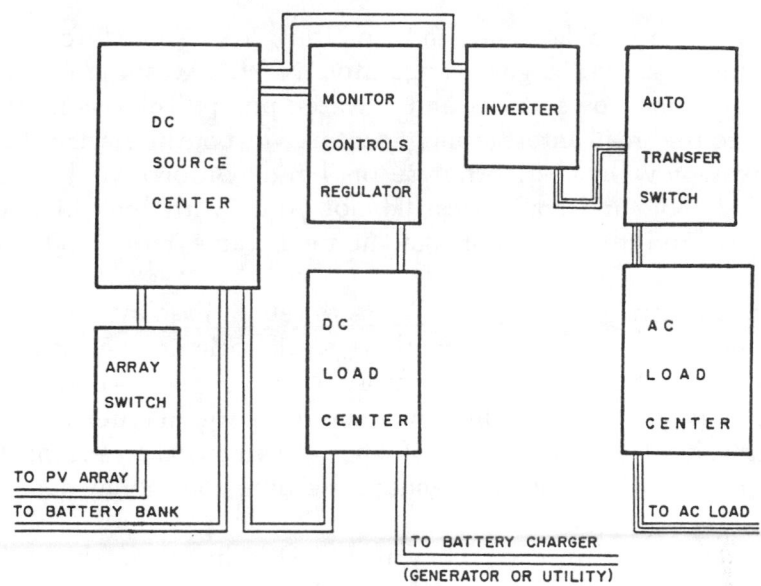

FIGURE 12.7: Block diagram of the Photron Powerboard.

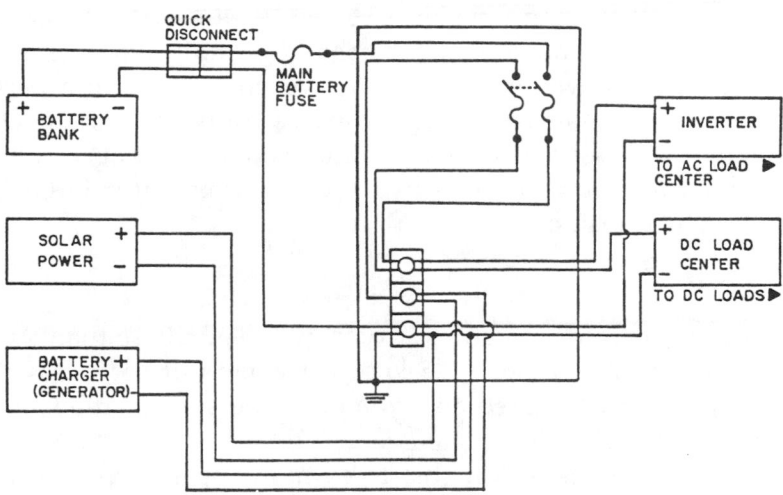

FIGURE 12.8: Schematic for Photron Powerline Source Center.

individuals got older and had children, they opted for more conveniences. First, a gas refrigerator. Next, a washing machine. Perhaps some power tools and a water pump. Televisions were added to the small entertainment center. Most of the electrical add-ons were low in cost, often secondhand. Unfortunately, these recycled tools and appliances did not come with built-in power supplies, and the energy-frugal home became more and more complicated and costly to run.

Being handy, some early PV users rebuilt gasoline generators for their big power loads. Still others, anticipating power needs for construction, bought new generators. Well, generators cost money to operate and require time and materials for maintenance and repairs. Needless to say, those who bought and used generators had to rationalize their use and validate the time and money spent on them.

Where did this lead? They were often in worse shape than before. Not only had they adopted the polluting and wasteful technology they had earlier rejected, they now had a fossil fuel plant in their backyard that they had to deal with regularly. Some of these generators were costing more than it cost to raise a child. Worse, the generator was stealing time better spent playing with the child.

By now you should conclude that I do not favor the use of generators. However, realizing that we do not live in a perfect world, and that most people go through a transition period before becoming 100% solar, I've included this section. It is hoped you will use this information not to justify the use of a generator, but to help you in your transition.

The Costs

Initially, a generator can be less expensive to purchase than a solar electric system. In 1986, a 2.5-kilowatt generator costs less than $2000. For a few hundred dollars more, a person could buy a fuel storage tank and wiring.

But how much does it cost to own and operate a generator? Depending on the circumstances, output from a 2.5-kilowatt generator—if efficiently operated at or near peak loading—will cost between $1.50 to over $2.00 per kilowatt hour. This is based on a life-cycle cost analysis that takes into consideration fuel, maintenance, repairs, and replacement over a 20-year period.

Terrance Paul of BEST Inverters, author of *How to Design an Independent Power System* (1981), lists the following costs:

2.5-kW generator	$1000.00
Gas and oil	0.52 per hour
Maintenance	0.25 per hour
Starting/handling	0.75 per hour
Depreciation	0.50 (2000-hour life)

Paul says it costs $2.02 per hour to operate a generator. He points out that the generator is loud, vibrates, pollutes with toxic fumes, wastes time (transporting fuel and changing oil), obligates the conscientious user (who wishes to save fuel and avoid noise) to turn the generator on and off frequently, and is hot and dangerous. It appears Mr. Paul doesn't like generators, either.

He compares this generator to his company's inverter system ($2200 for inverter, two batteries, alternator, and installation) which costs $0.35 per hour to operate. Listing battery recharging from an alternator at $0.16 per hour, $0.10 per hour for battery replacement, $0.01 per hour for maintenance, and $0.08 per hour depreciation over a 25,000-hour life, we can see why his company has sold so many inverters.

If PV is used to replace alternator, grid, or generator battery charging, you need a basic PV AC system: modules, mount, regulator, battery bank, inverter, and wiring. From this list, it's not hard to see where PV systems get most of their balance-of-systems (BOS) hardware. Generator, inverter, and uninterruptable power systems have been using similar components for decades.

Using a 20-year design life (175,000 hours), the eight-module solar electric system described in chapter 5 should require one battery replacement. Allowing for uncertainty, calculate that the cost of replacement batteries will increase 50%. There's no sense padding the figures to make it easier for the PV system to compete with a generator, even though battery costs in real dollars have remained relatively stable for the past 20 years. Allowing $25 per year for parts and servicing brings the simple 20-year life-cycle cost for the owner-installed PV system to $14,700.

If the solar electric system receives 4 peak hours of sun per day, it will produce about 1.6 kWh per day (440 peak watts x 4 peak hours of sun x 30% average increase from tracker = 2.28 kWh x 0.98 wire loss x 0.8 battery loss x 0.9 inverter loss x 365 days) or 580 kWh per year. The 11,700 kWh produced in 20 years will cost $1.26 per kWh.

Occasionally, an installer is asked to design a PV and generator power system when the exact power requirements are not known. This is often the case in new house construction. So it was with one of Don Loweburg's clients. Although the home is located in a utility-serviced suburban community north of Fresno, the homeowner wanted to produce his own power (see Figure 6.21). An 800-watt tracking solar array was mounted on the roof. Storage consists of an industrial-grade battery bank of 800 ampere hours. A Helionetics 2.5-kW inverter with load demand provides household power. For back-up power a 4-kW propane generator and battery charger was also installed (Figures 12.9 and 12.10).

FIGURE 12.9: The equipment room of a Fresno, California, home. From left to right: generator-powered battery charger, 2.5-kW inverter, PV regulator, DC panel, AC panel, and array safety switch. The twelve 2-volt industrial storage cells are located in the battery compartment below the AC breaker box. Designed and installed by Don Loweburg, Off Line, Independent Energy Systems.

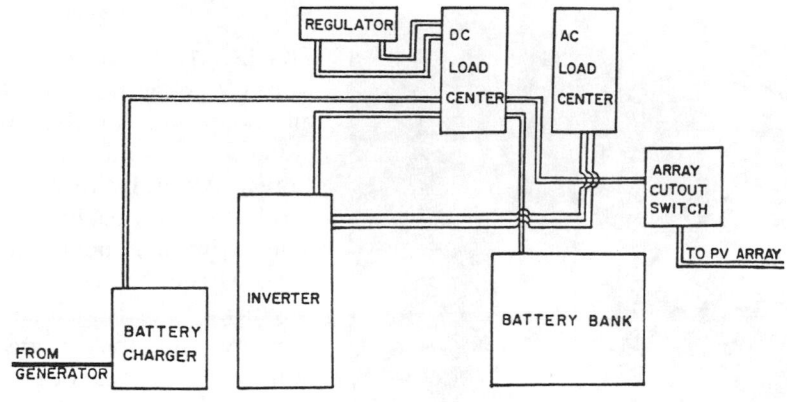

FIGURE 12.10: Block diagram of the equipment room shown in Figure 12.9.

As it turned out, the homeowner's power needs were almost within the system's production. From March to September, solar power provides all of the electrical needs. In December the generator is operated two hours per week to keep the batteries topped off. During January, February, October, and November, ony one hour per week of generator run-time is needed.

In this case, it may have been wiser for the homeowner to eliminate the generator and battery charger and use that money to expand the solar array and battery bank. This is possible when climatic conditions permit. In locations where winter cloud cover is extreme, a back-up generator can make sense.

The final point is to carefully consider your present and future power requirements. You may discover that the initial purchase of a fossil fuel generator is unnecessary. By adding a few more modules and paying close attention to your power requirements, you can go from energy user to energy producer and skip the fossil fuel generator transition period.

WIND AND SOLAR

When I first started using homemade electricity, there was much excitement about windchargers. Small homestead-sized units were

FIGURE 12.11: A combination PV/generator provides all the power for this spacious home and two cottages. All buildings have kitchens and bathrooms. The swimming pool pump runs three hours a day. Regular PV loads include all lighting and once a week laundry. Eighteen 66-watt modules; 32 each 220 ampere hour 6-volt batteries; 500-watt inverter. Designed and installed by James Whitcomb and Rob Hilbun. (Haleakala Resources, Inc., Kahului, HI)

beginning to reappear, similar to windchargers used in the U.S. before rural electrification. Although there have been improvements in controls and construction materials, they are basically the same as chargers made 40 years previous.

For a short time, I tried to use windpower. My experience (or lack of it) has biased me strongly against it. First of all, I don't like heights. Climbing a tower, even with safety straps, is not a welcome thrill.

Owning a windcharger means work. The good units are pretty dependable, the cheap chargers are pure aggravation. Regularly—the frequency depending on the quality of the windcharger and the weather—a windcharger needs servicing and repair.

If you are considering a windcharger, and if your budget is limited and you are handy, build your own or buy a used unit for under $1000. This will give you some experience. If you do like it, you can graduate to something better.

Before you decide on a windcharger, buy or lease wind monitoring equipment to evaluate your site. Windcharger dealers often provide this service at a nominal cost. Some state energy offices or colleges will assist you.

The following Arkansas PV/wind system belongs to Stephen and Laurie Cook. Stephen is a teacher, writer, amateur astronomer, and president of CompuSOLAR, a solar computer software and educational firm. While director of the North Arkansas Community College Energy Center at Harrison, Arkansas, Stephen installed and tested a PV/wind system at his remote mountain home. The following is adapted and updated from Stephen Cook's book *Achieving Self Reliance: Backyard Energy Lessons*, published by the Arkansas Energy Office.

Combining Solar and Wind—A First Report

If you are depending on a wind generator for electricity, summer breezes really do make you feel fine. That's because in many places the wind doesn't blow nearly as much then as at other times during the year. Back in 1979, I was pondering the relative lack of summer wind in the Ozarks. I wondered how we would meet our summer-time electrical needs with the 200-watt Winco wind machine we were planning to purchase. With costly excess battery storage? With a noisy, smelly gasoline back-up unit? How? Well, we managed by combining wind and solar electricity.

Besides the general observation that solar outputs will be high when wind outputs are low (summer) and vice versa in winter, there is another advantage. Wind generators, with their moving parts, need occasional maintenance and may be down for minor (or major) repairs once in a while. With the photovoltaics back-up, these intervals are less painful. Having both energy sources makes it easier on the batteries, too.

Let's look at how wind and solar inputs into a battery bank can be balanced by considering three separate systems.

System 1

Batteries will last much longer if there is no great seasonal variation in their state of charge. Photovoltaics companies consider this when they size systems for remote, constant-load applications (say for telecommunications equipment). ARCO Solar, for example, typically recommends a panel tilt of latitude plus 15° or 20° or more (a tilt equal to the latitude will maximize fixed-panel annual production). In other words, by pointing toward the winter sun, winter outputs will be increased, summer outputs cut, and a more uniform seasonal delivery of electricity achieved. By doing this, variations in the monthly kWh production can be kept within 10 to 15% of an average value—good for batteries, but wasteful of potentially available electricity.

System 2

Wind generators are hard on batteries. Consider a 200-watt Winco's output on a small tower at a good Ozark site (see Table 12.1). While 26 kWh/month is the average output, in July there is only 13.2 kWh available, yet 34.8 kWh is produced in April. The monthly production departs from the average value by nearly 50%. Without (power-wasting) voltage regulation and deep-cycle batteries, a constant remote load will be hard to power for long! And, of course, there is not much summer production.

Table 12.1: Wind Output

Typical Month	January	April	July	October	Average
Average wind speed at generator (mph)	11.1	12.0	8.5	10.0	10.5
kWh per month	30.4	34.8	13.2	23.4	26.0

How can we improve on these two situations? Consider solar first.

System 3

Suppose we want to maximize PV outputs without going to the additional complexity of tracking. By changing the tilt angle every so often, we can increase power output significantly for just minutes of work each year (if we mount our panels with this in mind). Consider a 200-watt PV array in the Ozarks (maybe six 35-watt modules). Suppose we adjust the tilt four times a year as follows:

Month	Tilt Angle
November, December, January	latitude plus 15°
February, March, April	latitude
May, June, July	latitude minus 15°
August, September, October	latitude

The average production is 9% greater than if we had kept the panels tilted at the latitude angle (here 37°), which gave greater production than the latitude plus 20° case! However, the seasonal variation is greater: about 25% departure from the average value with December–January low and July high. Certainly floating-type batteries will not last as long as they would in situation 1.

Table 12.2: PV Output

Typical Month	January	April	July	October	Average
kWh per month	18.7	23.3	29.1	24.5	23.7

Based on sunshine statistics for Springfield, Missouri.

Our next situation to consider should be clear: combining System 2 and System 3 and allowing wind and solar to charge the same battery bank. Summing from Tables 12.1 and 12.2, see Table 12.3. Our biggest variation from the average is now 17%—down from 50% wind only, 25% adjustable PV only—and almost as good as the throw-away energy PV situation of System 1.

Table 12.3: Wind and PV Output

Typical Month	January	April	July	October	Average
kWh per month (wind)	30.4	34.8	13.2	23.4	26.0
kWh per month (PV)	18.7	23.3	29.1	24.5	23.7
kWh per month (total)	49.1	58.2	42.3	47.9	49.7

Based on sunshine statistics for Springfield, Missouri.

In practice, 12-volt homes will not have constant year-round loads although I suspect some will roughly approach it. Certainly installations where 12-volt refrigeration is employed (as in my own) might be expected to show summer peaks, while smaller "for lights, mostly" systems might show winter peaks. If one has the proper understanding of seasonal variation of the photovoltaic (or wind) energy resource available, appliance use might be tailored to more closely match production. Of course, the kWh/month figures presented here are before losses due to transmission, voltage regulation, battery inefficiencies, inverters, and obstruction which block sunlight or hinder free wind flow.

Regarding the hook-up of combined wind and solar systems, the following points need to be made:

1. Use blocking diodes on both positive wind and solar legs not only to prevent battery discharge but to make sure the solar cells don't motor the wind generator.
2. Wind-only systems require deep-cycle batteries. However, if the generators are properly sized to minimize seasonal charging variation for a location, and if the (consumption) discharge pattern is (seasonally) nearly constant, floating-type batteries could be successfully employed.
3. Separate battery banks for wind and solar could be used, with the load being powered by one bank, while the other is kept (charging) in reserve. Of course this will be more costly.

Wind and Solar in Arkansas: A Follow-Up
(Courtesy Arkansas Energy Office)

For three years the Cooks lived without electricity. In July 1980, 105 watts of photovoltaics were installed, followed shortly by a 200-watt wind generator. With highly energy-efficient 12-volt appliances and meager needs, the Cooks have lived for six years on an average of 15 to 25 kilowatt hours per month. While their system configuration has changed (they now have more photovoltaics and no wind generator), they still use only renewable electricity with gasoline (vehicle charging) back-up. The Cooks' two-and-a-half year experience with wind electricity ended in March 1983 when high winds caused equipment damage.

FIGURE 12.12: Six 35-watt solar modules power the energy-efficient home of Stephen and Laurie Cook. Located near Jasper, Arkansas, the Cooks were among the first people in the Ozarks to use PV power. (Photo: Arkansas Energy Office)

Their 1000 square foot house is passively heated with 280 square feet of single glazing on the south wall. The house is built on and around a huge natural rock outcropping, in a low-cost way to earth-temper the structure and provide thermal mass. The house is insulated to R-20 in the walls, R-30 in the roof, has double-glazed east and west windows and employs movable insulation on the south glazing. Unique ventilation, earth-tempering, an overhang, and a forested location are summer cooling features. A wood stove is used for heating, and a combination wood/gas stove for cooking and hot water. A breadbox solar water heater heats the water which comes from a gravity-fed spring to a storage tank.

The cooling mode of the house works very well. Earth-tempering provides cool air which is pulled through the house by a naturally induced draft. A rooftop false ceiling is removed in summer exposing large peak vents where hot air can exit. During the record-setting heat wave of July 1980, the lower levels of the house never got above 79°F (61°C), and that's without the use of cooling fans.

In winter, an estimated 30% to 40% of the heating load is met by solar energy. However, winter shading of the south side of the house by the bare branches of deciduous trees has proven to be more a problem than expected. The Cooks are hesitant to cut too many of these trees because they shade the house in summer.

FIGURE 12.13: Alternate energy equipment at the North Arkansas Community College, Harrison, Arkansas. While director of the college's Energy Center, Stephen Cook's innovative projects such as a solar-powered learning and resource center and a mobile demonstration unit taught thousands of people about wind and solar power. A 700-watt roof-mounted solar array powers the college's learning center. The mobile unit has a solar module, 200-watt windcharger with an adjustable tower, a solar water heater, and attached greenhouse. (Photo: S. Cook)

Prior to March 1983, the energy system consisted of a 200-watt Winco wind machine on a 100-foot guyed tower, 400 feet of aluminum wire carrying 12 volt DC to the house, and three 35-watt solar modules. When the wind generator was taken out of service in 1983, two additional modules were added and the solar array moved to a sunnier location. A sixth module was added in 1985. The wind generator and the photovoltaics charged the same battery bank. Prior to January 1982, six 2-volt used telephone company 420 ampere-hour batteries hooked in series were used. Since then, six almost new 500 ampere-hour C&D batteries have been employed. There is no back-up generator, but a second battery has been installed in their vehicle. This auxiliary battery is charged by

FIGURE 12.14: This home located within sight of the state capitol at Little Rock, Arkansas, generates some of its power from wind and solar. (Winter winds are significant.) The solar array has a reflector to enhance collection. (Photo: S. Cook)

alternator-supplied electricity (which otherwise might be wasted by the car's voltage regulator).

The appliances are 12 volt DC. Lights are 20-watt quartz-iodine and 12-watt incandescent. A Koolatron 1.7 cubic foot refrigerator draws 24 watts at 70°F (21°C) ambient temperature. A car stereo cassette am/fm 20-watt radio and 12-inch black and white television (which draws 16 watts) are used. Other loads are an automobile vacuum cleaner (108 watts), kitchen blender (40 watts), and fan (14 watts). Additionally, there is occasional use of 12-volt tools such as a saw, sander, and drill. A 250-watt inverter is used to power a microcomputer and stereo turntable.

The Winco wind machine experienced difficulties because of excessive turbulence at its mountainside elevation. It was sited 1900 feet above sea level at the ground elevation with the tower adding another 100 feet. It was above all obstructions for two miles to the west but was within a couple hundred feet of obstructions to the east where trees and a mountaintop (elevation 2075 feet) hindered operation. With winds coming predominantly from the west, the site had plenty of wind but excessive turbulence.

Table 12.4: Electrical Consumption in kWh

Month	Overhead Lights	Directional Lights	TV	Refrig- erator	Vacuum	Radio/ Tape	Total
Dec 1980	7.4	1.1	3.0	0.7	0.2	0.3	12.7
Jan 1981	7.4	1.1	3.0	0.4	0.2	0.3	12.4
Feb 1981	5.6	0.7	2.2	1.3	0.6	0.2	10.5
Mar 1981	6.2	0.7	2.5	3.0	0.9	0.3	13.6
						Average:	12.3

The problems experienced included this power-robbing turbulence which at times reduced electrical production. The Winco voltage regulator, which sensed the solar array's higher daytime battery charging voltage would often cut out the wind generator's output. This would occur even when there was no real danger of battery overcharging. The regulator was finally taken out as it was wasting power. Battery voltage was regulated through appliance use, such as vacuuming during high wind periods. The leading-edge sheet-metal flap on the wind generator blades eventually tore off and grease and dirt got on the generator's collector ring copper brushes causing poor contact and reduced output.

In March 1983 winds estimated at 60 to 80 miles per hour slammed into the installation causing irreparable damage. The upper tower section was badly twisted, the generator was thrown to the ground, and the blades destroyed.

Cook concluded that the site was not appropriate for a wind machine. Before installation, spot monitoring had indicated promising wind speeds but turbulence was not observed. This could have been checked by tying a ribbon streamer to a pole and observing for uneven movement.

The photovoltaics have performed up to expectations and there have been no problems. After six years of operation, there has been no apparent reduction in output. Periodically, two mounting bolts are loosened and the panel tilt angle is adjusted to optimize output.

A Natural Power DC ampere-hour meter was used to record combined wind and solar production from December 1980 to March 1981. During the monitoring period the windcharger regulator (later removed) was in use and output was reduced. A 10% wire loss occurred in the wind generator operation because of the distance between it and the house. The PV array suffered a 15% reduced output from shading by obstructions. An average 20 kWh per month production is estimated to be only 60% of what the system would have produced without these problems. The 200-watt wind generator should have provided 20 kWh per month at this 10-mph average windspeed site.

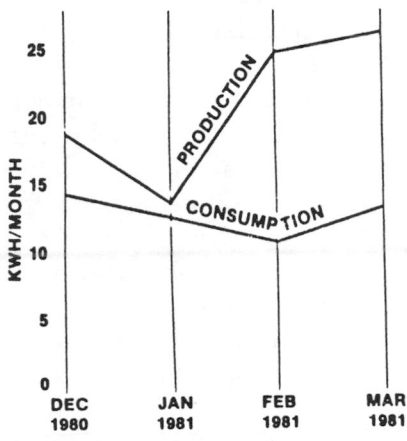

FIGURE 12.15: Electrical production vs consumption. (Courtesy Arkansas Energy Office; from *Achieving Self-Reliance* by Stephen Cook, page 97.)

The used batteries, 15-years-old when scrapped by the phone company, and supposedly with many good years remaining, were used one and a half years and replaced. They had an effective storage capacity of approximately 20% of their new 420 amp-hour rating and were clearly wasting power. As batteries age, their internal resistance increases and electricity is increasingly wasted as heat.

The inadequacy of these batteries was particularly evident when the new batteries were installed. The new battery bank storing six kilowatt hours of electricity could, by itself, run the house for up to ten days when fully charged. The Cook family, having no back-up generator, occasionally discharges these batteries to a lower level than they would like. After a few days of cloudy weather, they curtail appliance use when battery voltage drops below 12 volts. Below 11.2 volts, they would not use the batteries in an effort to prevent permanent damage, but so far, this has not been necessary. Previously, the Cooks relied on wind power during cloudy spells. Now, without the wind machine, cloudy-day diffuse-light energy production from the solar cells is critical.

The well-insulated portable refrigerator is the Cook's major consumer of summer electricity. In winter it is moved to an unheated porch and "charged" by cold nighttime temperatures. In summer, a variable thermostat controls refrigerator temperatures. In periods of low electricity production, the food is kept at a warmer temperature. The unit has performed well in six years of constant use, except for the plastic door latches which have been replaced three times. To help during summer periods of low electricity generation, Cook installed a 2.5 cubic foot icebox in the kitchen. They now occasionally augment the 12-volt Koolatron by purchasing ice blocks. They prefer this to running a generator.

High-efficiency quartz-iodine bulbs deliver light equivalent to a 60-watt incandescent bulb for only 20 watts of power. The Cooks say they like the light quality and view it as far superior to fluorescent lights. However, they disagree with the long-life claims. For them the bulbs provide 6 to 10 months of use, contrary to the advertised claims that they last three times as long as incandescents. This is a problem because replacement bulbs cost $3 to $4 each. Keeping these bulbs clean is important, as is careful handling since they are very fragile.

The chief expenditures for the Cooks' electricity generation system have been:

Photovoltaics	$1595
Wind generator	625
Wind tower	620
Transmission wire	355
Batteries (all)	400
Inverter	150
Miscellaneous	150

The house construction materials costs were $5000, a figure kept low by the use of recycled materials. The annual energy costs are estimated to be $120. This breaks down to $65 for LP gas, $40 for gasoline, oil, and maintenance for a chainsaw, and $15 for ice blocks. The Cooks live on 22 acres of forest in need of thinning so wood is free for the labor.

Stephen and Laurie Cook live an excellent example of frugal, open-minded, and intelligent energy production and use.

Chapter 13

PV in the North

Many of the systems in this book are located in the southwestern United States. The reasons for this are obvious. Anywhere you find a plentiful resource, people will use it. Where the wind blows, you find windmills and windchargers. Abundantly flowing water means hydropower. Countries rich in coal and oil will use these fossil fuels.

PV works anywhere the sun shines. Photovoltaic energy conversion does not require heat, just sunlight. In fact, cool and sunny weather results in more efficient solar cell performance. Therefore, the sunbelt has no monopoly on PV. In fact, some of the better-designed PV systems are located in the northeastern U.S.

Granted, the more sunshine, the more power you'll get from your PV power system. But don't let clouds, rain, and snow fool you. Unless it is nighttime, there are photons of energy pouring down waiting to be captured.

While Alaska in December, London in the rain, and Chicago in winter may seem poor candidates for solar set-ups, PV can make sense at these seemingly marginal locations. The combination of wind in winter and PV in summer has been used successfully at one

Alaskan site for almost ten years. Summer in Alaska is characterized by long days and short nights. That spells PV power.

As I've said, don't write off a location just because it doesn't "seem" a likely candidate for PV. Year-round climatic averages and microclimates can be deceiving. For example, solar radiation in Maine in summer is equal to that of Little Rock, Arkansas. And before jumping to conclusions, remember the importance of energy conservation and good design.

Sandra Dickson

Photovoltaics is a technology overwhelmingly peopled, but not dominated, by men. Perhaps this is because power production evolved in the "man's world" of gadgets and gears. But PV is not "man's work" and Sandra Dickson has been proving that for years. I had the honor of sharing a feature article with her in an 1982 *Christian Science Monitor* story about PV homes. Ms. Dickson is a PV and energy consultant (working under the name Solar Works, not to be confused with Solar Works! of Santa Fe, New Mexico). She has received state and national awards for her video documentary, "Yankee Independence: Solar Electricity on a Maine Island."

Sandra calls her system a PV/utility hybrid. Her Port Clyde home was already connected to the utilities when she bought it. The old 60-ampere service was renovated but not increased. Utility powers a 30-year old manual-defrost refrigerator, the furnace pump (which may someday be changed to 12 volt DC), some lights and a timer, a clock radio, hair dryer, coffee grinder, battery charger, and word processor.

The house has ample sun all day and until late in the afternoon for passive heating. Its open interior design, windows, and french doors provide natural lighting year round. Occasionally, the upstairs office on the southwest overheats in mid-winter. That doesn't sound like frigid Maine at all.

The kitchen range is propane. When the house was renovated, insulation and a vapor/air barrier was added. Now baking bread raises the house temperature in a half hour. Primary heat does not come from baking, however, but from a 100-year-old parlor stove that uses a cord and a half of wood per year. Oil back-up is used to keep the temperature above 58°F (14°C) at night.

FIGURE 13.1: James Schwarber designs and installs PV and windchargers. His log home in Alaska is a beautiful example of appropriate technology for northern climates. (Courtesy J. Schwarber, Bettles, Alaska)

The PV system consists of four 35-watt modules and six 110 ampere-hour deep-cycle batteries. These provide power for the lamps, a color television, stereo, and food blender. A utility-powered battery charger is used during the winter about one hour a night as back-up to the PV. Fluorescent lighting fixtures in the kitchen and bathroom were modified to 12 volts by David Sleeper. (David also designed and installed the PV system.)

The array was tilted to 60° for maximum winter collection and to shed snow. Only once in three years has the snow had to be swept off.

Average daily PV power consumption is about 100 watt hours. At a 50% depth of discharge, the battery bank can provide 5 (winter) to 12 (summer) days of autonomy. As stated, the home was not wired for 12 volts, and 12 AWG wiring was used during renovations to keep costs down. The electrician was reluctant to attempt to pull larger size wire through standard outlet boxes. No wire run is longer than 40 feet. However, an additional circuit using 10 AWG to the upstairs television was installed.

FIGURE 13.2: Sandra Dickson's Maine home. This small roof-mounted array provides part of the home's electrical needs. Installer: David Sleeper. (Photo: Robert Sardinsky, Rising Sun Enterprises)

Sandra's use of 60 to 90 kWh per month means her utility bills are a modest $6 per month. Utility rates in Port Clyde have risen to 8 cents per kWh since her system's installation. During the same period, the cost of PV modules has dropped 25%.

Sandra does not live very differently from her neighbors, although she does turn the lights and tv off when not in use. This artist, writer, PV pioneer likes the security of knowing that her lights will not go out during a winter storm. She enjoys contributing to the solar transition by example.

David Sleeper

David Sleeper, of Brook Farm, Inc., Falmouth, Maine, designs and installs PV systems. He has also done some outstanding pioneer work in PV. His installations are well thought out, meeting and exceeding federal, state, international, and Coast Guard specifications.

FIGURE 13.3: The Dickson home electric system combining PV and grid power. 12-volt DC powers lighting, stereo and television, while a grid-powered battery charger keeps the battery bank charged during cloudy periods. (Photo: Robert Sardinsky, Rising Sun Enterprises)

Peter Talmage

Talmage Engineering is another outstanding PV design and installation firm in Maine. Peter Talmage's Kennebunkport home will give you an idea of what is possible in a location where alternative energy has been considered impractical.

His 2300 square foot home has two solar arrays and a windcharger. DC voltages are 12 and 32 volts. The 12-volt battery bank is 786 ampere hour and the 32-volt bank is 840 ampere hour. A 3000-watt inverter changes 32 volts DC to 120 volts AC.

Lighting is both 12 and 32 volts. AC loads consist of shop tools, home appliances, and office equipment. The refrigerator, solar water heater pump, and baseboard hot water pump use 12 volts DC. The shop has an assortment of 32-volt tools, including a milling machine, lathe, drill press, grinder, and bandsaw.

Water is heated with wood, solar, and excess DC electricity. Cooking is done with propane and wood in winter. This passive

FIGURE 13.4: The first PV-powered United States Post Office. A 280-watt array charges a 48-volt battery bank on this PV/generator hybrid system. (David Sleeper, Brook Farm Inc.)

solar house has 425 square feet of south-facing windows with thermal night blinds and lots of insulation. The floors are R-26, walls are R-33, and ceiling is R-50. Power consumption for the home is two gallons of propane per month and two cords of soft wood per year. The PV system supplies all the summer electrical needs.

A well-designed Talmage Engineering PV system has been installed on an island home in Addison, Maine. The large house is all-electric except for a gas water heater and stove. All electricity is AC through an inverter and back-up generator. Power from the 1120 peak-watt 48-volt array is stored in 750 ampere-hour battery bank. Solar production is over 100 kWh per month year round. The largest load in the house is an Amana high-efficiency 16 cubic foot refrigerator/freezer (53 kWh per month). Lighting is the second largest load, approximately one-half that of the refrigerator. The balance of 20 kWh per month is consumed by the water pump, entertainment equipment, and miscellaneous appliances.

For this house, Peter recommended a small 1/3-hp water pump. He also recommended no continuous inverter loads, such as

FIGURE 13.5: Pilot Station, Monhegan Island, Maine, uses a PV/generator hybrid retrofit to power communications and lighting. (David Sleeper, Brook Farm Inc.)

nightlights or clocks, which would keep the inverter constantly on. By working with the client and explaining the best way to use PV power, designers like Peter Talmage are perfecting the custom system.

Ron LaPlace

Ron LaPlace of Colinton, Alberta, Canada, has been installing PV in what many people would consider a poor location for solar. Not only are Ron's installations very much to the north of the "sunbelt," the economics of that locale puts two strikes against it: no tax credits and an unfavorable exchange rate for the imported PV equipment.

Nonetheless, even with these odds, PV works. The Tarryall Resort, located on the shores of Catherine Lake 25 miles north of Kenora, Ontario (latitude 50°N) is approximately three miles from the nearest utility. Cost to run in a power line would be well over $80,000. Typically, residences and resorts there get their power from on-site diesel generators. These generators last 7 to 10 years. To provide for peak loads, the generator has to be relatively large, in

FIGURE 13.6: Another David Sleeper PV/generator hybrid retrofit. A 350-watt array provides 24-volt charging for this inverter system located at Blue Hill Bay, Maine. (David Sleeper, Brook Farm Inc.)

this case 7.5 kW. A 3-kW back-up generator is also on-site. Most of the time, though, much less power is needed. In fact, using 39 kWh per day, the generator was consuming fuel continuously while supplying an average of 22% of its potential rated output.

By installing a hybrid system, the generator operating time has been significantly reduced. Now, while running, the generator charges batteries and has a greater load which more efficiently utilizes fuel. Battery power can then be quietly consumed. A 564-watt PV array further reduces generator running time and provides a finish charge that greatly enhances battery life.

In the first month of operation, the hybrid system's average daily power consumption was 33.6 kWh. The consumption reduction is the result of Ron's suggestions for energy conservation. Before PV/hybrid, the generator ran 24 hours a day. Now the generator runs 27.2 hours *per week,* or an average of 3.9 hours per day. In U.S. dollars, the full value of this $25,500 PV/Hybrid system has a projected payback period of less than 5.3 years.

Before PV			*After PV*	
Annual fuel cost			Annual fuel cost	
1.7 liters/hour at $0.2625/liter			2.3 liters/hr	
$0.4462 x 8760 hours	$3909		$0.6037 x 1560 hours	$ 941
Maintenance			Maintenance	
oil change every 200 hours			8 oil changes	140
44 changes at $17.50 each	770		Major overhaul and repair	
Major overhaul and repair			every 15 years	210
$3150 every third year	1050		Generator depreciation 30 years	175
Generator depreciation			Battery depreciation 15 years	560
replacement every 8 years	656		Yearly operating costs	$2026
Yearly operating costs	$6385			

Year (5% inflation)	Before Hybrid	After Hybrid	Savings
1	$6385	$2026	$4359
2	6704	2127	4577
3	7039	2233	4806
4	7419	2345	5074
5	7761	2462	5299
6	8149	2585	5564
		Total:	$29679

FIGURE 13.7: The Talmage home has a hybrid system using wind and PV for all of its electrical needs. (Talmage Engineering, Kennebunkport, ME)

Paul Jeffrey Fowler

Paul Jeffrey Fowler of Worthington, Massachusetts, bought his first PV modules through one of my PV Network bulk purchases. Since

FIGURE 13.8: This attractive island home in Addison, Maine, gets all of its power from the sun. (Talmage Engineering, Kennebunkport, ME)

then Jeff has become a professional PV system designer and installer. He specializes in conservation and alternative energy equipment.

His own system has expanded and changed over the years. Now he uses 24 six-volt ribbon technology modules to produce 720 peak watts at 24 volts to power a 300-watt inverter. The system started as 12 volts which Jeff still uses to power his telephone answering machine, television, and stereo through a voltage equalizer (also known as a DC-to-DC converter). He also uses some 24-volt incandescent lights, although most of his lighting is high-efficiency 120 volts AC.

Appliances include gas refrigerator, wood and gas stove, tankless gas water heater, shop tools, one-third horsepower water pump, washing machine, computer, and the other items usually found in a modern home.

All PV system and house wiring is in compliance with the *National Electrical Code* and local ordinances. The solar array is fixed at 60° to optimize winter electrical production. Adjusting in summer to a lower tilt angle for more power is unnecessary. He has separately wired six parallel groups of four modules in series to a combiner box located in the cellar 25 feet from the array. Not only does this make testing each string easy, it keeps wire losses to a

FIGURE 13.9: The solar array for an Addison, Maine, home is mounted on the south-facing roof of the woodshed and grounded for lightning protection. (Talmage Engineering, Kennebunkport, ME)

FIGURE 13.10: Sixteen heavy-duty deep-cycle batteries in two series sets for 750 ampere hours at 48 volts get their charge from the shed-mounted PV array and a 45-amp charger tied into a back-up generator (Talmage Engineering, Kennebunkport, ME)

minimum. Array disconnect is a double pole switch rated at 40 amperes.

Each array string has its own diode and there is no diode in the single-stage regulator. Five strings go to the regulator and one string goes directly to the battery bank for constant trickle charging. There is also a fused regulator disconnect. The battery bank consists of 16 six-volt 200 ampere-hour batteries wired for 24 volts. The system has two DC distribution panels and an inverter disconnect.

As the home is located in a high lightning area, Jeff uses lightning suppressors (Transorbs) from the array positive to ground in the combiner box for each parallel array string. There is also a Transorb in the array disconnect box. In addition, fast acting canister-type lightning suppressors are used on the three wires to the well pump and at the AC distribution panel.

FIGURE 13.11: A 500-watt array for the La Place residence in Colinton, Alberta, Canada. Note the steep array tilt to maximize winter solar collection. (Courtesy Ron La Place, Sun Direct Energy Systems, Inc.)

FIGURE 13.12: Tarryall Resort, Kenora, Ontario, Canada, 560-watt peak PV array. (Photo: Sun Direct Energy Systems, Inc.)

FIGURE 13.13: The power equipment room for the Tarryall Resort (Figure 13.12). (Photo: Sun Direct Energy Systems, Inc.)

FIGURE 13.14: The Fowler residence in Massachusetts is an example of combining New England architecture with PV technology. The solar array located above the sunspace has been nicely integrated into the traditional design. (Photo: Fowler Solar Electric, Worthington, MA)

FIGURE 13.15: Located in Stamford, Vermont, this home has two ribbon technology modules producing 420 peak watts. The family uses a 3.5-kW generator for a portion of their power. The roof is braced to allow for the future additional solar modules. (Designed and installed by Fowler Solar Electric, Worthington, MA)

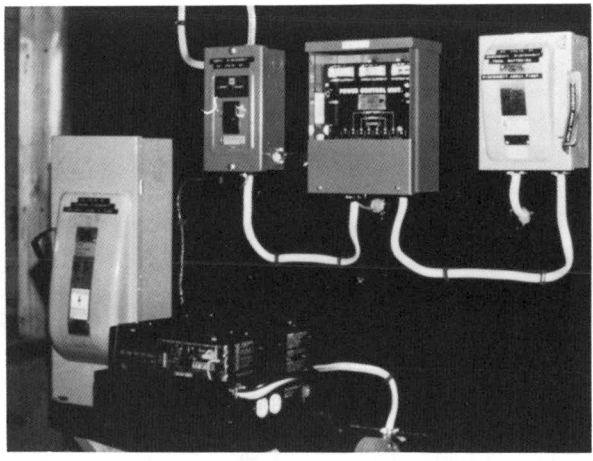

FIGURE 13.16: Code-approved wiring for the Stamford home shown in Figure 13.15. Pictured are array disconnect, BOSS controller, controller disconnect, 2000-watt inverter with built-in 50-ampere battery charger and automatic transfer switch. On the AC side is a disconnect switch and house wiring distribution panel. (Designed and installed by Fowler Solar Electric, Worthington, MA)

FIGURE 13.17: Engineers Robert Wills and Anitra Sorensen live in this passive solar home. The PV array has been moved to expand to 24 modules in three groups. The system uses a 2500-watt inverter and generator back-up. (Courtesy Skyline Engineering, Ely, VT)

FIGURE 13.18: Robert Cook's Worthington, Massachusetts, home has a PV/generator hybrid system. The ground-mounted array is above the snow and at a steep tilt. Sunlight reflected off the snow increases winter solar production. (Courtesy R. Cook)

FIGURE 13.19: The Cook residence (Figure 13.18) uses a battery equalizer for both 12 and 24 volts DC from the same battery bank. Pictured are the battery bank on a sturdy shelf, 2500-watt inverter, regulator, digital metering and alarm, equalizer and distribution panel. (Courtesy R. Cook)

FIGURE 13.20: South view of a PV-powered, stationary Airstream in New England. (Courtesy Jay Baldwin; Photo: Robert Sardinsky, Rising Sun Enterprises)

FIGURE 13.21: A small solar electric power system on an Idaho home. Even in locations where the winters are cold and cloudy, PV can provide power for the energy-conscious homeowner. (Courtesy Backwoods Cabin Electric Systems)

FIGURE 13.22: This 16-module low-profile adjustable solar array allows a clear sunpath while preventing visibility from the road. Power is used to pump water and for emergencies. (Courtesy Backwoods Cabin Electric Systems).

Chapter 14
Other PV Applications

By now, almost everyone has seen or used a solar-powered calculator. These thin lightweight, accurate, and low-cost devices have revolutionized the way we deal with daily computations.

And that's what technology is all about. If a technology can free us from drudgery, it will be adopted. If a technology does the job better, however we define it, that technology will be used. More and more we are finding that PV power does just that.

This short section is not an exhaustive survey of PV applications by any means. It is included to open your eyes to the virtually unlimited world of PV.

MARINE AND MOBILE

Photovoltaics, and its promise of energy independence, is a natural for the mobile. Sailors use PV to power lighting, navigation and communications equipment, sonar appliances, and tools. Land travelers, whether weekend campers or full-time motor coach or recreational vehicle (RV) residents, similarly use PV to meet their power requirements. The sailor is freed from the necessity of pricey and overpopulated moorings, RVers from costly and crowded campgrounds.

The space satellite, perhaps the ultimate in mobility, uses photovoltaics for its power source. Consequently it is restricted only by the imagination of its designers.

By replacing a combustion fuel generator with a PV system, the traveler is further freed—from noise, fumes, repairs, maintenance, fuel handling *and* fuel costs. Few things can destroy the mood of an idyllic spot faster than the sound of a running generator, except perhaps the awful silence of a broken one.

While some people choose to charge their on-board batteries by running the engine, unless they are travelling under power this is not very practical. Some boats and RVs have alternators and battery-charging equipment that puts out high current. However, most have small chargers in the 8 to 10 amp range. Running even a 100-horsepower engine to operate a 10-ampere battery charger wastes fuel, under-utilizes the engine, and will, eventually, carbon foul the cylinder head and valves.

Just look at the numbers. A 350-hp RV engine is capable of producing 130,000 watts at 50% efficiency. Most RV converter/battery chargers produce 10 amperes or 120 watts. In addition, there are some losses in the battery charger. Using nuclear energy

FIGURE 14.1: A neat installation of a marine PV module in a tight location. (Photo: Free Energy Systems, Inc.)

to produce 10,000 degrees to boil water to make steam-generated electricity is not appropriate technology, and neither is the under-utilization of a fossil fuel engine to charge batteries. It's just another example of "using a chainsaw to cut butter."

The characteristics that make PV so practical for fixed locations (rugged, dependable, quiet, low-maintenance) become especially important to the camper or boater. Proper selection of modules, wiring, and controls for mobile applications is important since vibration and salt air corrosion must be taken into consideration. For example, Steve Ackerson, inventor of the Heart inverter, did his design work while living off the coast of Costa Rica on a trimaran. He saw the need for a high-efficiency inverter and used his skills to design one. He also gave much thought to the effect of salt air on the inverter. Some other manufacturers also make quality equipment designed for harsh operating conditions.

There are 5-watt trickle chargers on the market that are ideal for keeping boat, RV, tractor, airplane, or automobile batteries charged. These small modules are connected directly to the battery with no regulator.

FIGURE 14.2: Barbara and Noel Kirkby and their children enjoy teaching recreational vehicle owners how to use PV power. Every year the Kirkbys hit the road, avoiding noisy campsites to stay at quiet, out of the way retreats. The Kirkbys have written a book on PV for the RVer, *The RVers' Guide to Solar Battery Charging*, available from **aatec publications.** (Courtesy Solar Electric Systems, Cave Creek, AZ)

FIGURE 14.3: Windy and Anne (holding a 12-volt grain mill) stand beside their 1964 Sunbeam powered by batteries and a 400-amp aircraft starter/generator. Their home's excess solar power is used to charge the car's battery bank for a 30 to 50 mile range. Although it costs nothing to run the car, battery replacement will equal the cost of gasoline. Like other solar vehicles, the project's goal was to increase public awareness and have fun. (Photo: Windy Dankoff, Windlight Workshop)

FIGURE 14.4: Solar Technologies, Key West, Florida, uses solar power to build PV homes. Twelve modules fitted to their service vehicle charges a battery bank and inverter. While building the home shown in Figures 3.6 and 3.7, a six-man crew had all the power they needed and avoided a $1200 temporary utility hook-up fee. (Courtesy Solar Technologies of the Florida Keys, Inc., Key West, FL)

SOLAR-POWER AND VEHICLES

Solar vehicles, while not a threat to the auto industry, do promote public awareness of PV's potential. For instance, in Switzerland, where there are 20 road-certified solar vehicles, there is an annual Tour de Sol rally for solarmobiles from around the world. But PV's uses are not limited to these unique conveyances. Thin-film technology is making possible a variety of more conventional applications: for example, an automobile sunroof that produces electricity. This sunroof will power on-board computers, security equipment, and a ventilation fan, plus keep the vehicle's battery trickle charged.

FIGURE 14.5: The SUN-RUNNER, built by Greg Johanson and Joel Davidson, set a world speed record of 24.7 mph in 700 watts per square meter sunshine and into a 12-mph wind. Capable of 40 mph in desert sun, this vehicle used no batteries. The solar array, if tied into a home battery bank, would produce over 4 kWh/day. (Photo: Solar Electrical Systems).

SOLAR-POWERED STRUCTURES
(Non-Residence)

PV is the answer in many unique setting situations. Stand-alone systems are ideal because they are less expensive, and less disruptive, than excavating and installing grid power via underground conduit. For powering lighting and ventilation in a restroom at a Palm Springs, California, golf course to a U.S. Forest Service fire station in Big Sur that uses a generator and PV for all its power, photovoltaics can be found almost everywhere.

FIGURE 14.6: Electrical Systems Consultants of Ft. Collins, Colorado, hired Architecture Plus to design their passive solar PV-powered offices. Twenty-four solar modules power lights. The system has automatic utility back-up. PV designed and installed by Home Energy Workshop, Ft. Collins. (Photo: Home Energy Workshop, Ft. Collins, CO)

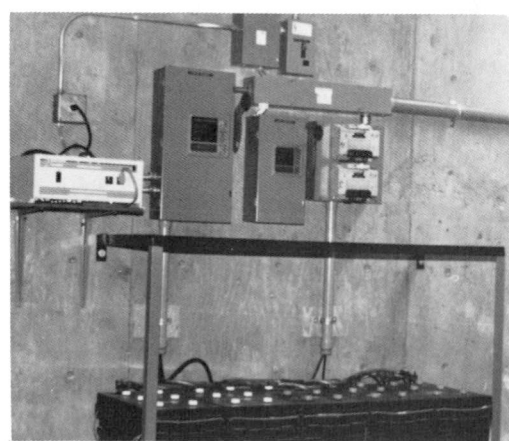

FIGURE 14.7: Battery bank, inverter, regulator, and distribution wiring for the office power system in Figure 14.8. Systems installed by qualified PV professionals are helping bring PV technology into the mainstream. (Photo: Home Energy Workshop, Ft. Collins, CO)

FIGURE 14.8: The Solviva Winter Garden, an extraordinary 400 square foot solar-heated and -electrified bioshelter built by Anna Edey on Martha's Vineyard, provides a nurturing environment for the year-round production of dozens of varieties of vegetables and herbs, eggs from 100 chickens, and fleece from 50 Angora rabbits. A 175-watt 12-volt PV array powers directly, or via a battery bank, two destratification fans, two pressurization fans (which keep the outer two of the structure's four thin polyester glazings taut during windy conditions), a circulating pump for solar water heating which is used to pre-heat irrigation water, and supplementary winter lighting for evening chores and to maintain egg production during shorter daylight hours. (Photo: Robert Sardinsky, Rising Sun Enterprises)

FIGURE 14.9: The city of Tucson, Arizona, uses PV-powered lights in their transit shelters. Special applications like this make sense. Virtually maintenance-free, these shelters can be set anywhere and moved as bus routes change. (Bill Cirrito, Electrasun Energy Systems, Tucson, AZ)

FIGURE 14.10: At White Sands, New Mexico, PV replaces a noisy generator to provide power for lighting the comfort station and a park service audio-visual trailer used for presentations to visitors. A Trombe wall passive solar system provides part of the heating requirements. Bright sunlight reflected off the sand increases this array output almost 20%. (Bill Cirrito, Electrasun Energy Systems, Tucson, AZ)

IRRIGATION AND WATER PUMPING

FIGURE 14.11: A 3-kW solar water-powered irrigation system located at the University of California, Davis. This solar array shown in the winter position has an adjustable mount. The tilt angle is lower in summer to get more power. (Courtesy California Energy Commission)

FIGURE 14.12: Crop irrigation using an efficient drip system. (Photo: ARCO Solar, Inc.)

FIGURE 14.13: Village water pumping is bringing hope to people in desert regions. (Photo: ARCO Solar, Inc.)

FIGURE 14.14 The use of DC motors and water tanks eliminates the need for battery storage. (Photo: ARCO Solar, Inc.)

COMMUNICATIONS

PV has performed reliably in communication systems for over 25 years. All the communications satellites are solar-powered. Photovoltaic-run radio repeaters for rural telephone systems are common worldwide. Lighthouses are electrified via photovoltaics. The possibilities are endless.

FIGURE 14.15: This mountaintop radio links geologists with ground and air crews. (Photo: ARCO Solar, Inc.)

Appendix A

A Short Course in Electricity

Electricity is electromotive force (voltage) and electron flow (current). If we think of a wire or conductor of electricity as a water hose, voltage would be the water pressure and current would be the rate of flow. Thus, 40 pounds per square inch at 4 gallons per minute is similar to 12 volts at 2 amperes. To expand this analogy, we can compare the size of the hose with the size of a wire. A large hose will let water flow with less restriction; a large wire will let electricity flow with less resistance.

We use two types of electricity: alternating current (AC) and direct current (DC). Our most common use of DC is in flashlights, portable radios, and cars and trucks. All of these devices are energized by the direct current from batteries. Most newer stereos and tvs use direct current but, with transformers and rectifiers, first change the 120-volt AC house current from the wall outlet into low-power direct current. Direct current is a one-way flow of electrons from minus to plus.

In our homes we normally use alternating current—the flow of electrons first in one direction and, then, reversed in the other direction. Typical house current is between 110 and 130 volts and alternates or cycles 60 times each second. The frequency of ordinary house current is therefore 60 hertz (Hz).

Alternating current is used almost universally because it can be transmitted over long distances of wire at high voltage with little power loss. Low-voltage DC wire loss is significant. That's why automobile jumper cables are short and thick.

At the present time, large-scale production of alternating current is more economical, although new developments in solid-state technology are making high-voltage DC transmission systems attractive. Small-scale DC power production can be economical, too. Years ago, many rural residences produced their own DC electricity. These home power plants were fueled by gasoline, diesel fuel, or the wind. Several firms manufactured appliances which could be powered by these home generation DC plants. With the spread of alternating current through the government's Rural Electrification Administration's programs, the nation's thousands of home power plants fell to disuse and were replaced by subsidized "cheap" alternating current from centralized large power plants.

With rising electric rates, more and more people are returning to home-grown DC power plants. In addition, new technology and the recreational vehicle boom have helped the small-scale electricity producer by providing a wide range of tools, equipment and appliances which use DC while consuming very little current (compared to those used in the 1930s). More efficient motors, DC fluorescent lamps, solid-state tvs, stereos, and radios are just some of the devices now available to the small-scale energy producer. New solid-state inverters which convert 12-volt DC to 120-volt AC make it possible for us to use all the appliances we have become accustomed to using with house current.

Many people have had just enough exposure to electricity in school to confuse and frighten them. The information you need to design and wire your own solar electric home is really quite simple. First the terms:

Voltage (E), measured in volts, indicates the electrical "pressure" in a system. The output of a flashlight cell is only 1.5 volts DC while the typical automobile battery puts out around 13 volts (even though it is called a 12-volt battery). The symbol for volts is V.

Current (I), measured in amperes (amps), describes how much electricity is flowing in a circuit. The symbol for amps is A. The current flowing through the starter when you start your car can be 200 amps or more, while that drawn by a portable radio may be only 0.1 amps or 100 milliamps (one milliamp, symbolized mA, is one-thousandth of an amp).

Resistance (R), measured in ohms, is just that—the resistance to the flow of electricity. It is a property of the electrical device or appliance. Sometimes the word **impedance** will be used instead of resistance. Impedance is a complex form of AC resistance that can change with the frequency. For a DC system, we will concern ourselves only with the resistance and not worry about the complex impedance of, for example, a fan motor. If a fan is designed to operate on 12 volts, and we are using 12 volts, we are concerned only about the current draw when the fan is operating properly. The symbol for ohms is Ω.

Power (W), measured in watts, indicates the rate at which electrical energy is being used. The power in watts is the voltage multiplied by the current used: $1 \text{ V} \times 1 \text{ A} = 1 \text{ W}$ (1 volt x 1 amp equals 1 watt of power). W is the symbol for both watts and power in general. Horsepower is the English measure for power; 746 watts is 1 horsepower. Sometimes you will see power stated in kilowatts (kW), a kilowatt being 1000 watts.

Energy is measured, for our purposes, in watt hours (Whr) or kilowatt hours (kWh). If you run a 25-watt light bulb for 1 hour, you have used 25 watt hours of electrical energy. If you left that bulb on for 100 hours, the total energy used would be $25 \times 100 = 2500$ Whr or 2.5 kWh. It is very important not to confuse energy and power. Power is the rate at which energy is used or changed.

Ohm's law is the relationship between volts, amps, and ohms. This law can be explained by Figure A.1. To make the wheel chart more complete and useful, power (W) is also included.

Don't let the chart scare you. It's really very simple and graphically shows the relationship between volts, amperes, ohms, and watts. We have already seen how volts and amperes are related and how resistance (ohms) can affect current flow. Watt is just another name for the product of volts times amperes.

This chart is a useful tool because we can refer to it to change values from watts to amperes as we size our PV power systems. We can also use it to determine the size wire we need to minimize resistance. Here are a few examples of the chart's use.

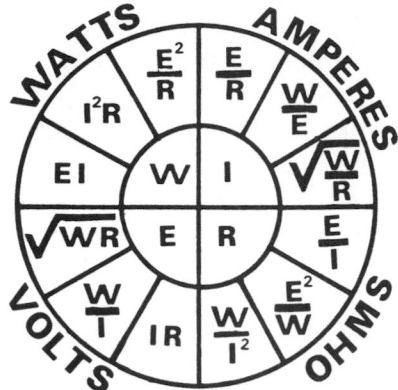

FIGURE A.1: Ohm's law.

CALCULATING RESISTANCE

Example 1. Fluorescent Lamp

Let's say we want to power a fluorescent lamp with a PV array. The lamp is 12 volt DC rated at 20 watts. Our PV module is rated at 35 peak watts. That seems to mean that the lamp can be powered by the PV module with some power to spare.

The same module may actually produce only 12 volts at 2 amps output. Thus, we find that the rated output (35 peak watts) is really the maximum output under perfect conditions. By multiplying volts by amperes, we find that the usable wattage is really 24 watts—still enough to power our lamp.

But we want to use the lamp at night, having produced power during the day and stored it in a 12-volt DC battery. If the sun usually shines 6 hours per day where you live, your PV module will produce 6 hours times 2 amperes or 12 ampere hours per day. Another way to describe this is to say that the PV module output is 144 watt hours per day (12 volts DC x 12 ampere hours = 144 watt hours).

If our lamp draws 20 watts, then we should be able to operate the lamp over 7 hours per night (144 watt hours ÷ 20 watts = 7.2). However, there is some internal loss in batteries (about 30%) and some wire loss (another 5%), so figure on about 5 hours of actual operation per night for that lamp. Of course, short winter days or cloudy skies will change the production output, so PV systems are generally sized with weekly or monthly solar output values.

Use the chart to change watts to amperes and back to watts. Batteries are rated by ampere hours of storage capacity, but most appliances are rated in watts. We now can see how easy it is to go back and forth from watts to amperes. Example: 20 watts at 12 volts DC = 1.6 amperes (I = W/E).

Volts divided by amperes will give you the resistance of a circuit. Resistance becomes an important factor with low-wattage (low voltage times amperage) power systems.

Example 2. Resistance Heater

Portable electric heaters are often used to take the chill off a small room. These inexpensive heaters typically have a wattage of 1200 watts and are designed to be used on 120 volts AC. If the heater has no fans or controls, only a resistance element, it will operate on DC just as well as on AC.

What is the current draw of this heater? From the chart, I = W/E. The current is the wattage divided by the voltage

$$I = \frac{1200\ W}{120\ V} = 10\ A$$

The resistance of the heater can be calculated from Ohm's law, R = E/I. Putting in the numbers:

$$R = \frac{120\ V}{10\ A} = 12\ \Omega$$

So the heater wire has a resistance of 12 ohms. You could connect the heater to a 12-volt DC electrical system without any problem. What would be the current draw in this situation? Taking the resistance of 12 Ω, which is a property of the wire essentially independent of the applied voltage, we use the chart again:

$$I = \frac{E}{R} = \frac{12\ V}{12\ \Omega} = 1\ A$$

What wattage would the heater put out? W = EI = 12 V x 1 A = 12 W. So on 12 volts, the 1200-watt heater would deliver only 12 watts— barely enough to keep warm. You could use an ordinary portable electric heater on 12 volts DC (provided you disconnect the fan, if it has one), but it will produce very little heat.

Example 3. Wire Resistance and Voltage Drop

A matter of concern in any electrical wiring job is the size wire to use. Chapter 10 examines wiring in detail, but we have included an example here to show how you can determine the loss expected when a particular size wire is used for a particular job.

#10 copper wire has a resistance of 0.1 Ω per 100 feet. If you hook up a battery to a 100-watt, 12-volt light bulb in the manner shown in Figure A.2, the total path length is 50 feet (25 feet to the bulb and 25 feet back). What voltage will be delivered to the bulb? First, let's calculate the resistance of the wire.

$$0.1\ \Omega\ \times\ \frac{50\ \text{ft}}{100\ \text{ft}}\ =\ 0.05\ \Omega\ \text{wire resistance}$$

That doesn't seem like much. The resistance of the bulb when it is operating normally is calculated from the Ohm's law chart ($R = E^2/W$ is the formula we use since we know the normal voltage is 12 volts and the design voltage is 100 watts):

$$R\ =\ \frac{12^2}{100}\ =\ \frac{144}{100}\ =\ 1.44\ \Omega$$

The total resistance of the system is the sum of the bulb plus the wire resistance, or 1.49 Ω. The total current drawn from the battery is given by the formula $I = E/R$ (notice we keep using the same formulas over and over):

$$I\ =\ \frac{12}{1.49}\ =\ 8.05\ \text{A}$$

which we can round off to 8 amps. The 8 amps flowing through the #10 wire will cause a voltage drop; the lamp will not get the full 12 volts the battery is delivering.

To calculate the voltage drop, simply use Ohm's law again: $E = IR$. The voltage drop, $E_w = 8\ \text{A} \times 0.05\ \Omega = 0.4\ \text{V}$, is determined by taking the current through the wire and multiplying it by the wire resistance. If 0.4 volts is lost in the wire, $12 - 0.4 = 11.6$ volts delivered to the bulb. The important thing to notice is that the voltage drop in the wire is directly proportional to the current the wire is called upon to carry and also directly proportional to the wire

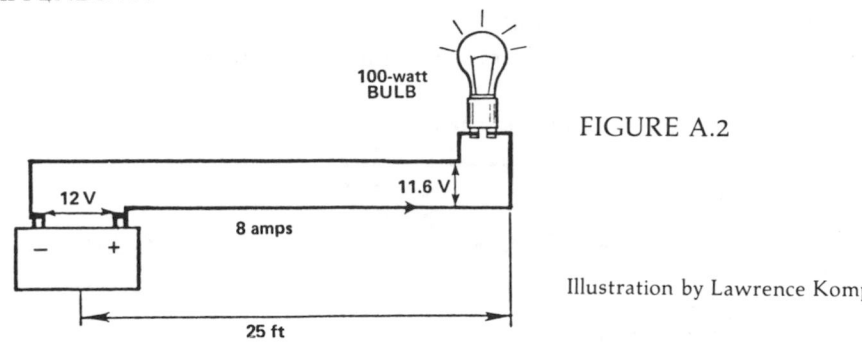

FIGURE A.2

Illustration by Lawrence Komp

resistance. The message is, if you want to carry a lot of current, you need short thick wires.

The last calculation in this example is the power loss in the wire. From the chart, the power $W = EI = 0.4\text{ V} \times 8\text{ A} = 3.2\text{ W}$. The power actually used by the nominal 100-watt bulb is $11.6\text{ V} = 8\text{ A} = 92.8\text{ W}$. This means the total power taken from the battery is $92.8\text{ W} + 3.2\text{ W} = 96\text{ W}$. Another way of arriving at the same conclusion is to multiply the total current by the total battery voltage ($12\text{ V} \times 8\text{ A} = 96\text{ W}$). Even the #10 wire, in this case, reduces considerably the actual power delivered by the light bulb. The bulb is cooler and will not be as bright, so the amount of light lost will be proportionally even larger. The only good thing about this situation is that the bulb will last longer.

Using the techniques in this example, you can now calculate the voltage drop and power loss in wires of various sizes in your own situations.

SERIES AND PARALLEL CIRCUITS

Series

Series circuits are probably the easiest for the beginner to understand. The wire resistance example had three resistances in series, as shown schematically in Figure A.3. There are two wires, each with its own resistance, R_1 and R_2, and there is the resistance of the load, R_3. The current from the battery has to go through all three resistances to get back to the negative terminal of the battery. The total resistance is simply the sum of the individual resistances.

$$R_{total} = R_1 + R_2 + R_3$$

This series circuit is like old-fashioned Christmas tree lights—if one light burns out, it breaks the circuit and no current flows at all.

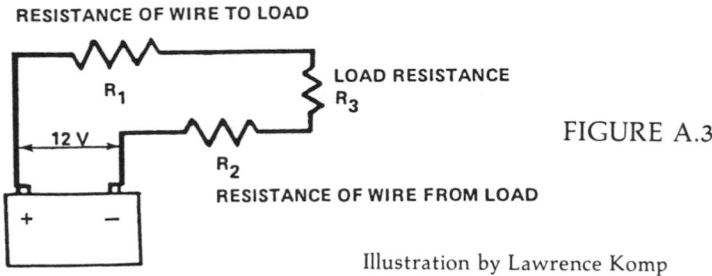

FIGURE A.3

Illustration by Lawrence Komp

Parallel

A parallel circuit is easy to understand if you think about the current going through each part (Figure A.4). Each load resistance is connected directly across the battery. R_1 might be a 12-volt light bulb, R_2 a 12-volt soldering iron, R_3 a portable radio. The parallel circuit is what you end up with when you wire your house. Each load is more or less independent of the others and draws its own current as calculated by Ohm's law ($I = E/R$). The total current will simply be the sum of all the individual currents. In the parallel circuit, since each device is connected separately to the battery, turning off one

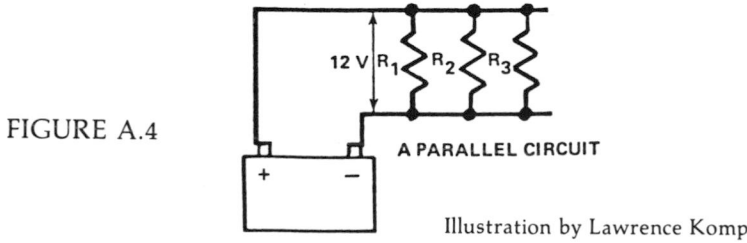

FIGURE A.4

Illustration by Lawrence Komp

device should have little influence on the others, provided that the connecting wires are large enough so that the wire resistance in series with the devices is small.

$$R_{total} = \frac{1}{\dfrac{1}{R_1} + \dfrac{1}{R_2} + \dfrac{1}{R_3}}$$

Appendix B
PV Sizing Data

PV SYSTEM SIZING WORKSHEET

STEP ONE: Calculate total daily load

Device	Quantity		Watts (E X I)		Hours per Day Used		Days per Week Used		Average Watt Hours per Day
_____	_____	X	_____	X	_____	X	_____	/ 7 =	_____
_____	_____	X	_____	X	_____	X	_____	/ 7 =	_____
_____	_____	X	_____	X	_____	X	_____	/ 7 =	_____
_____	_____	X	_____	X	_____	X	_____	/ 7 =	_____
_____	_____	X	_____	X	_____	X	_____	/ 7 =	_____

Maximum AC Surge _____   Watt Hours _____

STEP TWO: Allow for 2% wire loss (1.02) or actual loss
                        80% battery efficiency (1.25) or actual efficiency
                        90% inverter efficiency (1.11) or actual efficiency
                        OR skip inverter step for DC loads

Watt Hours x 1.02 x 1.25 x 1.11 = Daily Watt Hours

_____ **x 1.02 x 1.25 (x 1.11) =** _____ **Watt Hours**

STEP THREE: Calculate the required array peak watts using yearly average
                   or worse-month peak sun hours (see pages 347 to 380)

Daily Watt Hours / Peak Sun Hours = Array Peak Watts

_____ / _____ = _____ **Peak Watts**

STEP FOUR: Calculate the number of solar modules required. For 24-volt DC
                systems, round up to the next even number (i.e., 11 = 12)

Array Peak Watts / Module Wattage Rating = Number of Modules

_____ / _____ = _____ **Modules**

STEP FIVE: Calculate the battery bank size

Daily Watt Hours x Multiplier (see page 339, top map)
x 2.5 (for 40% depth of discharge)
x Temperature Correction Factor (see page 93) = Watt Hours

_____ x _____ x 2.5 x _____ = _____ **Watt Hours**

Notes:

BATTERY STORAGE REQUIREMENTS

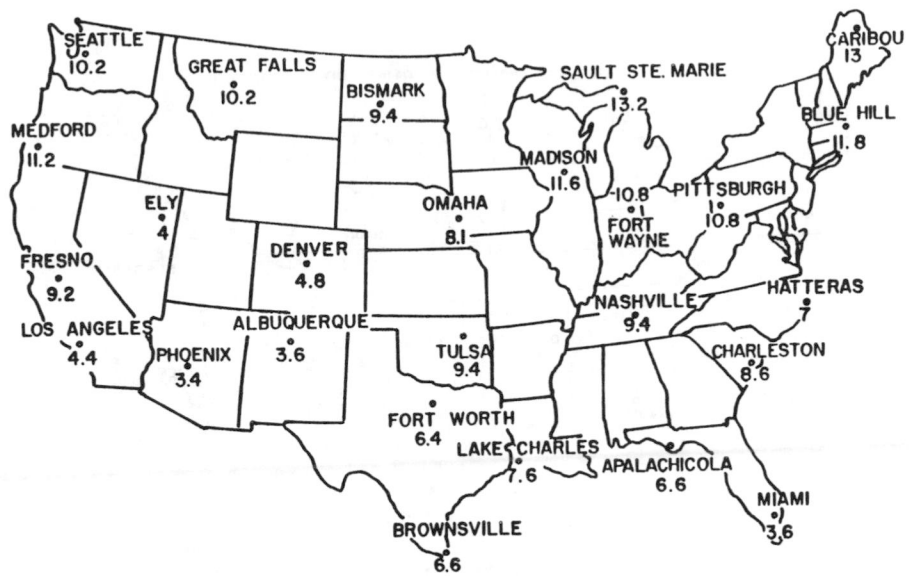

MULTIPLIER FACTORS MAP
(Continental United States)

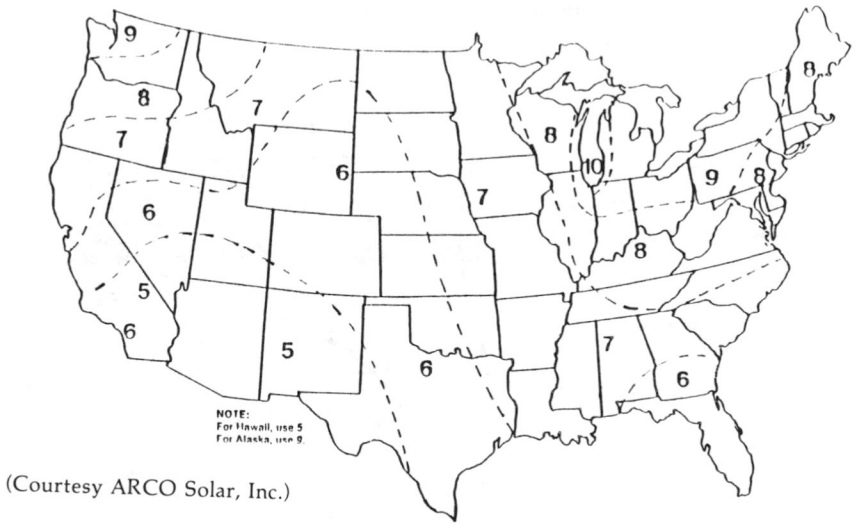

(Courtesy ARCO Solar, Inc.)

TOTAL INSOLATION ON A COLLECTOR TILTED 45°
ABOVE THE HORIZONTAL

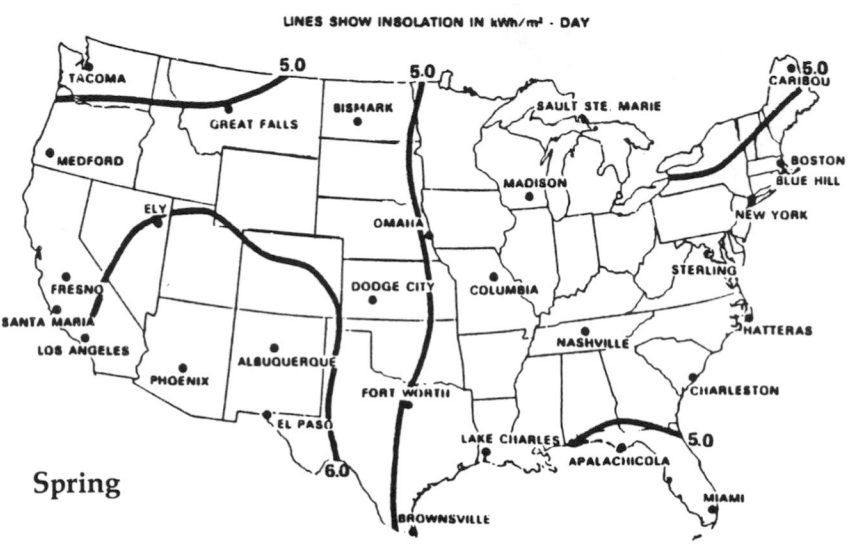

Spring

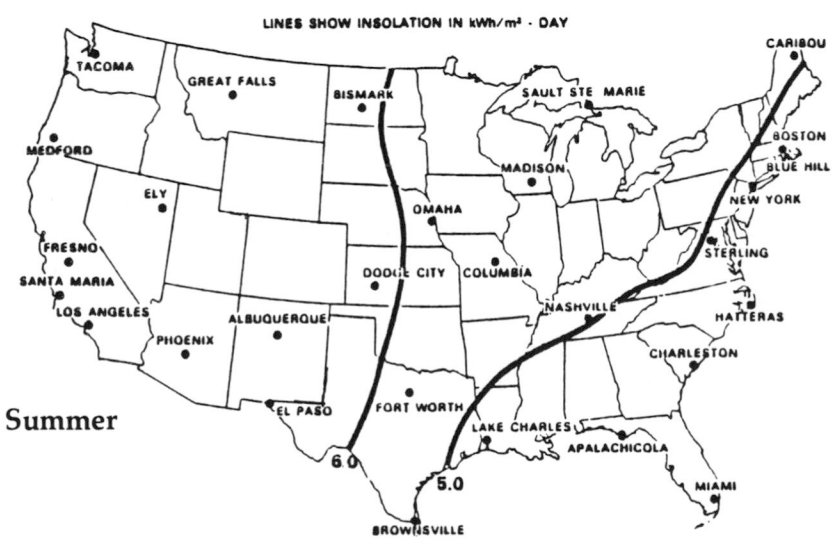

Summer

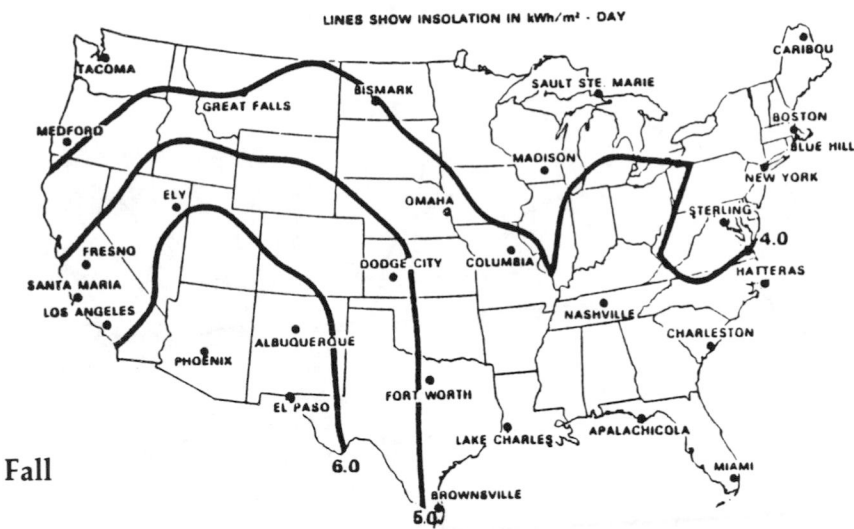

Fall

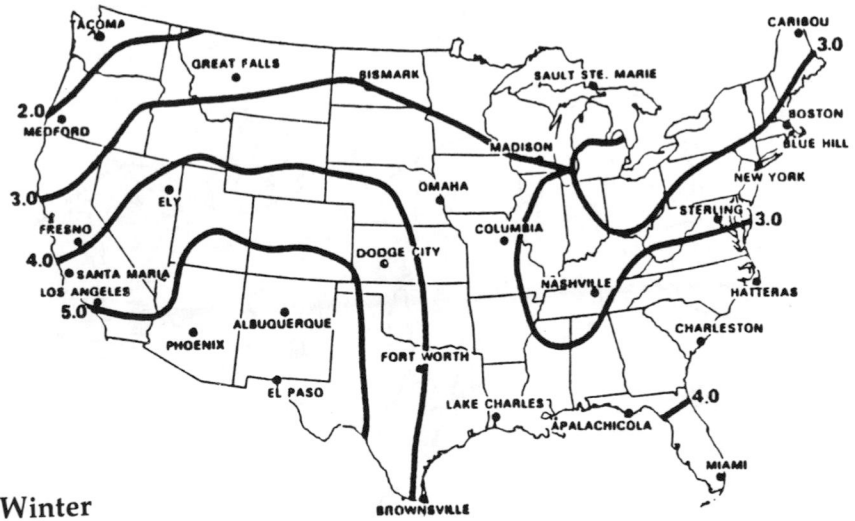

Winter

PEAK SUN HOURS PER DAY

Yearly Average

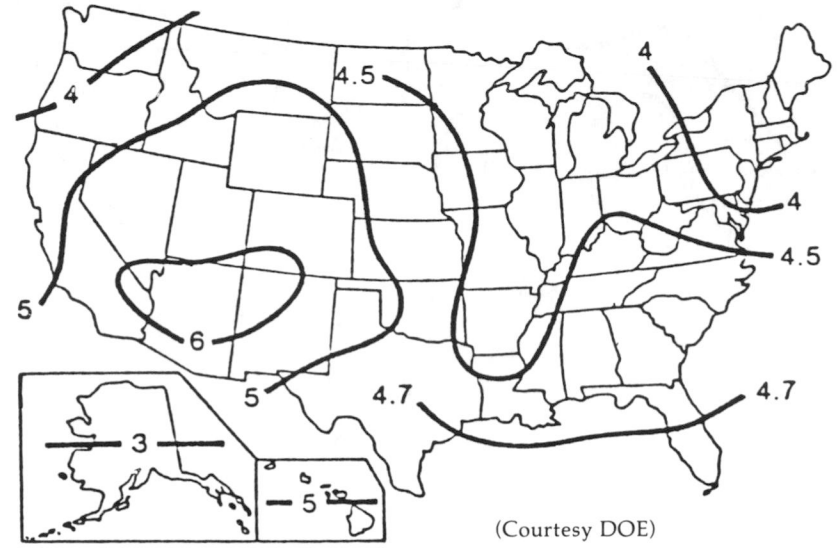

(Courtesy DOE)

Four-Week Average, 12/7–1/4

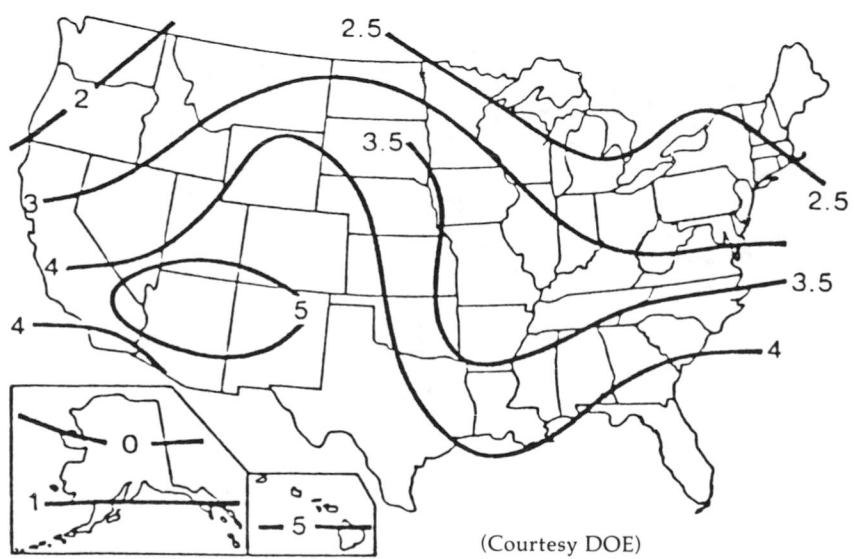

(Courtesy DOE)

PV PRODUCTION IN AMPERE HOURS

The following chart lists the average output of a 35-watt solar module for a 12-volt DC PV system. Battery efficiency of 80% and 5% wire losses are factored in. These figures can be used to determine the daily output of your solar electric system.

Example: What is the output of a solar array of 8 each 55-watt solar modules in a 12-volt PV system located in Omaha, Nebraska?

10.2 amp hours/day ÷ 35 watts = 0.2914 amp hours/day/watt
8 x 55 watts = 440 watt array
440 watts x 0.2914 = 128.2 amp hours/day
128.2 x 12 volts = 1538 watt hours/day

Note: These are average daily outputs based on the entire year's production. Tilt angle and time of year will make a big difference. If you are using an inverter, factor in inverter efficiency (i.e., 85% efficient inverter: 1538 x 0.85 = 1307 watt hours).

Lat.	Long.	Location	Ampere Hours Generated Per Day	Per Week	Tilt Angle
27N	1E	Algeria Aoulef	13.4	93.7	45S
30N	2W	Algeria, Beni-Abbes	13.1	91.5	45S
34N	1E	Algeria, Chottech Cherqui	12.2	85.6	45S
25N	1E	Algeria, Quallen	14.5	101.4	30S
23N	6E	Algeria, Tamanrasset	14.0	98.1	30S
30N	31E	Arab Rep. of Egypt, Giza	12.5	87.3	45S
41S	71W	Argentina, Bariloche	8.7	61.1	65N
35S	58W	Argentina, Buenos Aires	10.2	71.3	60N
28S	59W	Argentina, Corrientes	10.7	75.2	50N
31S	62W	Argentina, Rafaela	7.1	50.0	55N
32S	69W	Argentina, San Juan	11.2	78.5	45N
35S	139E	Australia, Adelaide	11.0	77.2	60N
35S	149E	Australia, Canberra	11.1	77.8	60N
38S	145E	Australia, Melbourne	9.6	67.3	60N
32S	116E	Australia, Perth	10.9	76.2	60N
34S	152E	Australia, Sydney	11.9	83.5	45N
19S	147E	Australia, Townsville	12.5	87.5	30N
48N	16E	Austria, Vienna	6.4	45.0	70S
17S	68W	Bolivia, La Paz	12.6	88.0	30N
20S	23E	Botswana, Maun	12.7	88.7	20N
43N	28E	Bulgaria, Varna	7.9	55.2	45S
59N	94W	Canada, Churchill	9.0	63.2	80S
45N	64W	Canada, Dartmouth	8.1	56.4	65S
54N	114W	Canada, Edmonton	9.2	64.2	75S

Lat.	Long.	Location	Ampere-Hours Generated		Tilt Angle
			Per Day	Per Week	
62N	121W	Canada, Fort Simpson	7.5	52.4	85S
46N	74W	Canada, Montreal	7.8	54.8	70S
44N	79W	Canada, Toronto	8.1	56.4	65S
49N	132W	Canada, Vancouver	6.8	47.4	70S
12N	15W	Chad, Fort Lamy	14.1	98.9	15S
23S	69W	Chile, Atacama	15.7	109.9	35N
34S	71W	Chile, Santiago	10.6	74.1	55N
44N	125E	China, Changchun	9.4	65.6	65S
38N	122E	China, Chefoo	9.8	68.4	55S
31N	121E	China, Shanghai	9.5	66.3	45S
5N	74W	Colombia, Bogota	10.0	69.8	15S
4S	15E	Congo Rep., Brazzaville	9.8	68.6	15N
14N	89E	El Salvador, San Salvador	13.2	92.7	15S
63N	26E	Finland, Luonetjarvi	5.9	41.5	85S
43N	5E	France, Marseilles	9.4	65.8	65S
49N	3E	France, Paris	6.9	48.4	70S
52N	7E	Germany, Bochum	4.8	33.7	70S
48N	12E	Germany, Munich	7.7	53.8	70S
6N	0E	Ghana, Accra	10.6	73.9	15S
38N	24E	Greece, Athens	9.7	68.2	55S
12N	16W	Guinea-Bissau	12.7	88.7	15S
22N	114E	Hong Kong	9.8	68.6	15S
19N	73E	India, Bombay	13.0	91.1	20S
23N	88E	India, Calcutta	11.6	81.1	25S
13N	80E	India, Madras	13.4	93.7	15S
6S	107E	Indonesia, Djakarta	9.8	68.6	15N
32N	46E	Iraq, Al-Kut	12.4	87.0	55S
32N	35E	Israel, Jerusalem	12.5	87.8	55S
41N	14E	Italy, Napoli	8.1	56.9	65S
46N	9E	Italy, Pallanza	8.8	61.3	65S
35N	137E	Japan, Nagoya	9.9	69.4	30S
43N	141E	Japan, Sapporo	7.6	53.3	65S
36N	140E	Japan, Tokyo	7.2	50.2	45S
1S	37E	Kenya, Nairobi	11.4	79.5	14N
35N	129E	Korea, Puzan	11.5	80.4	30S
38N	127E	Korea, Seoul	9.6	67.0	55S
34N	36E	Lebanon, Ksara	12.2	85.2	45S
15N	146E	Mariana Islands, Saipan	12.2	85.2	15S
17N	7W	Mauritania, Nema	12.2	85.7	20S
20N	99W	Mexico, Mexico City	11.8	82.4	15S
28N	107W	Mexico, Nonoava	13.7	96.0	35S
23N	110W	Mexico, Cabo San Lucas	12.1	84.5	30S
20N	106W	Mexico, Tomatlan	11.7	81.9	30N
18N	93W	Mexico, Tuxtla Gutierrez	9.6	67.0	15S
17N	100W	Mexico, Acapulco	12.1	85.0	15S
28N	13W	Morocco, Cabo Judy	11.1	77.6	35S

Lat.	Long.	Location	Ampere Hours Generated		Tilt Angle
			Per Day	Per Week	
23S	17E	Namibia (SWA), Windhoek	14.5	101.2	30N
30N	85E	Nepal, Saga	11.4	79.5	30S
4S	139E	New Guinea, Baliem	12.8	89.4	15N
9S	140E	New Guinea, Merauke	10.4	72.6	15N
4S	152E	New Guinea, Rabaul	11.1	77.9	15N
7S	147E	New Guinea, Bulolo	8.5	59.5	15N
18S	177E	New Zealand, Nandi	12.6	88.5	30N
29S	178E	New Zealand, Raoul Is.	10.7	74.8	50N
41S	175E	New Zealand, Wellington	8.7	61.1	65N
17N	8E	Niger, Agadez	15.1	105.4	15S
7N	6E	Nigeria, Genin City	8.8	61.3	15S
9N	12E	Nigeria, Yola	12.2	85.6	15S
60N	5E	Norway, Bergen	5.2	36.4	75S
25N	67E	Pakistan, Karachi	11.8	82.6	25S
30N	67E	Pakistan, Queta	13.2	92.3	50S
9N	80W	Panama, Panama	10.5	73.2	15S
12S	75W	Peru, Huancayo	14.4	100.7	15N
15N	121E	Philippines, Quezon City	9.7	67.7	15S
38N	9W	Portugal, Lisbon	11.3	78.8	60S
19N	66W	Puerto Rico, San Juan	13.6	95.5	20S
26N	50E	Saudi Arabia, Dhahran	12.6	88.3	45S
24N	50E	Saudi Arabia, Riyadh	13.2	92.5	30S
15N	17W	Senegal, Dakar	11.6	81.1	15S
9N	13W	Sierra Leone, Longo	10.4	72.7	15S
1N	104E	Singapore	9.2	64.2	15S
41N	4W	Spain, Madrid	9.6	67.2	65S
14N	25E	Sudan, El-Fasher	12.7	88.9	15S
20N	37E	Sudan, Port Sudan	12.7	89.0	30S
63N	14E	Sweden, Froson	6.9	48.4	75S
56N	13E	Sweden, Svalov	6.1	42.9	75S
25N	122E	Taiwan, Taipei	7.9	55.5	30S
23N	120E	Taiwan, Tainan	11.1	77.6	15S
24N	122E	Taiwan, Kwarenko	10.1	70.6	25S
14N	101E	Thailand, Bangkok	11.1	77.6	15S
3N	35E	Uganda, Moroto	14.2	99.1	15S
29S	17E	Republic, So. Africa, Alexander Bay	13.7	95.8	45N
30S	31E	Republic, So. Africa, Burban	10.7	74.8	35N
34S	18E	Republic, So. Africa, Capetown	12.1	85.0	45N
24S	29E	Republic, So. Africa, Pietersburg	13.2	92.3	30N
26S	28E	Republic, So. Africa, Pretoria	12.6	88.5	30N
29S	21E	Republic, So. Africa, Upington	13.3	93.0	40N
52N	0E	United Kingdom, Cambridge	6.7	42.9	70S
		United States of America			
32N	86W	AL, Montgomery	10.5	73.4	45S
61N	162W	AK, Bethel	8.1	56.9	75S
65N	148W	AK, Fairbanks	8.5	59.5	75S

Lat.	Long.	Location	Ampere Hours Generated		Tilt Angle
			Per Day	Per Week	
62N	149W	AK, Matanuska	7.6	52.9	75S
33N	112W	AZ, Phoenix	14.0	98.2	45S
35N	92W	AR, Little Rock	10.0	69.9	55S
39N	122W	CA, Davis	10.9	76.2	60S
37N	120W	CA, Fresno	11.9	83.5	45S
36N	118W	CA, Inyokern	15.4	107.8	45S
34N	118W	CA, Los Angeles	12.2	85.2	45S
34N	117W	CA, Riverside	12.7	89.0	45S
40N	105W	CO, Boulder	10.4	72.7	55S
40N	106W	CO, Grandby	12.1	84.9	55S
39N	74W	DC, Washington	9.0	63.2	60S
30N	82W	FL, Gainesville	11.2	78.6	45S
26N	80W	FL, Miami	12.0	83.8	25S
34N	84W	GA, Atlanta	10.5	73.6	45S
21N	158W	HI, Honolulu	12.8	89.9	35S
44N	116W	ID, Boise	10.5	73.4	65S
42N	87W	IL, Chicago	6.7	46.9	65S
40N	86W	IN, Indianapolis	9.0	62.8	60S
42N	94W	IA, Ames	9.4	65.6	65S
38N	100W	KN, Dodge City	12.3	86.4	55S
39N	97W	KN, Manhattan	9.7	68.2	60S
38N	85W	KY, Lexington	10.5	73.8	65S
30N	90W	LA, New Orleans	9.1	63.4	45S
32N	93W	LA, Shreveport	9.9	69.1	55S
44N	70W	ME, Portland	9.6	67.3	65S
42N	71W	MA, Boston	8.2	57.3	65S
43N	84W	MI, East Lansing	8.5	59.7	65S
46N	94W	MN, St. Cloud	9.9	69.4	65S
39N	90W	MO, St. Louis	9.3	65.4	60S
47N	111W	MT, Great Falls	10.5	73.6	65S
41N	96W	NB, N. Omaha	10.2	71.3	65S
36N	115W	NV, Las Vegas	14.2	99.3	45S
40N	75W	NJ, Sea Brook	9.0	62.8	60S
35N	107W	NM, Albuquerque	14.4	101.0	45S
41N	74W	NY, New York City	8.5	59.4	65S
39N	115W	NC, Ely	12.7	89.2	60S
36N	80W	NC, Greensboro	10.0	70.3	55S
47N	101W	ND, Bismarck	10.7	74.8	65S
41N	62W	OH, Cleveland	8.4	58.8	65S
35N	98W	OK, Oklahoma City	11.9	83.5	45S
42N	123W	OR, Medford	9.6	67.3	65S
40N	80W	PA, Pittsburgh	7.0	48.9	65S
41N	71W	RI, Newport	9.0	63.2	65S
33N	80W	SC, Charleston	10.8	75.5	45S
44N	102W	SD, Rapid City	11.2	78.1	65S
36N	87W	TN, Nashville	9.9	69.1	45S

Lat.	Long.	Location	Per Day	Per Week	Tilt Angle
			Ampere Hours Generated		
32N	101W	TX, Big Spring	11.8	82.8	45S
33N	97W	TX, Fort Worth	11.8	82.6	45S
41N	112W	UT, Salt Lake City	10.7	74.6	65S
38N	77W	VA, Richmond	8.8	61.6	60S
47N	122W	WA, Seattle	7.9	55.0	65S
48N	118W	WA, Spokane	9.5	66.8	70S
38N	82W	WV, Charleston	7.8	54.5	60S
43N	89W	WI, Madison	9.1	64.0	65S
43N	109W	WY, Lander	12.6	88.5	65S
46N	31E	USSR, Odessa	8.2	57.3	65N
43N	132E	USSR, Vladivostok	10.0	69.8	45S
60N	31E	USSR, Leningrad	6.0	42.3	75S
35S	56W	Uruguay, Montevideo	11.0	76.9	60N
10N	65W	Venezuela, Barcelona	11.9	83.5	15S
11N	72W	Venezuela, Maracaibo	11.6	82.6	15S
13N	45E	Yemen, Aden	13.0	91.1	15S
45N	21E	Yugoslavia, Beogard	8.2	57.5	65S
12S	28E	Zaire, Lubumbash	12.1	85.0	15N
1N	25E	Zaire, Kisangani	9.9	69.6	15S
20S	29E	Zimbabwe, Bulawayo	13.0	91.3	20N

SOLAR RADIATION CHARTS

The U.S. Department of Energy has been recording solar radiation for many locations for over 20 years. New data (see Seasonal Data) are said to be more accurate. Older data (Monthly Data) tend to underestimate actual available sunlight. It is recommended that you use the most conservative numbers when estimating your solar electric system. By doing so, you will be assured of at least that amount of production. The average daily amount of solar radiation in kilowatt hours per square meter is also the peak sun hours. The formula for calculating the performance of a solar electric system using these figures is:

solar module peak watts at insolation = 1000 watts/square meter
x
number of solar modules in array
x
peak sun hours = peak array kWh/day
x
0.98 (2%) wire losses
x
0.8 (80%) battery efficiency
x
0.9 (90%) inverter efficiency = usable kWh/day

SEASONAL DATA

AVERAGE DAILY INSOLATION AVAILABILITY
FOR A SURFACE TILTED AT LATITUDE - 15 DEGREES (KWH/M2)

SITE	WINTER D,J,F	SPRING M,A,M	SUMMER J,J,A	FALL S,O,N	ANNUAL
ALBUQUERQUE	4.92	7.59	7.90	6.35	6.70
ATLANTA	3.12	5.51	5.85	4.43	4.74
AUSTIN	3.78	5.35	6.42	4.84	5.10
BIRMINGHAM	3.08	5.35	5.58	4.47	4.62
BISMARCK	2.96	5.36	6.73	4.07	4.79
BOSTON	2.36	4.49	5.40	3.59	3.97
BROWNSVILLE	3.49	5.72	6.67	4.74	5.16
BRYCE CANYON	4.87	7.39	7.65	6.24	6.55
CARIBOU	2.46	5.09	5.35	2.97	3.98
COLUMBIA	2.90	5.39	6.34	4.38	4.77
DAGGETT	4.63	7.59	8.23	6.14	6.66
DALLAS-FORT WORTH	3.53	5.43	6.77	4.95	5.18
DENVER	4.52	6.75	7.16	5.75	6.06
DETROIT	2.13	4.65	5.72	3.57	4.03
ELKO	4.12	6.62	8.15	5.85	6.19
EL PASO	5.11	7.72	7.83	6.24	6.73
FAIRBANKS	0.54	5.03	4.83	1.96	3.11
FRESNO	3.17	7.12	8.33	5.90	6.14
GREAT FALLS	2.69	5.47	7.04	4.27	4.88
HONOLULU	4.19	5.76	6.27	5.11	5.34
LAS VEGAS	5.02	8.03	8.18	6.35	6.91
MADISON	2.68	5.04	5.95	3.75	4.36
MEDFORD	1.98	5.55	7.60	4.25	4.87
MIAMI	4.13	5.72	5.33	4.57	4.94
NASHVILLE	2.61	5.01	6.02	4.08	4.44
NEW ORLEANS	3.42	5.55	5.65	4.68	4.84
OKLAHOMA CITY	3.67	5.66	6.63	4.98	5.24
OMAHA	3.46	5.46	6.53	4.38	4.97
ORLANDO	4.02	6.00	5.51	4.68	5.06
PHOENIX	4.77	7.80	7.84	6.26	6.68
PITTSBURGH	1.90	4.33	5.39	3.41	3.76
RALEIGH-DURHAM	3.00	5.22	5.66	4.16	4.52
SACRAMENTO	3.07	6.76	8.32	5.62	5.96
SAN DIEGO	4.31	6.21	6.67	5.25	5.62
SAN JUAN	4.51	5.83	5.92	4.99	5.32
SEATTLE	1.37	4.46	5.92	2.89	3.67
SYRACUSE	1.71	4.29	5.41	3.02	3.62
WASHINGTON D.C.	2.80	4.95	5.60	3.89	4.32

AVERAGE DAILY INSOLATION AVAILABILITY
FOR A SOUTH FACING SURFACE TILTED AT LATITUDE DEGREES (KWH/M2)

SITE					
ALBUQUERQUE	5.70	7.57	7.49	6.92	6.92
ATLANTA	3.53	5.48	5.57	4.78	4.85
AUSTIN	4.32	5.34	6.08	5.21	5.24
BIRMINGHAM	3.47	5.32	5.32	4.80	4.73
BISMARCK	3.36	5.35	6.39	4.35	4.87
BOSTON	2.67	4.46	5.14	3.84	4.03
BROWNSVILLE	3.92	5.68	6.30	5.05	5.25
BRYCE CANYON	5.62	7.34	7.16	6.76	6.73
CARIBOU	2.77	5.09	5.10	3.15	4.04
COLUMBIA	3.28	5.37	6.01	4.71	4.85
DAGGETT	5.32	7.52	7.71	6.64	6.81
DALLAS-FORT WORTH	4.03	5.43	6.41	5.33	5.31
DENVER	5.22	6.73	6.76	6.22	6.24
DETROIT	2.39	4.64	5.43	3.78	4.07
ELKO	4.71	6.58	7.66	6.29	6.32
EL PASO	5.91	7.68	7.39	6.79	6.95
FAIRBANKS	0.60	5.02	4.62	2.05	3.09
FRESNO	3.54	7.05	7.82	6.14	6.20
GREAT FALLS	3.04	5.47	6.66	4.57	4.95
HONOLULU	4.74	5.73	5.93	5.49	5.48
LAS VEGAS	5.79	7.96	7.67	6.88	7.08
MADISON	3.02	5.05	5.66	3.99	4.14
MEDFORD	2.19	5.53	7.18	4.51	4.87
MIAMI	4.69	5.72	5.10	4.92	5.11
NASHVILLE	2.94	5.00	5.75	4.36	4.52
NEW ORLEANS	3.86	5.52	5.38	5.04	4.96
OKLAHOMA CITY	4.18	5.65	6.28	5.37	5.38
OMAHA	3.93	5.43	6.20	4.71	5.08
ORLANDO	4.57	5.96	5.26	5.05	5.21
PHOENIX	5.49	7.77	7.42	6.80	6.88
PITTSBURGH	2.11	4.29	5.15	3.62	3.80
RALEIGH-DURHAM	3.37	5.21	5.39	4.46	4.61
SACRAMENTO	3.47	6.70	7.80	6.03	6.01
SAN DIEGO	4.94	6.20	6.35	5.67	5.79
SAN JUAN	5.14	5.80	5.61	5.36	5.48
SEATTLE	1.51	4.43	5.66	3.03	3.67
SYRACUSE	1.90	4.27	5.16	3.20	3.64
WASHINGTON D.C.	3.17	4.93	5.33	4.16	4.41

AVERAGE DAILY INSOLATION AVAILABILITY
FOR A SOUTH FACING SURFACE TILTED AT LATITUDE + 15 DEGREES (KWH/M2)

SITE	WINTER D.J.F	SPRING M.A.M	SUMMER J.J.A	FALL S.O.N	ANNUAL
ALBUQUERQUE	6.14	7.18	6.72	7.12	6.79
ATLANTA	3.78	5.20	5.03	4.91	4.73
AUSTIN	4.63	5.11	5.45	5.33	5.13
BIRMINGHAM	3.70	5.05	4.84	4.91	4.63
BISMARCK	3.57	5.09	5.73	4.43	4.71
BOSTON	2.83	4.24	4.66	3.89	3.91
BROWNSVILLE	4.18	5.42	5.64	5.15	5.10
BRYCE CANYON	6.04	6.91	6.32	6.90	6.55
CARIBOU	2.92	4.88	4.62	3.20	3.91
COLUMBIA	3.49	5.10	5.41	4.80	4.71
DAGGETT	5.70	7.07	6.80	6.77	6.59
DALLAS-FORT WORTH	4.31	5.21	5.75	5.45	5.18
DENVER	5.60	6.37	6.01	6.35	6.09
DETROIT	2.53	4.42	4.91	3.82	3.93
ELKO	5.03	6.22	6.78	6.38	6.11
EL PASO	6.38	7.28	6.60	6.99	6.81
FAIRBANKS	0.62	4.77	4.23	2.07	2.93
FRESNO	3.76	6.64	6.90	6.44	5.95
GREAT FALLS	3.23	5.22	5.95	4.64	4.77
HONOLULU	5.07	5.45	5.34	5.60	5.37
LAS VEGAS	6.21	7.49	6.78	7.02	6.88
MADISON	3.20	4.84	5.12	4.04	4.31
MEDFORD	2.30	5.25	6.40	4.54	4.64
MIAMI	5.01	5.47	4.67	5.07	5.05
NASHVILLE	3.12	4.77	5.24	4.44	4.40
NEW ORLEANS	4.11	5.26	4.88	5.16	4.85
OKLAHOMA CITY	4.48	5.39	5.64	5.49	5.25
OMAHA	4.19	5.16	5.58	4.80	4.94
ORLANDO	4.88	5.66	4.79	5.19	5.13
PHOENIX	5.91	7.36	6.65	6.99	6.73
PITTSBURGH	2.23	4.09	4.68	3.66	3.68
RALEIGH-DURHAM	3.57	4.97	4.88	4.55	4.49
SACRAMENTO	3.68	6.30	6.89	6.12	5.76
SAN DIEGO	5.29	5.90	5.73	5.80	5.68
SAN JUAN	5.51	5.53	5.04	5.48	5.39
SEATTLE	1.59	4.24	5.13	3.05	3.52
SYRACUSE	2.00	4.08	4.70	3.24	3.51
WASHINGTON D.C.	3.37	4.70	4.84	4.24	4.29

AVERAGE DAILY INSOLATION AVAILABILITY
FOR A SURFACE TRACKING ABOUT A NORTH - SOUTH AXIS
TILTED AT LATITUDE - 15 DEGREES (KWH/M2)

SITE	WINTER	SPRING	SUMMER	FALL	ANNUAL
ALBUQUERQUE	6.34	10.13	10.60	8.32	8.86
ATLANTA	3.89	7.03	7.41	5.51	5.97
AUSTIN	4.80	6.75	8.27	6.14	6.50
BIRMINGHAM	3.80	6.78	7.00	5.57	5.80
BISMARCK	3.57	7.03	9.17	5.14	6.25
BOSTON	2.81	5.72	7.03	4.51	5.03
BROWNSVILLE	4.32	7.08	8.70	5.97	6.53
BRYCE CANYON	6.20	10.28	10.78	8.24	8.89
CARIBOU	2.93	6.61	7.14	3.63	5.09
COLUMBIA	3.44	6.84	8.55	5.55	6.12
DAGGETT	5.74	10.15	11.27	7.88	8.78
DALLAS-FORT WORTH	4.39	6.95	8.93	6.35	6.67
DENVER	5.76	9.09	9.73	7.49	8.03
DETROIT	2.54	5.88	7.36	4.33	5.04
ELKO	5.10	9.05	11.43	7.60	8.32
EL PASO	6.63	10.35	10.43	8.25	8.93
FAIRBANKS	0.60	6.67	6.85	2.33	4.14
FRESNO	3.80	9.55	11.55	7.57	8.15
GREAT FALLS	3.20	7.07	9.70	5.44	6.37
HONOLULU	5.21	7.12	7.91	6.48	6.69
LAS VEGAS	6.33	10.99	11.18	8.24	9.21
MADISON	3.17	6.49	7.67	4.62	5.50
MEDFORD	2.29	7.09	10.16	5.30	6.28
MIAMI	5.24	7.22	6.47	5.76	6.18
NASHVILLE	3.17	6.16	7.52	5.07	5.49
NEW ORLEANS	4.18	7.02	7.09	5.86	6.05
OKLAHOMA CITY	4.61	7.35	8.62	6.32	6.74
OMAHA	4.27	7.15	8.84	5.59	6.48
ORLANDO	4.99	7.66	6.82	5.82	6.33
PHOENIX	6.04	10.47	10.33	8.05	8.74
PITTSBURGH	2.23	5.37	6.80	4.12	4.65
RALEIGH-DURHAM	3.56	6.63	7.16	5.03	5.61
SACRAMENTO	3.67	8.96	11.46	7.11	7.82
SAN DIEGO	5.38	7.84	8.38	6.48	7.03
SAN JUAN	5.73	7.17	7.40	6.31	6.66
SEATTLE	1.57	5.61	7.82	3.49	4.64
SYRACUSE	2.00	5.41	6.88	3.63	4.50
WASHINGTON D.C.	3.40	6.33	7.20	4.80	5.45

AVERAGE DAILY INSOLATION AVAILABILITY
FOR A SURFACE TRACKING ABOUT A NORTH - SOUTH AXIS
TILTED AT LATITUDE DEGREES (KWH/M2)

SITE	WINTER D,J,F	SPRING M,A,M	SUMMER J,J,A	FALL S,O,N	ANNUAL
ALBUQUERQUE	6.94	10.13	10.35	8.73	9.05
ATLANTA	4.21	7.03	7.26	5.77	6.08
AUSTIN	5.22	6.76	8.08	6.43	6.63
BIRMINGHAM	4.11	6.78	6.85	5.82	5.90
BISMARCK	3.89	7.05	9.00	5.37	6.34
BOSTON	3.07	5.72	6.89	4.69	5.10
BROWNSVILLE	4.67	7.08	8.48	6.21	6.62
BRYCE CANYON	6.80	10.29	10.53	8.64	9.08
CARIBOU	3.18	6.63	7.01	3.78	5.17
COLUMBIA	3.77	6.85	8.38	5.80	6.21
DAGGETT	6.30	10.15	11.00	8.26	8.95
DALLAS-FORT WORTH	4.79	6.97	8.73	6.65	6.80
DENVER	6.31	9.10	9.51	7.86	8.21
DETROIT	2.76	5.88	7.21	4.49	5.10
ELKO	5.58	9.08	11.18	7.95	8.46
EL PASO	7.24	10.35	10.17	8.65	9.12
FAIRBANKS	0.64	6.70	6.75	2.41	4.15
FRESNO	4.12	9.55	11.28	7.91	8.24
GREAT FALLS	3.50	7.09	9.50	5.68	6.46
HONOLULU	5.67	7.11	7.72	6.77	6.82
LAS VEGAS	6.96	10.99	10.91	8.64	9.39
MADISON	3.46	6.52	7.52	4.81	5.59
MEDFORD	2.47	7.10	10.14	5.49	6.32
MIAMI	5.69	7.23	6.34	6.03	6.32
NASHVILLE	3.43	6.17	7.37	5.29	5.58
NEW ORLEANS	4.53	7.02	6.93	6.13	6.17
OKLAHOMA CITY	5.02	7.36	8.43	6.63	6.87
OMAHA	4.64	7.16	8.66	5.85	6.59
ORLANDO	5.43	7.66	6.66	6.12	6.47
PHOENIX	6.61	10.46	10.08	8.45	8.91
PITTSBURGH	2.41	5.37	6.67	4.29	4.70
RALEIGH-DURHAM	3.86	6.64	7.01	5.26	5.70
SACRAMENTO	3.99	8.96	11.20	7.43	7.91
SAN DIEGO	5.89	7.86	8.20	6.81	7.19
SAN JUAN	6.23	7.16	7.22	6.59	6.81
SEATTLE	1.69	5.61	7.67	3.60	4.66
SYRACUSE	2.16	5.41	6.75	3.77	4.53
WASHINGTON D.C.	3.70	6.34	7.05	5.02	5.54

AVERAGE DAILY INSOLATION AVAILABILITY
FOR A SURFACE TRACKING ABOUT A NORTH - SOUTH AXIS
TILTED AT LATITUDE + 15 DEGREES (KWH/M2)

SITE	WINTER	SPRING	SUMMER	FALL	ANNUAL
ALBUQUERQUE	7.24	9.85	9.78	8.85	8.94
ATLANTA	4.38	6.85	6.90	5.86	6.01
AUSTIN	5.44	6.61	7.64	6.51	6.56
BIRMINGHAM	4.28	6.61	6.52	5.90	5.83
BISMARCK	4.04	6.88	8.55	5.42	6.24
BOSTON	3.19	5.58	6.55	4.73	5.02
BROWNSVILLE	4.86	6.89	8.03	6.27	6.52
BRYCE CANYON	7.10	10.03	10.00	8.74	8.98
CARIBOU	3.30	6.50	6.71	3.80	5.09
COLUMBIA	3.92	6.66	7.99	5.86	6.12
DAGGETT	6.58	9.86	10.40	8.36	8.81
DALLAS-FORT WORTH	5.00	6.80	8.28	6.73	6.72
DENVER	6.59	8.87	9.02	7.95	8.12
DETROIT	2.86	5.74	6.86	4.53	5.01
ELKO	5.81	8.85	10.63	8.02	8.34
EL PASO	7.56	10.05	9.60	8.76	9.00
FAIRBANKS	0.66	6.53	6.51	2.42	4.05
FRESNO	4.28	9.29	10.70	7.99	8.08
GREAT FALLS	3.63	6.92	9.02	5.73	6.34
HONOLULU	5.89	6.91	7.29	6.86	6.74
LAS VEGAS	7.26	10.70	10.33	8.75	9.27
MADISON	3.58	6.37	7.15	4.86	5.50
MEDFORD	2.54	6.91	9.62	5.52	6.17
MIAMI	5.91	7.07	6.03	6.12	6.28
NASHVILLE	3.57	6.01	7.01	5.34	5.49
NEW ORLEANS	4.72	6.84	6.59	6.22	6.10
OKLAHOMA CITY	5.23	7.20	8.00	6.70	6.79
OMAHA	4.83	7.00	8.26	5.91	6.51
ORLANDO	5.66	7.47	6.34	6.21	6.42
PHOENIX	6.89	10.16	9.51	8.56	8.79
PITTSBURGH	2.50	5.24	6.35	4.32	4.61
RALEIGH-DURHAM	4.01	6.48	6.65	5.33	5.63
SACRAMENTO	4.14	8.71	10.61	7.49	7.76
SAN DIEGO	6.13	7.66	7.76	6.90	7.12
SAN JUAN	6.50	6.98	6.82	6.68	6.75
SEATTLE	1.74	5.47	7.32	3.63	4.56
SYRACUSE	2.23	5.28	6.42	3.80	4.45
WASHINGTON D.C.	3.84	6.18	6.71	5.08	5.46

AVERAGE DAILY INSOLATION AVAILABILITY
FOR A SURFACE TRACKING ABOUT A HORIZONTAL NORTH - SOUTH AXIS (KWH/M2)

SITE	WINTER D,J,F	SPRING M,A,M	SUMMER J,J,A	FALL S,O,N	ANNUAL
ALBUQUERQUE	5.32	9.76	10.59	7.42	8.29
ATLANTA	3.37	6.80	7.38	4.93	5.64
AUSTIN	4.24	6.58	8.27	5.67	6.20
BIRMINGHAM	3.29	6.57	6.97	5.02	5.47
BISMARCK	2.59	6.41	8.84	4.20	5.53
BOSTON	2.18	5.36	6.88	3.84	4.58
BROWNSVILLE	4.02	6.98	8.73	5.69	6.37
BRYCE CANYON	4.97	9.77	10.70	7.15	8.17
CARIBOU	2.17	6.00	6.91	2.99	4.53
COLUMBIA	2.76	6.47	8.45	4.84	5.65
DAGGETT	4.73	9.73	11.22	6.98	8.18
DALLAS-FORT WORTH	3.78	6.70	8.91	5.77	6.30
DENVER	4.52	8.55	9.59	6.42	7.29
DETROIT	2.02	5.51	7.20	3.70	4.63
ELKO	3.94	8.52	11.24	6.47	7.57
EL PASO	5.78	10.10	10.49	7.55	8.50
FAIRBANKS	0.33	5.45	6.20	1.62	3.42
FRESNO	3.18	9.14	11.47	6.66	7.64
GREAT FALLS	2.29	6.38	9.30	4.18	5.61
HONOLULU	4.98	7.08	7.93	6.31	6.59
LAS VEGAS	5.16	10.50	11.11	7.24	8.52
MADISON	2.41	6.00	7.48	3.90	4.96
MEDFORD	1.83	6.62	10.13	4.54	5.81
MIAMI	4.84	7.10	6.48	5.47	5.98
NASHVILLE	2.66	5.89	7.46	4.51	5.14
NEW ORLEANS	3.73	6.86	7.09	5.42	5.79
OKLAHOMA CITY	3.88	7.04	8.57	5.60	6.29
OMAHA	3.33	6.72	8.68	4.76	5.89
ORLANDO	4.48	7.51	6.82	5.42	6.07
PHOENIX	5.17	10.16	10.36	7.26	8.25
PITTSBURGH	1.82	5.10	6.66	3.58	4.31
RALEIGH-DURHAM	3.00	6.36	7.10	4.46	5.24
SACRAMENTO	2.97	8.51	11.34	6.16	7.27
SAN DIEGO	4.60	7.55	8.35	5.84	6.60
SAN JUAN	5.60	7.14	7.42	6.22	6.60
SEATTLE	1.20	5.14	7.50	2.90	4.21
SYRACUSE	1.60	5.08	6.71	3.09	4.13
WASHINGTON D.C.	2.76	6.00	7.10	4.19	5.02

Estimates of Available Solar Radiation and Photovoltaic Energy Production for Various Tilted and Tracking Surfaces throughout the U.S. Based on PVFORM, a Computerized Performance Model (Courtesy DOE)

MONTHLY DATA

SITE / ARRAY TILT	JAN	FEB	MAR	APR	MAY	JUN	JUL	AUG	SEP	OCT	NOV	DEC	ANNUAL TOTAL (KWH/SQ. M)	AVERAGE DAY (KWH/SQ. M)
JUNEAU AK LATITUDE: 58 DEGREES 22 MINUTES														
LATITUDE -15:	1.04	1.70	2.57	3.72	4.07	4.28	4.02	3.47	2.73	1.74	1.24	.68	951.8	2.6
LATITUDE :	1.16	1.81	2.57	3.53	3.71	3.86	3.67	3.28	2.72	1.83	1.39	.78	922.7	2.5
LATITUDE +15:	1.22	1.82	2.43	3.18	3.24	3.32	3.20	2.96	2.58	1.83	1.45	.83	853.6	2.3
KING SALMON AK LATITUDE: 58 DEGREES 41 MINUTES														
LATITUDE -15:	1.25	2.22	3.68	4.40	4.73	4.68	4.38	3.73	3.54	3.08	2.07	1.08	1183.5	3.2
LATITUDE :	1.39	2.36	3.72	4.19	4.32	4.22	4.00	3.54	3.57	3.32	2.35	1.25	1164.2	3.2
LATITUDE +15:	1.46	2.37	3.56	3.78	3.77	3.63	3.48	3.19	3.41	3.36	2.49	1.34	1090.5	3.0
KODIAK AK LATITUDE: 57 DEGREES 45 MINUTES														
LATITUDE -15:	1.13	1.92	3.50	4.38	4.35	4.65	4.46	4.20	3.55	3.06	1.90	.99	1160.0	3.2
LATITUDE :	1.26	2.03	3.53	4.16	3.98	4.20	4.08	3.99	3.58	3.29	2.15	1.13	1137.8	3.1
LATITUDE +15:	1.31	2.02	3.37	3.76	3.47	3.61	3.55	3.60	3.42	3.32	2.27	1.20	1062.9	2.9
NOME AK LATITUDE: 64 DEGREES 30 MINUTES														
LATITUDE -15:	.34	1.56	3.29	4.63	5.14	5.40	4.57	3.74	3.53	2.59	1.16	.01	1095.5	3.0
LATITUDE :	.38	1.65	3.32	4.39	4.69	4.85	4.15	3.54	3.58	2.79	1.32	.01	1055.8	2.9
LATITUDE +15:	.40	1.65	3.17	3.96	4.06	4.14	3.60	3.19	3.42	2.82	1.39	.01	968.4	2.7
BIRMINGHAM AL LATITUDE: 33 DEGREES 34 MINUTES														
LATITUDE -15:	2.75	3.53	4.44	5.39	5.72	5.83	5.58	5.57	5.01	4.54	3.44	2.70	1658.7	4.5
LATITUDE :	3.02	3.73	4.49	5.19	5.31	5.33	5.17	5.36	5.08	4.87	3.84	3.05	1657.0	4.5
LATITUDE +15:	3.14	3.74	4.31	4.75	4.68	4.63	4.57	4.91	4.89	4.94	4.03	3.23	1577.2	4.3
MOBILE AL LATITUDE: 30 DEGREES 41 MINUTES														
LATITUDE -15:	3.11	3.92	4.74	5.51	5.78	5.69	5.30	5.25	4.89	4.72	3.64	2.94	1688.3	4.6
LATITUDE :	3.43	4.16	4.80	5.31	5.37	5.22	4.93	5.06	4.95	5.06	4.07	3.32	1694.8	4.6
LATITUDE +15:	3.58	4.19	4.63	4.87	4.75	4.55	4.36	4.64	4.77	5.14	4.28	3.53	1620.9	4.4
MONTGOMERY AL LATITUDE: 32 DEGREES 18 MINUTES														
LATITUDE -15:	2.87	3.66	4.56	5.56	5.85	5.99	5.67	5.62	5.01	4.67	3.59	2.87	1703.1	4.7
LATITUDE :	3.17	3.88	4.61	5.36	5.43	5.47	5.26	5.42	5.08	5.01	4.02	3.25	1703.0	4.7
LATITUDE +15:	3.29	3.89	4.43	4.91	4.78	4.75	4.64	4.96	4.89	5.08	4.22	3.45	1622.1	4.4
FORT SMITH AR LATITUDE: 35 DEGREES 20 MINUTES														
LATITUDE -15:	3.03	3.75	4.56	5.23	5.90	6.33	6.37	6.11	5.27	4.64	3.55	2.94	1756.2	4.8
LATITUDE :	3.37	3.98	4.61	5.03	5.46	5.76	5.88	5.89	5.35	4.98	3.99	3.34	1754.8	4.8
LATITUDE +15:	3.52	4.00	4.43	4.60	4.81	4.98	5.16	5.38	5.15	5.05	4.20	3.56	1669.3	4.6

Stand-Alone Flat-Plate Photovoltaic Power Systems: System Sizing and Life-Cycle Costing Methodology for Federal Agencies (Courtesy DOE)

SITE ARRAY TILT		JAN	FEB	MAR	APR	MAY	JUN	JUL	AUG	SEP	OCT	NOV	DEC	ANNUAL TOTAL (KWH/SQ. M)	AVERAGE DAY (KWH/SQ. M)
LITTLE ROCK	AR	LATITUDE:		34 DEGREES	44 MINUTES										
LATITUDE -15:		2.93	3.73	4.54	5.21	5.95	6.38	6.27	6.04	5.30	4.71	3.47	2.84	1746.4	4.8
LATITUDE :		3.24	3.96	4.59	5.01	5.51	5.82	5.80	5.82	5.38	5.06	3.89	3.22	1744.7	4.8
LATITUDE +15:		3.39	3.98	4.41	4.59	4.85	5.03	5.09	5.32	5.19	5.14	4.09	3.43	1659.4	4.5
PHOENIX	AZ	LATITUDE:		33 DEGREES	26 MINUTES										
LATITUDE -15:		4.19	5.21	6.36	7.67	8.26	8.28	7.66	7.46	7.13	6.11	4.83	4.02	2348.9	6.4
LATITUDE :		4.74	5.62	6.51	7.40	7.60	7.48	7.06	7.20	7.32	6.66	5.52	4.66	2366.3	6.5
LATITUDE +15:		5.02	5.72	6.31	6.76	6.60	6.36	6.15	6.57	7.10	6.83	5.89	5.03	2260.9	6.2
PRESCOTT	AZ	LATITUDE:		34 DEGREES	39 MINUTES										
LATITUDE -15:		4.30	5.14	6.30	7.44	8.11	8.34	7.12	6.82	6.97	6.09	4.94	4.15	2304.2	6.3
LATITUDE :		4.88	5.54	6.45	7.18	7.47	7.53	6.57	6.58	7.16	6.64	5.66	4.83	2327.1	6.4
LATITUDE +15:		5.18	5.64	6.24	6.56	6.50	6.41	5.74	6.01	6.94	6.81	6.05	5.23	2229.6	6.1
TUCSON	AZ	LATITUDE:		32 DEGREES	7 MINUTES										
LATITUDE -15:		4.42	5.35	6.47	7.66	8.23	8.24	7.21	7.06	6.90	6.08	4.94	4.18	2335.1	6.4
LATITUDE :		5.00	5.77	6.63	7.39	7.57	7.44	6.65	6.81	7.07	6.62	5.65	4.85	2356.1	6.5
LATITUDE +15:		5.30	5.87	6.42	6.74	6.57	6.33	5.81	6.22	6.84	6.79	6.02	5.25	2255.1	6.2
WINSLOW	AZ	LATITUDE:		35 DEGREES	1 MINUTES										
LATITUDE -15:		4.19	5.15	6.34	7.47	8.01	8.19	7.24	7.00	6.90	6.01	4.90	4.04	2295.2	6.3
LATITUDE :		4.75	5.55	6.49	7.21	7.37	7.39	6.67	6.74	7.07	6.54	5.61	4.69	2314.0	6.3
LATITUDE +15:		5.03	5.64	6.28	6.57	6.40	6.29	5.81	6.15	6.84	6.70	5.99	5.07	2213.3	6.1
YUMA	AZ	LATITUDE:		32 DEGREES	40 MINUTES										
LATITUDE -15:		4.45	5.43	6.70	7.84	8.42	8.50	7.56	7.56	7.20	6.23	5.04	4.26	2410.2	6.6
LATITUDE :		5.05	5.87	6.88	7.57	7.75	7.68	6.97	7.31	7.40	6.80	5.77	4.97	2434.4	6.7
LATITUDE +15:		5.37	5.98	6.67	6.91	6.73	6.53	6.08	6.67	7.18	6.98	6.17	5.38	2331.3	6.4
BAKERSFIELD	CA	LATITUDE:		35 DEGREES	25 MINUTES										
LATITUDE -15:		3.15	4.19	5.63	6.85	7.75	8.31	8.29	7.97	7.19	5.79	4.03	2.92	2194.8	6.0
LATITUDE :		3.50	4.46	5.74	6.60	7.14	7.50	7.62	7.69	7.39	6.30	4.56	3.33	2187.2	6.0
LATITUDE +15:		3.67	4.50	5.54	6.03	6.22	6.38	6.61	7.01	7.16	6.45	4.83	3.55	2068.6	5.7
CHINA LAKE	CA	LATITUDE:		35 DEGREES	41 MINUTES										
LATITUDE -15:		3.88	4.77	6.19	7.33	7.88	8.30	8.07	8.63	7.16	5.88	4.53	3.82	2328.1	6.4
LATITUDE :		4.38	5.13	6.34	7.07	7.26	7.50	7.43	8.34	7.36	6.41	5.17	4.43	2338.3	6.4
LATITUDE +15:		4.63	5.20	6.13	6.46	6.32	6.38	6.45	7.61	7.14	6.57	5.51	4.78	2227.5	6.1

SITE ARRAY TILT		JAN	FEB	MAR	APR	MAY	JUN	JUL	AUG	SEP	OCT	NOV	DEC	ANNUAL TOTAL (KWH/SQ. M)	AVERAGE DAY (KWH/SQ. M)
DAGGETT	CA	LATITUDE:		34 DEGREES	52 MINUTES										
LATITUDE -15:		4.02	4.92	6.29	7.44	8.01	8.36	8.03	7.82	7.20	5.98	4.68	3.90	2334.1	6.4
LATITUDE :		4.55	5.30	6.44	7.18	7.38	7.55	7.40	7.56	7.41	6.53	5.35	4.52	2348.6	6.4
LATITUDE +15:		4.82	5.38	6.24	6.56	6.42	6.43	6.43	6.90	7.19	6.69	5.71	4.89	2241.3	6.1
EL TORO	CA	LATITUDE:		33 DEGREES	40 MINUTES										
LATITUDE -15:		3.85	4.65	5.60	6.25	6.39	6.65	7.29	7.00	6.07	5.17	4.24	3.73	2036.4	5.6
LATITUDE :		4.33	4.98	5.71	6.02	5.91	6.05	6.73	6.76	6.20	5.59	4.81	4.30	2051.4	5.6
LATITUDE +15:		4.57	5.05	5.52	5.51	5.20	5.22	5.87	6.17	5.99	5.69	5.10	4.63	1963.7	5.4
FRESNO	CA	LATITUDE:		36 DEGREES	46 MINUTES										
LATITUDE -15:		2.72	3.89	5.59	6.89	7.69	8.26	8.31	8.04	7.27	5.81	3.90	2.50	2158.6	5.9
LATITUDE :		3.01	4.14	5.70	6.64	7.08	7.46	7.64	7.76	7.48	6.33	4.41	2.83	2146.5	5.9
LATITUDE +15:		3.14	4.17	5.51	6.07	6.18	6.36	6.63	7.08	7.26	6.48	4.68	3.00	2025.7	5.5
LONG BEACH	CA	LATITUDE:		33 DEGREES	49 MINUTES										
LATITUDE -15:		3.77	4.56	5.61	6.28	6.37	6.48	7.09	6.83	5.94	5.05	4.14	3.62	2001.8	5.5
LATITUDE :		4.24	4.89	5.72	6.06	5.90	5.91	6.55	6.59	6.06	5.45	4.69	4.18	2015.9	5.5
LATITUDE +15:		4.48	4.95	5.53	5.54	5.18	5.11	5.73	6.02	5.86	5.55	4.98	4.49	1929.4	5.3
LOS ANGELES	CA	LATITUDE:		33 DEGREES	56 MINUTES										
LATITUDE -15:		3.78	4.57	5.65	6.32	6.36	6.43	7.12	6.76	5.88	5.02	4.16	3.65	1999.6	5.5
LATITUDE :		4.25	4.89	5.76	6.10	5.89	5.86	6.57	6.52	5.99	5.41	4.72	4.21	2014.3	5.5
LATITUDE +15:		4.48	4.96	5.57	5.58	5.18	5.07	5.75	5.96	5.79	5.51	5.00	4.53	1928.6	5.3
NEEDLES	CA	LATITUDE:		34 DEGREES	46 MINUTES										
LATITUDE -15:		4.15	5.23	6.49	7.59	8.19	8.44	7.84	7.45	7.21	6.07	4.87	4.09	2362.5	6.5
LATITUDE :		4.70	5.65	6.65	7.33	7.54	7.62	7.22	7.19	7.42	6.62	5.58	4.76	2381.9	6.5
LATITUDE +15:		4.98	5.76	6.45	6.69	6.56	6.48	6.29	6.57	7.20	6.79	5.96	5.16	2277.5	6.2
OAKLAND	CA	LATITUDE:		37 DEGREES	44 MINUTES										
LATITUDE -15:		3.05	3.97	5.22	6.32	6.85	7.12	7.19	6.78	6.19	4.89	3.66	2.98	1955.4	5.4
LATITUDE :		3.40	4.23	5.31	6.09	6.32	6.46	6.62	6.53	6.33	5.28	4.13	3.40	1951.7	5.3
LATITUDE +15:		3.56	4.27	5.12	5.56	5.54	5.55	5.78	5.96	6.12	5.37	4.37	3.64	1851.7	5.1
RED BLUFF	CA	LATITUDE:		40 DEGREES	9 MINUTES										
LATITUDE -15:		2.53	3.58	4.95	6.36	7.39	7.87	8.31	7.79	7.00	5.24	3.29	2.44	2033.9	5.6
LATITUDE :		2.80	3.79	5.02	6.11	6.80	7.11	7.63	7.51	7.19	5.68	3.70	2.77	2013.7	5.5
LATITUDE +15:		2.92	3.81	4.83	5.57	5.93	6.07	6.61	6.85	6.96	5.79	3.89	2.94	1893.3	5.2

SITE / ARRAY TILT		JAN	FEB	MAR	APR	MAY	JUN	JUL	AUG	SEP	OCT	NOV	DEC	ANNUAL TOTAL (KWH/SQ. M)	AVERAGE DAY (KWH/SQ. M)
SACRAMENTO	CA	LATITUDE: 38 DEGREES 31 MINUTES													
LATITUDE -15:		2.55	3.69	5.27	6.63	7.55	8.11	8.35	7.92	7.11	5.47	3.53	2.45	2090.9	5.7
LATITUDE :		2.81	3.91	5.36	6.38	6.95	7.33	7.67	7.65	7.31	5.94	3.98	2.77	2073.2	5.7
LATITUDE +15:		2.93	3.94	5.17	5.83	6.06	6.25	6.65	6.97	7.09	6.07	4.20	2.94	1951.6	5.3
SAN DIEGO	CA	LATITUDE: 32 DEGREES 44 MINUTES													
LATITUDE -15:		3.90	4.70	5.64	6.25	6.18	6.26	6.74	6.65	5.95	5.16	4.31	3.79	1994.8	5.5
LATITUDE :		4.39	5.04	5.75	6.03	5.72	5.72	6.23	6.42	6.07	5.58	4.89	4.37	2014.9	5.5
LATITUDE +15:		4.63	5.11	5.56	5.52	5.04	4.95	5.46	5.87	5.87	5.68	5.19	4.71	1934.5	5.3
SAN FRANCISCO	CA	LATITUDE: 37 DEGREES 37 MINUTES													
LATITUDE -15:		3.04	3.93	5.22	6.31	6.90	7.19	7.41	7.00	6.35	4.94	3.64	2.94	1975.5	5.4
LATITUDE :		3.39	4.19	5.30	6.08	6.36	6.52	6.82	6.75	6.49	5.33	4.11	3.36	1970.1	5.4
LATITUDE +15:		3.55	4.22	5.11	5.55	5.57	5.60	5.95	6.16	6.28	5.43	4.34	3.59	1867.3	5.1
SANTA MARIA	CA	LATITUDE: 34 DEGREES 54 MINUTES													
LATITUDE -15:		3.51	4.31	5.56	6.25	6.60	7.12	7.22	6.88	6.11	5.26	4.11	3.52	2023.8	5.5
LATITUDE :		3.93	4.61	5.66	6.03	6.11	6.46	6.67	6.64	6.25	5.70	4.66	4.06	2032.8	5.6
LATITUDE +15:		4.14	4.66	5.47	5.51	5.36	5.56	5.82	6.06	6.04	5.82	4.95	4.36	1940.4	5.3
SUNNYVALE	CA	LATITUDE: 37 DEGREES 25 MINUTES													
LATITUDE -15:		3.17	4.04	5.32	6.39	7.05	7.42	7.55	7.17	6.41	5.03	3.74	3.02	2019.1	5.5
LATITUDE :		3.54	4.31	5.41	6.15	6.50	6.72	6.95	6.91	6.55	5.43	4.22	3.45	2013.9	5.5
LATITUDE +15:		3.72	4.34	5.21	5.62	5.68	5.76	6.05	6.30	6.34	5.53	4.46	3.69	1908.7	5.2
COLORADO SPRINGS	CO	LATITUDE: 38 DEGREES 49 MINUTES													
LATITUDE -15:		4.17	4.81	5.66	6.39	6.60	7.17	6.86	6.72	6.51	5.72	4.50	3.92	2100.9	5.8
LATITUDE :		4.74	5.18	5.77	6.15	6.09	6.50	6.32	6.48	6.67	6.23	5.16	4.57	2126.1	5.8
LATITUDE +15:		5.04	5.27	5.57	5.62	5.34	5.59	5.53	5.91	6.45	6.38	5.50	4.95	2042.8	5.6
DENVER	CO	LATITUDE: 39 DEGREES 45 MINUTES													
LATITUDE -15:		4.01	4.65	5.65	6.24	6.62	7.12	7.05	6.82	6.45	5.55	4.29	3.75	2076.2	5.7
LATITUDE :		4.56	5.01	5.76	6.00	6.11	6.46	6.50	6.57	6.60	6.04	4.90	4.36	2095.9	5.7
LATITUDE +15:		4.84	5.08	5.56	5.48	5.35	5.55	5.67	6.00	6.39	6.17	5.22	4.72	2009.0	5.5
GRAND JUNCTION	CO	LATITUDE: 39 DEGREES 7 MINUTES													
LATITUDE -15:		3.65	4.56	5.70	6.59	7.38	7.86	7.65	7.28	6.86	5.70	4.41	3.66	2171.0	5.9
LATITUDE :		4.12	4.89	5.81	6.34	6.79	7.10	7.03	7.02	7.03	6.20	5.03	4.25	2180.1	6.0
LATITUDE +15:		4.35	4.96	5.61	5.78	5.92	6.06	6.12	6.39	6.81	6.34	5.36	4.58	2078.0	5.7

SITE ARRAY TILT		JAN	FEB	MAR	APR	MAY	JUN	JUL	AUG	SEP	OCT	NOV	DEC	ANNUAL TOTAL (KWH/SQ. M)	AVERAGE DAY (KWH/SQ. M)
PUEBLO	CO	LATITUDE:		38 DEGREES	17 MINUTES										
LATITUDE -15:		4.12	4.73	5.68	6.46	6.69	7.36	7.15	6.97	6.55	5.67	4.49	3.86	2122.9	5.8
LATITUDE :		4.68	5.09	5.79	6.22	6.17	6.67	6.59	6.72	6.71	6.16	5.12	4.48	2142.6	5.9
LATITUDE +15:		4.97	5.16	5.59	5.68	5.40	5.71	5.74	6.13	6.49	6.30	5.46	4.85	2053.1	5.6
HARTFORD	CT	LATITUDE:		41 DEGREES	56 MINUTES										
LATITUDE -15:		2.14	2.85	3.51	4.33	4.86	5.11	5.11	4.70	4.22	3.53	2.27	1.81	1352.9	3.7
LATITUDE :		2.35	3.00	3.53	4.15	4.49	4.66	4.72	4.51	4.26	3.76	2.51	2.02	1337.8	3.7
LATITUDE +15:		2.44	2.99	3.37	3.78	3.96	4.05	4.16	4.11	4.08	3.79	2.61	2.13	1262.1	3.5
GUANTANAMO BAY	CU	LATITUDE:		19 DEGREES	54 MINUTES										
LATITUDE -15:		4.66	5.38	6.17	6.69	6.36	6.10	6.50	6.31	5.84	5.17	4.76	4.39	2079.2	5.7
LATITUDE :		5.27	5.79	6.31	6.48	5.93	5.62	6.06	6.12	5.96	5.56	5.40	5.07	2115.9	5.8
LATITUDE +15:		5.60	5.90	6.12	5.95	5.23	4.89	5.33	5.62	5.77	5.67	5.74	5.47	2046.1	5.6
WASHINGTON-STERLINDC		LATITUDE:		38 DEGREES	57 MINUTES										
LATITUDE -15:		2.44	3.15	3.98	4.77	5.32	5.76	5.62	5.30	4.81	4.02	2.86	2.16	1528.7	4.2
LATITUDE :		2.69	3.32	4.01	4.58	4.93	5.26	5.20	5.10	4.87	4.29	3.19	2.43	1518.5	4.2
LATITUDE +15:		2.80	3.33	3.85	4.19	4.35	4.56	4.58	4.66	4.68	4.34	3.34	2.57	1437.8	3.9
WILMINGTON	DE	LATITUDE:		39 DEGREES	40 MINUTES										
LATITUDE -15:		2.49	3.25	4.11	4.85	5.30	5.71	5.65	5.32	4.76	3.97	2.90	2.26	1539.3	4.2
LATITUDE :		2.76	3.43	4.14	4.65	4.90	5.21	5.22	5.11	4.82	4.25	3.23	2.55	1530.0	4.2
LATITUDE +15:		2.87	3.44	3.97	4.25	4.32	4.51	4.59	4.66	4.63	4.29	3.39	2.70	1449.3	4.0
APALACHICOLA	FL	LATITUDE:		29 DEGREES	44 MINUTES										
LATITUDE -15:		3.16	3.98	4.94	6.02	6.45	6.08	5.60	5.39	5.16	4.93	3.94	3.12	1788.9	4.9
LATITUDE :		3.49	4.22	5.01	5.81	5.98	5.56	5.21	5.21	5.24	5.31	4.42	3.54	1795.8	4.9
LATITUDE +15:		3.65	4.25	4.84	5.32	5.26	4.83	4.61	4.78	5.05	5.40	4.67	3.77	1716.4	4.7
DAYTONA BEACH	FL	LATITUDE:		29 DEGREES	11 MINUTES										
LATITUDE -15:		3.56	4.29	5.19	6.02	6.07	5.57	5.51	5.36	4.94	4.45	3.89	3.32	1769.5	4.8
LATITUDE :		3.96	4.56	5.28	5.81	5.63	5.11	5.12	5.18	5.00	4.75	4.35	3.77	1780.2	4.9
LATITUDE +15:		4.15	4.60	5.09	5.32	4.96	4.46	4.53	4.75	4.81	4.80	4.58	4.02	1705.4	4.7
JACKSONVILLE	FL	LATITUDE:		30 DEGREES	30 MINUTES										
LATITUDE -15:		3.40	4.16	5.14	5.95	6.03	5.73	5.56	5.42	4.85	4.40	3.81	3.18	1753.7	4.8
LATITUDE :		3.77	4.42	5.22	5.74	5.60	5.25	5.17	5.23	4.91	4.71	4.27	3.61	1761.6	4.8
LATITUDE +15:		3.95	4.46	5.03	5.26	4.94	4.57	4.57	4.80	4.73	4.76	4.50	3.85	1685.2	4.6

SITE ARRAY TILT		JAN	FEB	MAR	APR	MAY	JUN	JUL	AUG	SEP	OCT	NOV	DEC	ANNUAL TOTAL (KWH/SQ. M)	AVERAGE DAY (KWH/SQ. M)
MIAMI	FL	LATITUDE:		25 DEGREES		48 MINUTES									
LATITUDE -15:		3.74	4.50	5.27	5.90	5.71	5.25	5.46	5.17	4.76	4.46	3.99	3.69	1761.3	4.8
LATITUDE :		4.17	4.80	5.36	5.70	5.33	4.84	5.09	5.00	4.83	4.76	4.47	4.21	1781.2	4.9
LATITUDE +15:		4.38	4.85	5.18	5.24	4.72	4.24	4.52	4.60	4.65	4.82	4.71	4.50	1715.6	4.7
ORLANDO	FL	LATITUDE:		28 DEGREES		33 MINUTES									
LATITUDE -15:		3.68	4.37	5.29	6.06	6.14	5.60	5.56	5.33	4.99	4.61	4.08	3.50	1801.8	4.9
LATITUDE :		4.10	4.65	5.38	5.86	5.71	5.14	5.18	5.15	5.06	4.94	4.59	3.99	1817.2	5.0
LATITUDE +15:		4.31	4.70	5.19	5.37	5.03	4.49	4.58	4.73	4.87	5.01	4.85	4.26	1744.9	4.8
TALLAHASSEE	FL	LATITUDE:		30 DEGREES		23 MINUTES									
LATITUDE -15:		3.30	4.05	4.98	5.84	5.98	5.73	5.39	5.36	5.03	4.78	3.86	3.16	1748.3	4.8
LATITUDE :		3.66	4.31	5.06	5.63	5.55	5.25	5.02	5.17	5.10	5.13	4.33	3.58	1757.6	4.8
LATITUDE +15:		3.82	4.34	4.87	5.16	4.89	4.57	4.44	4.74	4.91	5.21	4.56	3.81	1682.6	4.6
TAMPA	FL	LATITUDE:		27 DEGREES		58 MINUTES									
LATITUDE -15:		3.68	4.39	5.30	6.09	6.18	5.65	5.42	5.26	4.95	4.73	4.08	3.49	1802.0	4.9
LATITUDE :		4.10	4.68	5.40	5.88	5.74	5.19	5.05	5.09	5.02	5.07	4.59	3.98	1819.3	5.0
LATITUDE +15:		4.31	4.73	5.22	5.40	5.07	4.53	4.48	4.67	4.84	5.15	4.85	4.25	1748.7	4.8
ATLANTA	GA	LATITUDE:		33 DEGREES		39 MINUTES									
LATITUDE -15:		2.80	3.55	4.47	5.43	5.72	5.81	5.59	5.51	4.89	4.50	3.55	2.76	1661.6	4.6
LATITUDE :		3.08	3.75	4.52	5.23	5.31	5.31	5.19	5.31	4.96	4.82	3.98	3.12	1661.3	4.6
LATITUDE +15:		3.21	3.76	4.34	4.79	4.68	4.62	4.58	4.86	4.77	4.89	4.18	3.32	1582.5	4.3
AUGUSTA	GA	LATITUDE:		33 DEGREES		22 MINUTES									
LATITUDE -15:		2.94	3.72	4.59	5.58	5.75	5.78	5.55	5.37	4.84	4.57	3.69	2.97	1684.7	4.6
LATITUDE :		3.25	3.94	4.64	5.37	5.34	5.29	5.15	5.17	4.89	4.90	4.14	3.37	1687.5	4.6
LATITUDE +15:		3.38	3.96	4.46	4.92	4.71	4.59	4.55	4.74	4.71	4.97	4.36	3.58	1610.3	4.4
MACON	GA	LATITUDE:		32 DEGREES		42 MINUTES									
LATITUDE -15:		2.97	3.70	4.65	5.60	5.81	5.83	5.50	5.53	4.92	4.63	3.74	2.94	1699.3	4.7
LATITUDE :		3.28	3.92	4.71	5.39	5.39	5.34	5.11	5.33	4.99	4.97	4.20	3.34	1703.3	4.7
LATITUDE +15:		3.42	3.94	4.53	4.94	4.76	4.64	4.51	4.88	4.80	5.05	4.42	3.55	1626.3	4.5
SAVANNAH	GA	LATITUDE:		32 DEGREES		8 MINUTES									
LATITUDE -15:		3.06	3.77	4.75	5.66	5.72	5.60	5.49	5.21	4.63	4.48	3.70	3.02	1677.2	4.6
LATITUDE :		3.38	3.99	4.81	5.46	5.31	5.13	5.10	5.02	4.68	4.79	4.15	3.43	1681.1	4.6
LATITUDE +15:		3.52	4.01	4.63	5.00	4.68	4.47	4.50	4.59	4.49	4.85	4.36	3.65	1605.4	4.4

SITE ARRAY TILT		JAN	FEB	MAR	APR	MAY	JUN	JUL	AUG	SEP	OCT	NOV	DEC	ANNUAL TOTAL (KWH/SQ. M)	AVERAGE DAY (KWH/SQ. M)
BARBERS POINT	HI	LATITUDE:	21 DEGREES	19 MINUTES											
LATITUDE -15:		4.06	4.75	5.29	5.79	6.13	6.26	6.28	6.22	5.85	5.15	4.39	3.97	1951.9	5.3
LATITUDE :		4.54	5.07	5.38	5.61	5.72	5.74	5.85	6.02	5.96	5.53	4.94	4.54	1974.8	5.4
LATITUDE +15:		4.78	5.14	5.21	5.16	5.05	4.98	5.15	5.53	5.77	5.63	5.23	4.86	1900.3	5.2
HILO	HI	LATITUDE:	19 DEGREES	43 MINUTES											
LATITUDE -15:		3.69	4.04	4.29	4.52	4.85	5.17	5.07	5.02	4.94	4.46	3.64	3.37	1614.9	4.4
LATITUDE :		4.09	4.29	4.35	4.39	4.56	4.79	4.76	4.88	5.01	4.76	4.03	3.81	1634.2	4.5
LATITUDE +15:		4.29	4.32	4.21	4.06	4.08	4.21	4.25	4.50	4.85	4.81	4.23	4.04	1577.0	4.3
HONOLULU	HI	LATITUDE:	21 DEGREES	20 MINUTES											
LATITUDE -15:		3.97	4.60	5.22	5.67	6.07	6.21	6.22	6.21	5.84	5.10	4.27	3.85	1924.4	5.3
LATITUDE :		4.43	4.90	5.31	5.49	5.66	5.70	5.80	6.01	5.95	5.48	4.80	4.39	1944.7	5.3
LATITUDE +15:		4.65	4.96	5.13	5.05	5.00	4.95	5.11	5.52	5.76	5.57	5.07	4.69	1869.8	5.1
BURLINGTON	IA	LATITUDE:	40 DEGREES	47 MINUTES											
LATITUDE -15:		2.63	3.46	4.21	5.07	5.82	6.42	6.48	6.09	5.23	4.45	3.11	2.31	1684.2	4.6
LATITUDE :		2.92	3.67	4.26	4.86	5.38	5.84	5.98	5.86	5.32	4.79	3.49	2.62	1674.1	4.6
LATITUDE +15:		3.06	3.68	4.08	4.44	4.73	5.03	5.23	5.35	5.12	4.86	3.68	2.78	1584.1	4.3
DES MOINES	IA	LATITUDE:	41 DEGREES	32 MINUTES											
LATITUDE -15:		2.71	3.52	4.31	5.16	5.80	6.43	6.51	6.12	5.36	4.57	3.16	2.42	1707.7	4.7
LATITUDE :		3.02	3.74	4.35	4.94	5.35	5.84	6.00	5.88	5.45	4.92	3.55	2.75	1699.3	4.7
LATITUDE +15:		3.16	3.76	4.18	4.51	4.70	5.04	5.25	5.36	5.25	5.00	3.74	2.93	1609.3	4.4
MASON CITY	IA	LATITUDE:	43 DEGREES	9 MINUTES											
LATITUDE -15:		2.71	3.53	4.34	5.06	5.90	6.40	6.49	6.19	5.34	4.46	3.00	2.31	1697.1	4.6
LATITUDE :		3.02	3.75	4.39	4.84	5.44	5.80	5.97	5.95	5.43	4.80	3.37	2.62	1685.9	4.6
LATITUDE +15:		3.17	3.77	4.21	4.41	4.76	5.00	5.22	5.42	5.23	4.86	3.54	2.79	1593.7	4.4
SIOUX CITY	IA	LATITUDE:	42 DEGREES	24 MINUTES											
LATITUDE -15:		2.73	3.49	4.31	5.24	5.91	6.44	6.60	6.20	5.36	4.51	3.17	2.40	1716.6	4.7
LATITUDE :		3.04	3.71	4.35	5.03	5.45	5.84	6.08	5.96	5.45	4.86	3.57	2.73	1707.0	4.7
LATITUDE +15:		3.18	3.72	4.17	4.58	4.78	5.03	5.31	5.43	5.25	4.93	3.76	2.90	1615.2	4.4
BOISE	ID	LATITUDE:	43 DEGREES	34 MINUTES											
LATITUDE -15:		2.33	3.57	4.95	6.18	7.13	7.48	8.20	7.55	6.87	5.20	3.23	2.30	1980.0	5.4
LATITUDE :		2.58	3.80	5.03	5.94	6.56	6.76	7.53	7.28	7.06	5.65	3.64	2.62	1962.5	5.4
LATITUDE +15:		2.69	3.82	4.84	5.41	5.73	5.79	6.53	6.63	6.84	5.77	3.84	2.78	1847.2	5.1

SITE / ARRAY TILT	JAN	FEB	MAR	APR	MAY	JUN	JUL	AUG	SEP	OCT	NOV	DEC	ANNUAL TOTAL (KWH/SQ. M)	AVERAGE DAY (KWH/SQ. M)
LEWISTON ID														
LATITUDE: 46 DEGREES 23 MINUTES														
LATITUDE -15:	1.62	2.59	3.88	4.84	5.77	6.11	7.36	6.70	5.76	3.98	2.13	1.50	1593.0	4.4
LATITUDE :	1.76	2.71	3.91	4.63	5.31	5.54	6.76	6.44	5.87	4.27	2.37	1.68	1562.4	4.3
LATITUDE +15:	1.81	2.70	3.74	4.21	4.65	4.77	5.88	5.86	5.66	4.33	2.47	1.76	1458.1	4.0
POCATELLO ID														
LATITUDE: 42 DEGREES 55 MINUTES														
LATITUDE -15:	2.59	3.73	5.19	6.14	7.13	7.53	8.15	7.67	6.95	5.47	3.54	2.49	2028.8	5.6
LATITUDE :	2.89	3.97	5.28	5.90	6.57	6.81	7.49	7.40	7.14	5.95	4.01	2.85	2018.7	5.5
LATITUDE +15:	3.02	4.00	5.09	5.38	5.74	5.84	6.51	6.75	6.93	6.09	4.25	3.04	1907.2	5.2
CHICAGO IL														
LATITUDE: 41 DEGREES 47 MINUTES														
LATITUDE -15:	2.30	3.05	4.03	4.82	5.55	6.09	6.04	5.74	5.03	4.09	2.65	1.91	1562.9	4.3
LATITUDE :	2.54	3.22	4.06	4.62	5.13	5.54	5.57	5.51	5.11	4.39	2.96	2.14	1546.6	4.2
LATITUDE +15:	2.64	3.22	3.89	4.22	4.51	4.78	4.88	5.03	4.91	4.44	3.10	2.26	1457.9	4.0
SPRINGFIELD IL														
LATITUDE: 39 DEGREES 50 MINUTES														
LATITUDE -15:	2.58	3.41	4.08	4.97	5.78	6.35	6.38	5.98	5.33	4.40	3.09	2.28	1663.6	4.6
LATITUDE :	2.86	3.61	4.12	4.77	5.34	5.78	5.88	5.75	5.42	4.73	3.46	2.58	1653.3	4.5
LATITUDE +15:	2.98	3.63	3.95	4.36	4.70	4.99	5.15	5.25	5.22	4.80	3.63	2.74	1564.5	4.3
FORT WAYNE IN														
LATITUDE: 41 DEGREES 0 MINUTES														
LATITUDE -15:	1.95	2.73	3.50	4.46	5.18	5.57	5.53	5.27	4.65	3.80	2.31	1.65	1419.2	3.9
LATITUDE :	2.13	2.85	3.51	4.27	4.78	5.07	5.10	5.05	4.70	4.05	2.54	1.83	1397.2	3.8
LATITUDE +15:	2.19	2.84	3.35	3.89	4.21	4.40	4.48	4.60	4.51	4.09	2.64	1.91	1311.9	3.6
INDIANAPOLIS IN														
LATITUDE: 39 DEGREES 44 MINUTES														
LATITUDE -15:	2.09	2.88	3.67	4.58	5.23	5.66	5.59	5.42	4.79	3.95	2.55	1.84	1469.6	4.0
LATITUDE :	2.28	3.02	3.69	4.39	4.84	5.17	5.17	5.20	4.85	4.22	2.82	2.05	1452.7	4.0
LATITUDE +15:	2.36	3.02	3.53	4.01	4.27	4.48	4.54	4.75	4.66	4.26	2.95	2.15	1369.2	3.8
SOUTH BEND IN														
LATITUDE: 41 DEGREES 42 MINUTES														
LATITUDE -15:	1.77	2.57	3.56	4.57	5.34	5.83	5.75	5.55	4.76	3.78	2.25	1.52	1439.3	3.9
LATITUDE :	1.92	2.69	3.57	4.38	4.93	5.30	5.30	5.33	4.82	4.03	2.48	1.67	1414.3	3.9
LATITUDE +15:	1.97	2.67	3.42	4.00	4.34	4.59	4.66	4.86	4.63	4.07	2.58	1.73	1324.9	3.6
DODGE CITY KS														
LATITUDE: 37 DEGREES 46 MINUTES														
LATITUDE -15:	3.67	4.46	5.31	6.20	6.47	7.14	7.11	6.79	6.14	5.31	4.05	3.47	2012.6	5.5
LATITUDE :	4.15	4.78	5.40	5.97	5.98	6.48	6.55	6.54	6.27	5.76	4.60	4.01	2023.5	5.5
LATITUDE +15:	4.38	4.84	5.21	5.45	5.24	5.57	5.72	5.97	6.07	5.88	4.88	4.31	1933.0	5.3

SITE ARRAY TILT		JAN	FEB	MAR	APR	MAY	JUN	JUL	AUG	SEP	OCT	NOV	DEC	ANNUAL TOTAL (KWH/SQ. M)	AVERAGE DAY (KWH/SQ. M)
GOODLAND	KS	LATITUDE:	39 DEGREES	22 MINUTES											
LATITUDE -15:		3.67	4.28	5.19	6.05	6.39	7.13	7.19	6.81	6.07	5.34	4.07	3.47	1999.4	5.5
LATITUDE :		4.15	4.58	5.27	5.82	5.90	6.46	6.62	6.56	6.20	5.80	4.63	4.01	2008.8	5.5
LATITUDE +15:		4.38	4.63	5.08	5.31	5.17	5.54	5.77	5.98	5.98	5.92	4.92	4.32	1917.5	5.3
TOPEKA	KS	LATITUDE:	39 DEGREES	4 MINUTES											
LATITUDE -15:		3.03	3.72	4.50	5.39	5.93	6.44	6.59	6.32	5.54	4.72	3.55	2.76	1781.4	4.9
LATITUDE :		3.38	3.95	4.55	5.18	5.48	5.85	6.08	6.08	5.64	5.08	4.00	3.15	1778.0	4.9
LATITUDE +15:		3.55	3.98	4.37	4.73	4.81	5.04	5.31	5.54	5.43	5.16	4.22	3.36	1688.6	4.6
WICHITA	KS	LATITUDE:	37 DEGREES	39 MINUTES											
LATITUDE -15:		3.43	4.15	5.01	5.84	6.30	6.86	6.93	6.71	5.84	5.06	3.92	3.21	1926.1	5.3
LATITUDE :		3.85	4.43	5.09	5.62	5.82	6.23	6.39	6.47	5.96	5.47	4.44	3.69	1931.4	5.3
LATITUDE +15:		4.05	4.48	4.90	5.14	5.11	5.36	5.58	5.90	5.75	5.57	4.70	3.95	1841.3	5.0
LEXINGTON	KY	LATITUDE:	38 DEGREES	2 MINUTES											
LATITUDE -15:		2.25	2.95	3.85	4.81	5.39	5.74	5.72	5.52	4.85	4.15	2.83	2.12	1528.7	4.2
LATITUDE :		2.46	3.10	3.88	4.62	4.99	5.23	5.28	5.30	4.91	4.43	3.14	2.37	1513.9	4.1
LATITUDE +15:		2.54	3.09	3.71	4.22	4.40	4.54	4.64	4.84	4.72	4.47	3.29	2.50	1429.3	3.9
LOUISVILLE	KY	LATITUDE:	38 DEGREES	11 MINUTES											
LATITUDE -15:		2.26	3.00	3.87	4.78	5.31	5.76	5.67	5.51	4.86	4.14	2.82	2.14	1526.7	4.2
LATITUDE :		2.47	3.15	3.89	4.59	4.92	5.26	5.24	5.29	4.92	4.42	3.13	2.40	1512.8	4.1
LATITUDE +15:		2.55	3.14	3.73	4.19	4.34	4.56	4.61	4.83	4.73	4.47	3.27	2.53	1429.0	3.9
BATON ROUGE	LA	LATITUDE:	30 DEGREES	32 MINUTES											
LATITUDE -15:		2.92	3.74	4.64	5.38	5.77	5.86	5.38	5.37	4.93	4.71	3.48	2.83	1675.1	4.6
LATITUDE :		3.22	3.96	4.70	5.19	5.36	5.37	5.01	5.18	5.00	5.06	3.88	3.19	1677.2	4.6
LATITUDE +15:		3.35	3.98	4.53	4.76	4.74	4.67	4.43	4.75	4.81	5.13	4.07	3.39	1600.1	4.4
LAKE CHARLES	LA	LATITUDE:	30 DEGREES	7 MINUTES											
LATITUDE -15:		2.67	3.56	4.39	5.01	5.71	5.99	5.51	5.29	4.99	5.01	3.45	2.67	1651.9	4.5
LATITUDE :		2.91	3.75	4.44	4.83	5.31	5.48	5.12	5.11	5.06	5.39	3.84	3.00	1650.6	4.5
LATITUDE +15:		3.01	3.76	4.27	4.43	4.69	4.75	4.53	4.68	4.87	5.47	4.02	3.16	1571.8	4.3
NEW ORLEANS	LA	LATITUDE:	29 DEGREES	59 MINUTES											
LATITUDE -15:		3.09	3.93	4.74	5.70	6.08	6.11	5.60	5.49	5.09	4.81	3.66	2.98	1743.3	4.8
LATITUDE :		3.41	4.17	4.81	5.50	5.65	5.59	5.21	5.30	5.17	5.17	4.09	3.37	1747.9	4.8
LATITUDE +15:		3.56	4.20	4.63	5.05	4.98	4.85	4.61	4.86	4.99	5.25	4.30	3.58	1669.2	4.6

SITE ARRAY TILT		JAN	FEB	MAR	APR	MAY	JUN	JUL	AUG	SEP	OCT	NOV	DEC	ANNUAL TOTAL (KWH/SQ. M)	AVERAGE DAY (KWH/SQ. M)
SHREVEPORT	LA	LATITUDE:	32 DEGREES	28 MINUTES											
LATITUDE -15:		2.93	3.77	4.56	5.18	5.82	6.27	6.20	6.05	5.33	4.85	3.67	2.94	1753.2	4.8
LATITUDE :		3.23	3.99	4.62	4.99	5.40	5.72	5.74	5.84	5.42	5.22	4.11	3.33	1753.8	4.8
LATITUDE +15:		3.37	4.01	4.44	4.57	4.76	4.95	5.05	5.34	5.22	5.30	4.32	3.54	1670.6	4.6
BOSTON	MA	LATITUDE:	42 DEGREES	22 MINUTES											
LATITUDE -15:		2.17	2.86	3.69	4.37	5.02	5.50	5.43	4.93	4.67	3.74	2.34	1.97	1421.2	3.9
LATITUDE :		2.38	3.00	3.71	4.18	4.64	5.01	5.01	4.73	4.72	3.99	2.59	2.21	1405.3	3.9
LATITUDE +15:		2.47	2.99	3.54	3.81	4.08	4.34	4.40	4.31	4.53	4.02	2.70	2.33	1325.0	3.6
BALTIMORE	MD	LATITUDE:	39 DEGREES	11 MINUTES											
LATITUDE -15:		2.55	3.28	4.13	4.87	5.29	5.69	5.63	5.25	4.79	4.01	2.94	2.29	1544.2	4.2
LATITUDE :		2.81	3.47	4.17	4.67	4.90	5.19	5.20	5.04	4.84	4.28	3.28	2.58	1534.8	4.2
LATITUDE +15:		2.93	3.47	3.99	4.26	4.31	4.50	4.57	4.60	4.65	4.32	3.44	2.73	1453.9	4.0
PATUXENT RIVER	MD	LATITUDE:	38 DEGREES	17 MINUTES											
LATITUDE -15:		2.59	3.33	4.17	5.02	5.45	5.73	5.61	5.33	4.85	4.05	3.10	2.42	1572.7	4.3
LATITUDE :		2.86	3.52	4.21	4.82	5.04	5.23	5.19	5.12	4.91	4.32	3.47	2.74	1565.0	4.3
LATITUDE +15:		2.98	3.53	4.04	4.40	4.44	4.53	4.56	4.67	4.72	4.36	3.64	2.91	1484.2	4.1
BANGOR	ME	LATITUDE:	44 DEGREES	48 MINUTES											
LATITUDE -15:		2.25	3.09	4.11	4.83	5.39	5.63	5.80	5.45	4.80	3.69	2.36	2.02	1504.8	4.1
LATITUDE :		2.49	3.27	4.15	4.62	4.97	5.12	5.34	5.22	4.86	3.95	2.63	2.29	1489.3	4.1
LATITUDE +15:		2.60	3.27	3.98	4.21	4.36	4.43	4.68	4.76	4.67	3.98	2.75	2.43	1404.1	3.8
CARIBOU	ME	LATITUDE:	46 DEGREES	52 MINUTES											
LATITUDE -15:		2.21	3.24	4.40	4.78	4.93	5.33	5.51	5.10	4.26	3.07	1.85	1.74	1413.0	3.9
LATITUDE :		2.46	3.44	4.46	4.58	4.54	4.84	5.07	4.88	4.30	3.27	2.03	1.97	1394.9	3.8
LATITUDE +15:		2.57	3.46	4.28	4.16	3.99	4.19	4.44	4.44	4.12	3.28	2.11	2.08	1312.1	3.6
PORTLAND	ME	LATITUDE:	43 DEGREES	39 MINUTES											
LATITUDE -15:		2.12	2.79	3.54	4.31	4.86	5.18	5.15	4.87	4.32	3.51	2.18	1.81	1359.8	3.7
LATITUDE :		2.33	2.93	3.56	4.13	4.49	4.72	4.75	4.67	4.36	3.74	2.41	2.03	1343.3	3.7
LATITUDE +15:		2.43	2.93	3.40	3.76	3.95	4.10	4.17	4.25	4.18	3.77	2.51	2.14	1265.6	3.5
ALPENA	MI	LATITUDE:	45 DEGREES	4 MINUTES											
LATITUDE -15:		1.67	2.55	3.85	4.71	5.36	5.69	5.87	5.35	4.39	3.23	1.81	1.49	1400.8	3.8
LATITUDE :		1.82	2.66	3.88	4.50	4.94	5.16	5.40	5.12	4.43	3.42	1.98	1.70	1371.6	3.8
LATITUDE +15:		1.87	2.65	3.71	4.10	4.33	4.46	4.73	4.66	4.24	3.44	2.04	1.81	1280.1	3.5

SITE ARRAY TILT		JAN	FEB	MAR	APR	MAY	JUN	JUL	AUG	SEP	OCT	NOV	DEC	ANNUAL TOTAL (KWH/SQ. M)	AVERAGE DAY (KWH/SQ. M)
DETROIT	MI	LATITUDE:		42 DEGREES		25 MINUTES									
LATITUDE -15:		1.83	2.71	3.61	4.63	5.32	5.66	5.70	5.24	4.65	3.69	2.19	1.58	1425.7	3.9
LATITUDE :		1.99	2.84	3.63	4.43	4.92	5.15	5.26	5.02	4.70	3.93	2.42	1.75	1401.6	3.8
LATITUDE +15:		2.05	2.83	3.47	4.04	4.32	4.46	4.61	4.58	4.51	3.96	2.51	1.83	1313.9	3.6
FLINT	MI	LATITUDE:		42 DEGREES		58 MINUTES									
LATITUDE -15:		1.65	2.53	3.46	4.42	5.15	5.49	5.59	5.19	4.43	3.49	1.94	1.60	1369.4	3.8
LATITUDE :		1.79	2.64	3.47	4.24	4.76	5.00	5.16	4.98	4.48	3.72	2.13	1.82	1345.7	3.7
LATITUDE +15:		1.83	2.62	3.31	3.86	4.19	4.34	4.53	4.54	4.30	3.75	2.20	1.94	1261.1	3.5
GRAND RAPIDS	MI	LATITUDE:		42 DEGREES		53 MINUTES									
LATITUDE -15:		1.78	2.57	3.69	4.68	5.46	5.93	5.95	5.62	4.71	3.63	2.03	1.60	1452.3	4.0
LATITUDE :		1.98	2.69	3.71	4.48	5.04	5.40	5.49	5.39	4.77	3.87	2.23	1.83	1428.6	3.9
LATITUDE +15:		2.08	2.68	3.55	4.09	4.43	4.67	4.81	4.92	4.58	3.90	2.32	1.95	1339.2	3.7
HOUGHTON	MI	LATITUDE:		47 DEGREES		10 MINUTES									
LATITUDE -15:		1.23	1.96	3.54	4.61	5.18	5.56	5.76	5.19	3.88	3.01	1.58	1.05	1297.0	3.6
LATITUDE :		1.35	2.02	3.56	4.40	4.77	5.04	5.29	4.96	3.90	3.19	1.76	1.17	1262.7	3.5
LATITUDE +15:		1.40	1.98	3.40	4.00	4.18	4.36	4.62	4.51	3.73	3.20	1.84	1.23	1172.0	3.2
SAULT STE. MARIE	MI	LATITUDE:		46 DEGREES		28 MINUTES									
LATITUDE -15:		1.52	2.56	3.93	4.66	5.26	5.49	5.74	5.16	4.00	2.96	1.58	1.48	1351.2	3.7
LATITUDE :		1.64	2.69	3.96	4.45	4.85	4.98	5.28	4.94	4.03	3.14	1.72	1.69	1321.4	3.6
LATITUDE +15:		1.69	2.68	3.79	4.05	4.25	4.31	4.62	4.49	3.86	3.15	1.76	1.80	1231.9	3.4
TRAVERSE CITY	MI	LATITUDE:		44 DEGREES		44 MINUTES									
LATITUDE -15:		1.53	2.27	3.72	4.69	5.39	5.80	5.96	5.44	4.40	3.25	1.74	1.37	1388.7	3.8
LATITUDE :		1.70	2.36	3.74	4.49	4.97	5.27	5.49	5.21	4.45	3.46	1.90	1.55	1358.7	3.7
LATITUDE +15:		1.77	2.33	3.58	4.09	4.36	4.55	4.81	4.75	4.26	3.47	1.96	1.64	1267.1	3.5
DULUTH	MN	LATITUDE:		46 DEGREES		50 MINUTES									
LATITUDE -15:		1.99	2.97	3.97	4.63	5.13	5.35	5.81	5.27	4.23	3.27	1.95	1.59	1405.7	3.9
LATITUDE :		2.20	3.14	4.00	4.43	4.73	4.86	5.34	5.05	4.27	3.49	2.15	1.79	1383.8	3.8
LATITUDE +15:		2.29	3.14	3.83	4.03	4.15	4.21	4.68	4.59	4.09	3.51	2.24	1.89	1298.1	3.6
INTERNATIONAL FALL	MN	LATITUDE:		48 DEGREES		34 MINUTES									
LATITUDE -15:		1.93	3.05	4.13	4.94	5.38	5.62	6.04	5.60	4.45	3.32	1.86	1.63	1460.5	4.0
LATITUDE :		2.13	3.24	4.17	4.73	4.95	5.10	5.55	5.36	4.51	3.55	2.06	1.84	1436.4	3.9
LATITUDE +15:		2.22	3.25	3.99	4.29	4.33	4.40	4.85	4.88	4.32	3.57	2.14	1.95	1345.2	3.7

SITE ARRAY TILT	JAN	FEB	MAR	APR	MAY	JUN	JUL	AUG	SEP	OCT	NOV	DEC	ANNUAL TOTAL (KWH/SQ. M)	AVERAGE DAY (KWH/SQ. M)
MINNEAPOLIS-ST.PAUL MN			LATITUDE:	44 DEGREES	53 MINUTES									
LATITUDE -15:	2.31	3.29	4.16	4.84	5.41	5.84	6.16	5.73	4.81	3.81	2.43	1.85	1541.8	4.2
LATITUDE :	2.57	3.49	4.20	4.63	4.99	5.31	5.67	5.50	4.87	4.08	2.71	2.09	1525.4	4.2
LATITUDE +15:	2.69	3.50	4.03	4.22	4.38	4.59	4.96	5.01	4.68	4.12	2.83	2.21	1437.2	3.9
ROCHESTER MN			LATITUDE:	43 DEGREES	55 MINUTES									
LATITUDE -15:	2.31	3.16	4.02	4.70	5.28	5.76	5.95	5.60	4.72	3.77	2.42	1.88	1510.0	4.1
LATITUDE :	2.56	3.35	4.06	4.50	4.87	5.24	5.49	5.38	4.79	4.03	2.70	2.12	1494.0	4.1
LATITUDE +15:	2.67	3.36	3.89	4.10	4.28	4.54	4.81	4.90	4.60	4.07	2.82	2.24	1408.3	3.9
COLUMBIA MO			LATITUDE:	38 DEGREES	49 MINUTES									
LATITUDE -15:	2.64	3.41	4.19	4.99	5.82	6.33	6.55	6.21	5.24	4.47	3.13	2.38	1686.8	4.6
LATITUDE :	2.93	3.61	4.23	4.79	5.39	5.76	6.05	5.98	5.33	4.80	3.50	2.69	1676.9	4.6
LATITUDE +15:	3.05	3.62	4.06	4.38	4.74	4.98	5.30	5.46	5.13	4.87	3.68	2.86	1587.2	4.3
KANSAS CITY MO			LATITUDE:	39 DEGREES	18 MINUTES									
LATITUDE -15:	2.88	3.53	4.30	5.17	5.80	6.30	6.51	6.16	5.29	4.48	3.37	2.66	1718.9	4.7
LATITUDE :	3.21	3.74	4.34	4.96	5.36	5.72	6.00	5.93	5.37	4.81	3.78	3.02	1712.9	4.7
LATITUDE +15:	3.36	3.76	4.17	4.53	4.71	4.94	5.25	5.41	5.17	4.88	3.99	3.22	1624.9	4.5
SPRINGFIELD MO			LATITUDE:	37 DEGREES	14 MINUTES									
LATITUDE -15:	2.89	3.55	4.34	5.23	5.82	6.28	6.37	6.14	5.28	4.53	3.36	2.70	1720.5	4.7
LATITUDE :	3.21	3.76	4.38	5.02	5.38	5.72	5.88	5.91	5.36	4.87	3.76	3.06	1714.8	4.7
LATITUDE +15:	3.36	3.77	4.21	4.59	4.73	4.94	5.15	5.40	5.16	4.94	3.95	3.26	1627.2	4.5
ST. LOUIS MO			LATITUDE:	38 DEGREES	45 MINUTES									
LATITUDE -15:	2.72	3.46	4.28	5.12	5.79	6.34	6.34	5.99	5.28	4.45	3.20	2.42	1687.1	4.6
LATITUDE :	3.02	3.66	4.33	4.92	5.36	5.77	5.85	5.76	5.37	4.79	3.59	2.74	1679.1	4.6
LATITUDE +15:	3.16	3.68	4.15	4.50	4.71	4.99	5.13	5.26	5.17	4.85	3.77	2.91	1591.1	4.4
JACKSON MS			LATITUDE:	32 DEGREES	19 MINUTES									
LATITUDE -15:	2.88	3.71	4.66	5.49	5.98	6.14	5.88	5.73	5.16	4.72	3.54	2.83	1726.7	4.7
LATITUDE :	3.17	3.93	4.72	5.29	5.55	5.61	5.45	5.52	5.23	5.06	3.95	3.20	1725.1	4.7
LATITUDE +15:	3.30	3.94	4.54	4.84	4.89	4.86	4.80	5.05	5.04	5.14	4.14	3.39	1641.5	4.5
MERIDIAN MS			LATITUDE:	32 DEGREES	20 MINUTES									
LATITUDE -15:	2.84	3.66	4.51	5.33	5.73	5.96	5.62	5.60	4.96	4.66	3.51	2.79	1679.7	4.6
LATITUDE :	3.13	3.88	4.57	5.14	5.32	5.45	5.21	5.39	5.03	5.00	3.92	3.15	1679.2	4.6
LATITUDE +15:	3.25	3.89	4.39	4.71	4.69	4.73	4.60	4.94	4.84	5.07	4.11	3.34	1599.3	4.4

SITE ARRAY TILT	JAN	FEB	MAR	APR	MAY	JUN	JUL	AUG	SEP	OCT	NOV	DEC	ANNUAL TOTAL (KWH/SQ. M)	AVERAGE DAY (KWH/SQ. M)
BILLINGS MT LATITUDE: 45 DEGREES 48 MINUTES														
LATITUDE -15:	2.55	3.36	4.59	5.15	5.99	6.61	7.51	7.02	5.87	4.65	3.08	2.45	1792.7	4.9
LATITUDE :	2.85	3.57	4.65	4.93	5.52	5.99	6.90	6.76	5.99	5.02	3.48	2.81	1781.6	4.9
LATITUDE +15:	2.99	3.59	4.47	4.49	4.83	5.15	6.01	6.16	5.79	5.11	3.67	3.00	1683.0	4.6
CUT BANK MT LATITUDE: 48 DEGREES 36 MINUTES														
LATITUDE -15:	2.28	3.20	4.50	5.10	5.93	6.22	7.25	6.68	5.57	4.36	2.89	2.17	1711.5	4.7
LATITUDE :	2.55	3.40	4.56	4.87	5.45	5.63	6.65	6.42	5.68	4.71	3.26	2.49	1696.8	4.6
LATITUDE +15:	2.67	3.42	4.38	4.43	4.77	4.84	5.79	5.85	5.47	4.78	3.44	2.66	1599.3	4.4
DILLON MT LATITUDE: 45 DEGREES 15 MINUTES														
LATITUDE -15:	2.76	3.75	4.96	5.56	6.23	6.50	7.53	6.99	6.05	4.79	3.30	2.59	1858.4	5.1
LATITUDE :	3.09	4.00	5.04	5.32	5.73	5.89	6.91	6.72	6.18	5.19	3.73	2.97	1850.8	5.1
LATITUDE +15:	3.25	4.03	4.85	4.85	5.01	5.06	6.01	6.12	5.97	5.28	3.94	3.18	1751.6	4.8
GLASGOW MT LATITUDE: 48 DEGREES 13 MINUTES														
LATITUDE -15:	2.13	3.08	4.37	5.09	5.73	6.22	6.93	6.53	5.48	4.34	2.82	2.12	1671.4	4.6
LATITUDE :	2.37	3.26	4.42	4.87	5.27	5.62	6.36	6.27	5.58	4.68	3.17	2.43	1654.7	4.5
LATITUDE +15:	2.48	3.27	4.24	4.42	4.61	4.84	5.54	5.71	5.38	4.75	3.34	2.59	1557.9	4.3
GREAT FALLS MT LATITUDE: 47 DEGREES 29 MINUTES														
LATITUDE -15:	2.28	3.28	4.62	5.07	5.80	6.39	7.36	6.76	5.60	4.52	2.85	2.02	1723.5	4.7
LATITUDE :	2.55	3.48	4.68	4.85	5.33	5.79	6.76	6.50	5.71	4.88	3.21	2.31	1707.5	4.7
LATITUDE +15:	2.66	3.50	4.50	4.41	4.67	4.97	5.88	5.92	5.50	4.96	3.38	2.46	1608.4	4.4
HELENA MT LATITUDE: 46 DEGREES 36 MINUTES														
LATITUDE -15:	2.18	3.13	4.44	5.04	5.82	6.20	7.36	6.70	5.67	4.41	2.90	2.13	1706.7	4.7
LATITUDE :	2.43	3.32	4.50	4.82	5.36	5.62	6.76	6.44	5.79	4.76	3.26	2.43	1690.7	4.6
LATITUDE +15:	2.53	3.33	4.32	4.39	4.69	4.84	5.88	5.87	5.58	4.83	3.44	2.59	1592.8	4.4
LEWISTOWN MT LATITUDE: 47 DEGREES 3 MINUTES														
LATITUDE -15:	2.24	3.09	4.40	4.90	5.66	6.25	7.22	6.62	5.52	4.35	2.83	2.19	1684.1	4.6
LATITUDE :	2.49	3.27	4.45	4.68	5.20	5.66	6.62	6.35	5.62	4.69	3.18	2.50	1667.2	4.6
LATITUDE +15:	2.60	3.28	4.27	4.26	4.55	4.86	5.76	5.78	5.41	4.76	3.35	2.66	1569.7	4.3
MILES CITY MT LATITUDE: 46 DEGREES 26 MINUTES														
LATITUDE -15:	2.43	3.32	4.61	5.23	5.94	6.52	7.22	6.88	5.80	4.59	3.10	2.38	1767.4	4.8
LATITUDE :	2.71	3.53	4.67	5.01	5.47	5.90	6.63	6.61	5.92	4.96	3.51	2.72	1755.3	4.8
LATITUDE +15:	2.84	3.54	4.48	4.56	4.78	5.07	5.77	6.02	5.71	5.04	3.70	2.91	1657.2	4.5

SITE / ARRAY TILT	JAN	FEB	MAR	APR	MAY	JUN	JUL	AUG	SEP	OCT	NOV	DEC	ANNUAL TOTAL (KWH/SQ. M)	AVERAGE DAY (KWH/SQ. M)
MISSOULA — MT — LATITUDE: 46 DEGREES 55 MINUTES														
LATITUDE −15:	1.46	2.43	3.74	4.67	5.33	5.87	7.35	6.54	5.43	3.77	2.16	1.41	1537.4	4.2
LATITUDE:	1.58	2.55	3.77	4.47	5.14	5.33	6.76	6.28	5.54	4.04	2.40	1.57	1506.6	4.1
LATITUDE +15:	1.62	2.53	3.60	4.06	4.51	4.60	5.88	5.72	5.34	4.08	2.51	1.64	1405.2	3.8
ASHEVILLE — NC — LATITUDE: 35 DEGREES 26 MINUTES														
LATITUDE −15:	2.93	3.65	4.54	5.41	5.57	5.63	5.47	5.27	4.73	4.40	3.56	2.82	1643.1	4.5
LATITUDE:	3.25	3.86	4.59	5.20	5.16	5.15	5.07	5.07	4.79	4.72	3.99	3.20	1645.1	4.5
LATITUDE +15:	3.40	3.88	4.42	4.76	4.55	4.48	4.47	4.64	4.60	4.78	4.21	3.41	1569.4	4.3
CAPE HATTERAS — NC — LATITUDE: 35 DEGREES 16 MINUTES														
LATITUDE −15:	2.75	3.55	4.61	5.76	6.05	6.17	5.92	5.53	5.14	4.35	3.66	2.82	1714.8	4.7
LATITUDE:	3.04	3.76	4.66	5.55	5.60	5.62	5.48	5.32	5.22	4.66	4.11	3.20	1711.1	4.7
LATITUDE +15:	3.17	3.77	4.48	5.07	4.93	4.87	4.82	4.87	5.02	4.72	4.34	3.40	1625.9	4.5
CHARLOTTE — NC — LATITUDE: 35 DEGREES 13 MINUTES														
LATITUDE −15:	2.91	3.64	4.57	5.50	5.72	5.83	5.64	5.50	4.93	4.51	3.62	2.88	1681.8	4.6
LATITUDE:	3.22	3.85	4.62	5.29	5.30	5.32	5.23	5.29	5.00	4.83	4.06	3.28	1682.8	4.6
LATITUDE +15:	3.36	3.87	4.44	4.84	4.67	4.62	4.60	4.84	4.81	4.90	4.28	3.49	1603.7	4.4
CHERRY POINT — NC — LATITUDE: 34 DEGREES 54 MINUTES														
LATITUDE −15:	3.06	3.83	4.82	5.82	5.94	5.89	5.64	5.29	4.96	4.46	3.78	3.07	1721.2	4.7
LATITUDE:	3.40	4.07	4.88	5.61	5.50	5.38	5.23	5.09	5.03	4.79	4.26	3.51	1726.8	4.7
LATITUDE +15:	3.56	4.10	4.70	5.13	4.85	4.67	4.61	4.66	4.84	4.85	4.51	3.75	1649.7	4.5
GREENSBORO — NC — LATITUDE: 36 DEGREES 5 MINUTES														
LATITUDE −15:	2.95	3.67	4.59	5.47	5.76	5.92	5.75	5.52	4.98	4.43	3.58	2.89	1689.5	4.6
LATITUDE:	3.28	3.89	4.64	5.26	5.33	5.40	5.32	5.30	5.05	4.75	4.02	3.28	1689.4	4.6
LATITUDE +15:	3.42	3.91	4.46	4.80	4.69	4.68	4.67	4.84	4.85	4.81	4.23	3.50	1608.9	4.4
RALEIGH-DURHAM — NC — LATITUDE: 35 DEGREES 52 MINUTES														
LATITUDE −15:	2.83	3.55	4.44	5.34	5.58	5.66	5.47	5.22	4.81	4.25	3.40	2.74	1622.3	4.4
LATITUDE:	3.13	3.75	4.49	5.14	5.18	5.17	5.07	5.03	4.87	4.55	3.81	3.11	1622.5	4.4
LATITUDE +15:	3.27	3.77	4.32	4.70	4.57	4.50	4.48	4.60	4.69	4.60	4.01	3.31	1546.1	4.2
BISMARCK — ND — LATITUDE: 46 DEGREES 46 MINUTES														
LATITUDE −15:	2.53	3.51	4.55	4.94	5.79	6.26	6.88	6.52	5.40	4.31	2.82	2.22	1697.8	4.7
LATITUDE:	2.83	3.75	4.61	4.73	5.33	5.67	6.32	6.26	5.51	4.65	3.18	2.54	1686.3	4.6
LATITUDE +15:	2.97	3.77	4.43	4.30	4.67	4.88	5.51	5.70	5.30	4.72	3.35	2.71	1592.8	4.4

SITE / ARRAY TILT		JAN	FEB	MAR	APR	MAY	JUN	JUL	AUG	SEP	OCT	NOV	DEC	ANNUAL TOTAL (KWH/SQ. M)	AVERAGE DAY (KWH/SQ. M)
FARGO	ND	LATITUDE: 46 DEGREES 54 MINUTES													
LATITUDE -15:		2.18	3.15	4.25	5.00	5.75	6.05	6.67	6.32	5.19	4.13	2.49	1.95	1618.6	4.4
LATITUDE :		2.42	3.34	4.30	4.79	5.29	5.49	6.13	6.07	5.28	4.44	2.79	2.22	1601.1	4.4
LATITUDE +15:		2.53	3.35	4.12	4.36	4.64	4.73	5.35	5.53	5.08	4.51	2.93	2.36	1507.0	4.1
MINOT	ND	LATITUDE: 48 DEGREES 16 MINUTES													
LATITUDE -15:		2.11	2.99	4.10	4.99	5.80	6.00	6.62	6.29	5.18	4.18	2.52	1.92	1606.1	4.4
LATITUDE :		2.34	3.17	4.14	4.77	5.34	5.43	6.07	6.03	5.26	4.50	2.83	2.19	1586.7	4.3
LATITUDE +15:		2.45	3.18	3.97	4.33	4.66	4.68	5.29	5.49	5.06	4.56	2.97	2.33	1491.4	4.1
GRAND ISLAND	NE	LATITUDE: 40 DEGREES 58 MINUTES													
LATITUDE -15:		3.10	3.75	4.62	5.62	6.13	6.80	6.89	6.50	5.63	4.86	3.58	2.86	1837.6	5.0
LATITUDE :		3.48	3.99	4.68	5.40	5.66	6.17	6.35	6.26	5.74	5.26	4.05	3.28	1836.5	5.0
LATITUDE +15:		3.66	4.02	4.50	4.93	4.97	5.31	5.55	5.71	5.54	5.35	4.29	3.52	1745.5	4.8
NORTH OMAHA	NE	LATITUDE: 41 DEGREES 22 MINUTES													
LATITUDE -15:		3.00	3.66	4.48	5.16	5.81	6.43	6.55	6.22	5.08	4.46	3.06	2.56	1719.8	4.7
LATITUDE :		3.36	3.89	4.53	4.95	5.36	5.84	6.03	5.98	5.16	4.80	3.43	2.91	1712.3	4.7
LATITUDE +15:		3.53	3.91	4.35	4.51	4.71	5.03	5.27	5.45	4.96	4.87	3.61	3.11	1622.5	4.4
NORTH PLATTE	NE	LATITUDE: 41 DEGREES 8 MINUTES													
LATITUDE -15:		3.31	3.95	4.91	5.73	6.17	6.87	7.09	6.67	5.88	5.09	3.74	3.13	1905.3	5.2
LATITUDE :		3.72	4.22	4.98	5.50	5.69	6.22	6.52	6.42	6.00	5.51	4.24	3.61	1907.1	5.2
LATITUDE +15:		3.92	4.25	4.79	5.01	4.98	5.35	5.68	5.85	5.79	5.62	4.49	3.87	1814.5	5.0
SCOTTSBLUFF	NE	LATITUDE: 41 DEGREES 52 MINUTES													
LATITUDE -15:		3.29	3.98	4.85	5.55	6.02	6.78	7.12	6.75	6.08	5.01	3.61	3.02	1889.5	5.2
LATITUDE :		3.71	4.25	4.92	5.33	5.55	6.15	6.55	6.50	6.22	5.42	4.09	3.48	1892.9	5.2
LATITUDE +15:		3.91	4.29	4.74	4.87	4.87	5.29	5.72	5.93	6.01	5.53	4.33	3.73	1802.4	4.9
CONCORD	NH	LATITUDE: 43 DEGREES 12 MINUTES													
LATITUDE -15:		2.14	2.78	3.54	4.35	4.90	5.16	5.19	4.83	4.22	3.45	2.17	1.77	1355.8	3.7
LATITUDE :		2.36	2.92	3.56	4.16	4.53	4.70	4.79	4.63	4.25	3.67	2.39	1.98	1337.6	3.7
LATITUDE +15:		2.45	2.91	3.40	3.79	3.98	4.08	4.20	4.21	4.07	3.69	2.49	2.08	1258.8	3.4
LAKEHURST	NJ	LATITUDE: 40 DEGREES 2 MINUTES													
LATITUDE -15:		2.47	3.13	3.96	4.78	5.17	5.38	5.26	5.04	4.55	3.88	2.81	2.22	1480.9	4.1
LATITUDE :		2.73	3.30	3.99	4.58	4.78	4.90	4.86	4.83	4.60	4.13	3.13	2.50	1470.9	4.0
LATITUDE +15:		2.84	3.30	3.82	4.18	4.21	4.26	4.28	4.41	4.41	4.17	3.28	2.65	1392.8	3.8

SITE ARRAY TILT		JAN	FEB	MAR	APR	MAY	JUN	JUL	AUG	SEP	OCT	NOV	DEC	ANNUAL TOTAL (KWH/SQ. M)	AVERAGE DAY (KWH/SQ. M)
NEWARK	NJ	LATITUDE:		40 DEGREES		42 MINUTES									
LATITUDE -15:		2.47	3.15	3.98	4.76	5.23	5.44	5.46	5.16	4.63	3.90	2.72	2.14	1493.2	4.1
LATITUDE :		2.73	3.32	4.01	4.57	4.84	4.96	5.04	4.95	4.69	4.16	3.03	2.41	1483.1	4.1
LATITUDE +15:		2.85	3.33	3.85	4.17	4.26	4.31	4.43	4.52	4.50	4.20	3.18	2.55	1404.2	3.8
ALBUQUERQUE	NM	LATITUDE:		35 DEGREES		3 MINUTES									
LATITUDE -15:		4.35	5.22	6.28	7.29	7.83	8.09	7.67	7.50	7.07	6.17	4.98	4.23	2334.2	6.4
LATITUDE :		4.94	5.62	6.43	7.03	7.21	7.30	7.06	7.24	7.26	6.73	5.70	4.92	2356.8	6.5
LATITUDE +15:		5.24	5.72	6.22	6.42	6.27	6.22	6.14	6.60	7.03	6.90	6.09	5.33	2257.0	6.2
CLAYTON	NM	LATITUDE:		36 DEGREES		27 MINUTES									
LATITUDE -15:		4.24	4.88	5.92	6.69	6.86	7.31	7.05	6.89	6.51	5.80	4.62	4.04	2155.5	5.9
LATITUDE :		4.81	5.25	6.05	6.44	6.34	6.63	6.50	6.64	6.66	6.31	5.28	4.71	2179.6	6.0
LATITUDE +15:		5.11	5.33	5.84	5.88	5.55	5.69	5.68	6.06	6.45	6.46	5.63	5.10	2092.3	5.7
FARMINGTON	NM	LATITUDE:		36 DEGREES		45 MINUTES									
LATITUDE -15:		4.19	5.09	6.10	7.02	7.59	8.06	7.66	7.44	7.05	6.04	4.76	3.94	2280.6	6.2
LATITUDE :		4.75	5.49	6.24	6.77	6.99	7.29	7.05	7.18	7.25	6.59	5.45	4.59	2301.7	6.3
LATITUDE +15:		5.05	5.59	6.03	6.18	6.10	6.22	6.14	6.55	7.03	6.76	5.83	4.96	2203.8	6.0
ROSWELL	NM	LATITUDE:		33 DEGREES		24 MINUTES									
LATITUDE -15:		4.30	5.20	6.33	7.21	7.58	7.90	7.52	7.29	6.73	5.88	4.73	4.12	2276.3	6.2
LATITUDE :		4.87	5.61	6.48	6.96	6.99	7.14	6.93	7.03	6.89	6.40	5.40	4.78	2297.1	6.3
LATITUDE +15:		5.16	5.71	6.27	6.35	6.10	6.10	6.04	6.42	6.67	6.55	5.76	5.17	2199.6	6.0
TRUTH OR CONSEQUEN	NM	LATITUDE:		33 DEGREES		14 MINUTES									
LATITUDE -15:		4.64	5.53	6.62	7.60	7.89	8.01	7.28	7.20	6.82	6.10	5.13	4.37	2348.5	6.4
LATITUDE :		5.28	5.98	6.79	7.33	7.27	7.24	6.72	6.95	6.98	6.64	5.88	5.09	2377.3	6.5
LATITUDE +15:		5.61	6.09	6.58	6.69	6.32	6.18	5.86	6.34	6.76	6.81	6.29	5.52	2282.5	6.3
TUCUMCARI	NM	LATITUDE:		35 DEGREES		11 MINUTES									
LATITUDE -15:		4.33	5.02	6.08	6.86	7.14	7.51	7.24	7.08	6.53	5.70	4.67	4.14	2201.1	6.0
LATITUDE :		4.91	5.41	6.21	6.61	6.59	6.80	6.67	6.83	6.68	6.20	5.34	4.82	2223.5	6.1
LATITUDE +15:		5.21	5.49	6.00	6.04	5.76	5.82	5.82	6.23	6.46	6.34	5.69	5.22	2132.1	5.8
ZUNI	NM	LATITUDE:		35 DEGREES		6 MINUTES									
LATITUDE -15:		4.20	5.02	5.98	7.08	7.63	7.86	6.98	6.79	6.78	5.93	4.75	4.04	2223.0	6.1
LATITUDE :		4.76	5.40	6.11	6.83	7.03	7.11	6.44	6.54	6.94	6.46	5.43	4.69	2243.3	6.1
LATITUDE +15:		5.05	5.49	5.90	6.23	6.13	6.07	5.62	5.97	6.72	6.61	5.79	5.07	2148.6	5.9

SITE / ARRAY TILT		JAN	FEB	MAR	APR	MAY	JUN	JUL	AUG	SEP	OCT	NOV	DEC	ANNUAL TOTAL (KWH/SQ. M)	AVERAGE DAY (KWH/SQ. M)
ELKO	NV					LATITUDE: 40 DEGREES 50 MINUTES									
LATITUDE -15:		3.25	4.31	5.44	6.34	7.17	7.67	8.18	7.84	7.28	5.80	4.01	3.16	2145.9	5.9
LATITUDE :		3.65	4.62	5.55	6.10	6.61	6.94	7.52	7.57	7.50	6.33	4.57	3.65	2149.4	5.9
LATITUDE +15:		3.85	4.67	5.35	5.57	5.77	5.94	6.53	6.90	7.28	6.49	4.86	3.92	2043.2	5.6
ELY	NV					LATITUDE: 39 DEGREES 17 MINUTES									
LATITUDE -15:		3.84	4.69	5.93	6.68	7.17	7.61	7.60	7.46	7.31	6.04	4.48	3.63	2204.7	6.0
LATITUDE :		4.34	5.04	6.06	6.43	6.60	6.88	6.99	7.19	7.52	6.59	5.12	4.21	2220.7	6.1
LATITUDE +15:		4.60	5.11	5.85	5.87	5.76	5.89	6.08	6.56	7.29	6.75	5.46	4.54	2122.5	5.8
LAS VEGAS	NV					LATITUDE: 36 DEGREES 5 MINUTES									
LATITUDE -15:		4.28	5.30	6.58	7.64	8.18	8.39	7.99	7.76	7.43	6.27	4.88	4.11	2398.8	6.6
LATITUDE :		4.86	5.72	6.74	7.37	7.52	7.56	7.35	7.49	7.64	6.85	5.59	4.79	2418.3	6.6
LATITUDE +15:		5.16	5.83	6.53	6.72	6.53	6.43	6.38	6.83	7.41	7.03	5.97	5.18	2311.4	6.3
LOVELOCK	NV					LATITUDE: 40 DEGREES 4 MINUTES									
LATITUDE -15:		3.86	4.88	6.20	7.26	7.95	8.32	8.67	8.40	7.76	6.41	4.64	3.70	2376.4	6.5
LATITUDE :		4.37	5.26	6.34	6.99	7.31	7.49	7.95	8.10	7.99	7.02	5.31	4.29	2387.9	6.5
LATITUDE +15:		4.62	5.34	6.13	6.38	6.35	6.38	6.89	7.38	7.75	7.21	5.67	4.63	2274.7	6.2
RENO	NV					LATITUDE: 39 DEGREES 30 MINUTES									
LATITUDE -15:		3.75	4.74	6.12	7.21	7.85	8.17	8.37	8.10	7.58	6.19	4.43	3.55	2316.2	6.3
LATITUDE :		4.24	5.10	6.26	6.95	7.22	7.37	7.68	7.82	7.80	6.78	5.06	4.12	2326.3	6.4
LATITUDE +15:		4.49	5.18	6.06	6.34	6.29	6.29	6.66	7.13	7.58	6.96	5.40	4.44	2215.6	6.1
TONOPAH	NV					LATITUDE: 38 DEGREES 4 MINUTES									
LATITUDE -15:		4.23	5.20	6.54	7.47	7.99	8.41	8.38	8.14	7.63	6.45	4.90	4.10	2419.3	6.6
LATITUDE :		4.81	5.62	6.71	7.20	7.35	7.58	7.69	7.86	7.85	7.06	5.62	4.78	2439.2	6.7
LATITUDE +15:		5.10	5.71	6.49	6.57	6.39	6.44	6.67	7.16	7.62	7.26	6.01	5.18	2331.1	6.4
WINNEMUCCA	NV					LATITUDE: 40 DEGREES 54 MINUTES									
LATITUDE -15:		3.26	4.27	5.48	6.59	7.37	7.79	8.36	7.97	7.35	5.81	4.01	3.18	2175.9	6.0
LATITUDE :		3.67	4.58	5.59	6.34	6.78	7.04	7.68	7.69	7.57	6.34	4.56	3.67	2178.2	6.0
LATITUDE +15:		3.87	4.63	5.40	5.79	5.92	6.02	6.67	7.02	7.35	6.50	4.85	3.95	2069.0	5.7
YUCCA FLATS	NV					LATITUDE: 36 DEGREES 57 MINUTES									
LATITUDE -15:		4.25	5.07	6.39	7.42	7.98	8.27	8.21	7.90	7.44	6.25	4.75	4.07	2375.0	6.5
LATITUDE :		4.84	5.47	6.55	7.17	7.35	7.47	7.55	7.63	7.66	6.84	5.45	4.75	2396.5	6.6
LATITUDE +15:		5.14	5.57	6.35	6.54	6.40	6.37	6.56	6.96	7.45	7.03	5.83	5.14	2292.6	6.3

SITE / ARRAY TILT		JAN	FEB	MAR	APR	MAY	JUN	JUL	AUG	SEP	OCT	NOV	DEC	ANNUAL TOTAL (KWH/SQ. M)	AVERAGE DAY (KWH/SQ. M)
ALBANY NY	LATITUDE: 42 DEGREES 45 MINUTES														
LATITUDE -15:		2.08	2.77	3.57	4.40	4.86	5.24	5.35	4.98	4.32	3.41	2.09	1.68	1363.8	3.7
LATITUDE :		2.28	2.91	3.59	4.22	4.50	4.94	4.94	4.36	4.36	3.63	2.30	1.88	1344.6	3.7
LATITUDE +15:		2.37	2.90	3.43	3.84	3.96	4.15	4.34	4.35	4.18	3.65	2.39	1.97	1264.5	3.5
BINGHAMTON NY	LATITUDE: 42 DEGREES 13 MINUTES														
LATITUDE -15:		1.84	2.20	3.05	4.07	4.63	5.09	5.14	4.71	4.13	3.20	1.80	1.48	1259.3	3.5
LATITUDE :		2.05	2.27	3.04	3.89	4.28	4.64	4.74	4.51	4.16	3.39	1.96	1.67	1236.8	3.4
LATITUDE +15:		2.15	2.23	2.89	3.55	3.77	4.03	4.17	4.11	3.98	3.40	2.01	1.77	1159.4	3.2
BUFFALO NY	LATITUDE: 42 DEGREES 56 MINUTES														
LATITUDE -15:		1.65	2.27	3.18	4.34	4.96	5.47	5.52	5.04	4.25	3.26	1.78	1.43	1315.2	3.6
LATITUDE :		1.83	2.41	3.18	4.15	4.58	4.98	5.10	4.83	4.29	3.47	1.94	1.61	1291.3	3.5
LATITUDE +15:		1.92	2.42	3.03	3.78	4.03	4.32	4.48	4.40	4.12	3.48	2.00	1.71	1209.3	3.3
CENTRAL PARK NY	LATITUDE: 40 DEGREES 47 MINUTES														
LATITUDE -15:		2.19	2.82	3.70	4.48	5.07	5.18	5.22	4.89	4.40	3.65	2.38	1.84	1395.3	3.8
LATITUDE :		2.40	2.96	3.72	4.29	4.69	4.73	4.83	4.69	4.44	3.88	2.63	2.05	1380.0	3.8
LATITUDE +15:		2.49	2.95	3.56	3.92	4.14	4.12	4.25	4.28	4.26	3.91	2.74	2.16	1302.0	3.6
LA GUARDIA NY	LATITUDE: 40 DEGREES 46 MINUTES														
LATITUDE -15:		2.45	3.16	4.02	4.79	5.24	5.47	5.52	5.23	4.66	3.90	2.71	2.16	1501.8	4.1
LATITUDE :		2.71	3.34	4.05	4.60	4.85	4.98	5.10	5.02	4.72	4.17	3.02	2.44	1491.7	4.1
LATITUDE +15:		2.82	3.34	3.89	4.20	4.27	4.33	4.49	4.58	4.53	4.21	3.16	2.59	1412.3	3.9
MASSENA NY	LATITUDE: 44 DEGREES 56 MINUTES														
LATITUDE -15:		1.84	2.56	3.63	4.48	5.03	5.39	5.45	5.06	4.23	3.17	1.83	1.43	1341.4	3.7
LATITUDE :		2.02	2.68	3.65	4.29	4.64	4.90	5.02	4.85	4.28	3.37	2.00	1.58	1316.2	3.6
LATITUDE +15:		2.09	2.67	3.49	3.91	4.08	4.25	4.41	4.42	4.10	3.39	2.07	1.66	1231.7	3.4
ROCHESTER NY	LATITUDE: 43 DEGREES 7 MINUTES														
LATITUDE -15:		1.77	2.16	3.25	4.43	4.98	5.50	5.53	5.01	4.29	3.27	1.80	1.43	1325.1	3.6
LATITUDE :		1.97	2.23	3.25	4.24	4.60	5.00	5.09	4.80	4.33	3.47	1.96	1.61	1298.4	3.6
LATITUDE +15:		2.06	2.19	3.09	3.86	4.05	4.33	4.47	4.37	4.15	3.48	2.02	1.71	1213.6	3.3
SYRACUSE NY	LATITUDE: 43 DEGREES 7 MINUTES														
LATITUDE -15:		1.68	2.21	3.19	4.38	4.89	5.38	5.46	5.01	4.32	3.26	1.77	1.45	1310.7	3.6
LATITUDE :		1.82	2.29	3.19	4.18	4.52	4.90	5.03	4.80	4.36	3.45	1.93	1.64	1283.0	3.5
LATITUDE +15:		1.86	2.25	3.03	3.81	3.97	4.24	4.42	4.37	4.18	3.46	1.98	1.74	1198.3	3.3

SITE ARRAY TILT		JAN	FEB	MAR	APR	MAY	JUN	JUL	AUG	SEP	OCT	NOV	DEC	ANNUAL TOTAL (KWH/SQ. M)	AVERAGE DAY (KWH/SQ. M)
AKRON-CANTON	OH	LATITUDE:		40 DEGREES		55 MINUTES									
LATITUDE -15:		1.79	2.49	3.41	4.45	5.17	5.58	5.54	5.28	4.64	3.71	2.23	1.54	1396.3	3.8
LATITUDE :		1.94	2.59	3.42	4.27	4.79	5.09	5.12	5.07	4.69	3.96	2.45	1.69	1372.9	3.8
LATITUDE +15:		1.99	2.57	3.27	3.90	4.22	4.42	4.50	4.63	4.50	3.99	2.55	1.76	1287.7	3.5
CINCINNATI	OH	LATITUDE:		39 DEGREES		4 MINUTES									
LATITUDE -15:		2.08	2.82	3.60	4.56	5.17	5.57	5.47	5.37	4.70	3.97	2.55	1.90	1454.5	4.0
LATITUDE :		2.26	2.95	3.62	4.37	4.78	5.08	5.06	5.15	4.76	4.23	2.82	2.11	1436.8	3.9
LATITUDE +15:		2.33	2.93	3.46	3.99	4.21	4.40	4.45	4.70	4.56	4.27	2.94	2.21	1353.6	3.7
CLEVELAND	OH	LATITUDE:		41 DEGREES		24 MINUTES									
LATITUDE -15:		1.80	2.29	3.26	4.43	5.21	5.59	5.67	5.24	4.54	3.56	2.05	1.55	1377.5	3.8
LATITUDE :		2.00	2.37	3.26	4.24	4.82	5.09	5.23	5.03	4.58	3.79	2.25	1.76	1353.7	3.7
LATITUDE +15:		2.10	2.33	3.11	3.87	4.24	4.41	4.59	4.58	4.40	3.82	2.32	1.87	1268.8	3.5
COLUMBUS	OH	LATITUDE:		40 DEGREES		0 MINUTES									
LATITUDE -15:		1.91	2.58	3.45	4.42	5.09	5.49	5.43	5.41	4.63	3.83	2.35	1.70	1410.4	3.9
LATITUDE :		2.07	2.68	3.46	4.23	4.71	5.00	5.01	5.19	4.68	4.08	2.59	1.87	1388.7	3.8
LATITUDE +15:		2.13	2.66	3.30	3.86	4.15	4.34	4.41	4.73	4.49	4.12	2.69	1.95	1304.3	3.6
DAYTON	OH	LATITUDE:		39 DEGREES		54 MINUTES									
LATITUDE -15:		2.07	2.79	3.62	4.60	5.26	5.68	5.60	5.43	4.77	3.93	2.48	1.80	1462.8	4.0
LATITUDE :		2.26	2.93	3.64	4.41	4.87	5.18	5.18	5.21	4.83	4.19	2.75	2.00	1444.8	4.0
LATITUDE +15:		2.33	2.92	3.48	4.03	4.30	4.49	4.56	4.76	4.64	4.23	2.86	2.10	1360.8	3.7
TOLEDO	OH	LATITUDE:		41 DEGREES		36 MINUTES									
LATITUDE -15:		1.87	2.67	3.57	4.55	5.33	5.69	5.74	5.36	4.69	3.79	2.25	1.60	1434.9	3.9
LATITUDE :		2.04	2.79	3.59	4.36	4.92	5.18	5.29	5.15	4.74	4.04	2.48	1.77	1411.6	3.9
LATITUDE +15:		2.10	2.77	3.43	3.98	4.33	4.49	4.65	4.69	4.55	4.07	2.58	1.85	1324.2	3.6
YOUNGSTOWN	OH	LATITUDE:		41 DEGREES		16 MINUTES									
LATITUDE -15:		1.77	2.21	3.13	4.18	4.91	5.33	5.37	4.97	4.35	3.48	1.99	1.53	1317.5	3.6
LATITUDE :		1.97	2.28	3.13	4.00	4.54	4.86	4.96	4.76	4.39	3.70	2.17	1.73	1294.8	3.5
LATITUDE +15:		2.06	2.25	2.98	3.65	4.00	4.21	4.36	4.34	4.20	3.72	2.24	1.84	1214.2	3.3
OKLAHOMA CITY	OK	LATITUDE:		35 DEGREES		24 MINUTES									
LATITUDE -15:		3.30	3.99	4.89	5.60	5.91	6.49	6.57	6.37	5.47	4.78	3.82	3.16	1837.9	5.0
LATITUDE :		3.69	4.25	4.96	5.39	5.47	5.91	6.07	6.13	5.57	5.15	4.31	3.61	1841.6	5.0
LATITUDE +15:		3.87	4.28	4.78	4.93	4.82	5.10	5.31	5.60	5.37	5.23	4.55	3.86	1755.8	4.8

THE NEW SOLAR ELECTRIC HOME

SITE / ARRAY TILT		JAN	FEB	MAR	APR	MAY	JUN	JUL	AUG	SEP	OCT	NOV	DEC	ANNUAL TOTAL (KWH/SQ. M)	AVERAGE DAY (KWH/SQ. M)
TULSA	OK	LATITUDE: 36 DEGREES 12 MINUTES													
LATITUDE −15:		3.04	3.71	4.57	5.20	5.62	6.12	6.26	6.10	5.20	4.54	3.52	2.90	1729.3	4.7
LATITUDE :		3.38	3.94	4.62	5.00	5.21	5.58	5.78	5.87	5.28	4.87	3.95	3.30	1728.5	4.7
LATITUDE +15:		3.54	3.96	4.44	4.57	4.59	4.83	5.07	5.36	5.08	4.94	4.16	3.52	1645.0	4.5
ASTORIA	OR	LATITUDE: 46 DEGREES 9 MINUTES													
LATITUDE −15:		1.66	2.23	3.20	4.19	5.00	4.91	5.45	5.07	4.57	3.16	1.93	1.29	1300.5	3.6
LATITUDE :		1.85	2.32	3.20	3.99	4.61	4.47	5.01	4.85	4.63	3.35	2.13	1.42	1274.8	3.5
LATITUDE +15:		1.94	2.29	3.05	3.63	4.04	3.87	4.39	4.41	4.43	3.37	2.21	1.48	1191.7	3.3
BURNS	OR	LATITUDE: 43 DEGREES 35 MINUTES													
LATITUDE −15:		2.35	3.34	4.44	5.53	6.41	6.92	7.70	7.13	6.33	4.67	3.00	2.26	1831.1	5.0
LATITUDE :		2.61	3.54	4.49	5.31	5.91	6.27	7.08	6.86	6.48	5.04	3.38	2.57	1813.6	5.0
LATITUDE +15:		2.72	3.55	4.31	4.84	5.17	5.38	6.16	6.25	6.27	5.13	3.55	2.73	1707.5	4.7
MEDFORD	OR	LATITUDE: 42 DEGREES 22 MINUTES													
LATITUDE −15:		1.76	2.99	4.16	5.47	6.33	6.90	7.73	7.20	6.08	4.22	2.35	1.54	1728.7	4.7
LATITUDE :		1.91	3.15	4.19	5.25	5.83	6.25	7.10	6.93	6.22	4.53	2.60	1.70	1695.6	4.6
LATITUDE +15:		1.96	3.15	4.02	4.78	5.11	5.37	6.18	6.32	6.00	4.58	2.71	1.77	1581.9	4.3
NORTH BEND	OR	LATITUDE: 43 DEGREES 25 MINUTES													
LATITUDE −15:		2.03	2.89	3.90	5.03	5.79	6.04	6.57	6.03	5.24	3.85	2.57	1.91	1579.9	4.3
LATITUDE :		2.23	3.04	3.92	4.82	5.33	5.49	6.05	5.80	5.33	4.12	2.86	2.15	1557.5	4.3
LATITUDE +15:		2.31	3.04	3.75	4.39	4.68	4.74	5.28	5.28	5.12	4.16	3.00	2.27	1462.2	4.0
PENDLETON	OR	LATITUDE: 45 DEGREES 41 MINUTES													
LATITUDE −15:		1.62	2.57	3.95	5.07	6.02	6.52	7.54	6.91	6.00	4.18	2.23	1.49	1649.4	4.5
LATITUDE :		1.76	2.69	3.98	4.85	5.55	5.91	6.93	6.64	6.14	4.50	2.48	1.66	1618.1	4.4
LATITUDE +15:		1.81	2.68	3.81	4.41	4.86	5.08	6.03	6.05	5.93	4.56	2.59	1.74	1510.0	4.1
PORTLAND	OR	LATITUDE: 45 DEGREES 36 MINUTES													
LATITUDE −15:		1.59	2.25	3.31	4.37	5.18	5.37	6.38	5.70	4.68	3.16	1.88	1.46	1382.1	3.8
LATITUDE :		1.76	2.34	3.31	4.18	4.77	4.89	5.87	5.46	4.74	3.35	2.07	1.66	1353.9	3.7
LATITUDE +15:		1.84	2.31	3.16	3.80	4.19	4.23	5.13	4.97	4.55	3.37	2.15	1.77	1264.0	3.5
REDMOND	OR	LATITUDE: 44 DEGREES 16 MINUTES													
LATITUDE −15:		2.43	3.31	4.50	5.68	6.50	6.94	7.68	7.11	6.25	4.52	2.95	2.30	1833.4	5.0
LATITUDE :		2.70	3.50	4.55	5.45	5.98	6.28	7.05	6.84	6.39	4.88	3.31	2.62	1814.4	5.0
LATITUDE +15:		2.82	3.52	4.36	4.96	5.23	5.39	6.13	6.23	6.17	4.95	3.49	2.78	1706.8	4.7

SITE ARRAY TILT		JAN	FEB	MAR	APR	MAY	JUN	JUL	AUG	SEP	OCT	NOV	DEC	ANNUAL TOTAL (KWH/SQ. M)	AVERAGE DAY (KWH/SQ. M)
SALEM	OR	LATITUDE:	44 DEGREES	55 MINUTES											
LATITUDE -15:		1.68	2.39	3.50	4.58	5.42	5.61	6.72	6.04	5.14	3.34	1.98	1.53	1461.0	4.0
LATITUDE :		1.88	2.49	3.51	4.38	5.00	5.10	6.18	5.81	5.22	3.55	2.18	1.74	1434.1	3.9
LATITUDE +15:		1.97	2.47	3.36	3.99	4.39	4.41	5.40	5.29	5.03	3.58	2.26	1.86	1340.9	3.7
ALLENTOWN	PA	LATITUDE:	40 DEGREES	39 MINUTES											
LATITUDE -15:		2.33	3.01	3.86	4.63	5.07	5.38	5.47	5.10	4.49	3.79	2.57	2.00	1452.4	4.0
LATITUDE :		2.57	3.17	3.89	4.44	4.69	4.91	5.05	4.89	4.54	4.04	2.85	2.24	1439.4	3.9
LATITUDE +15:		2.67	3.17	3.72	4.05	4.14	4.27	4.44	4.46	4.36	4.07	2.98	2.36	1360.1	3.7
ERIE	PA	LATITUDE:	42 DEGREES	5 MINUTES											
LATITUDE -15:		1.59	2.20	3.28	4.48	5.10	5.58	5.69	4.82	4.41	3.43	1.81	1.34	1332.2	3.6
LATITUDE :		1.75	2.27	3.28	4.28	4.71	5.08	5.24	4.61	4.45	3.64	1.96	1.50	1303.8	3.6
LATITUDE +15:		1.82	2.24	3.12	3.90	4.14	4.40	4.60	4.20	4.26	3.66	2.02	1.59	1217.3	3.3
HARRISBURG	PA	LATITUDE:	40 DEGREES	13 MINUTES											
LATITUDE -15:		2.34	3.02	3.86	4.62	5.11	5.47	5.45	5.10	4.59	3.80	2.60	2.07	1462.7	4.0
LATITUDE :		2.58	3.18	3.89	4.43	4.73	4.99	5.04	4.89	4.63	4.05	2.88	2.32	1449.2	4.0
LATITUDE +15:		2.68	3.18	3.72	4.04	4.17	4.33	4.43	4.46	4.44	4.08	3.01	2.45	1369.0	3.8
PHILADELPHIA	PA	LATITUDE:	39 DEGREES	53 MINUTES											
LATITUDE -15:		2.42	3.11	3.95	4.69	5.14	5.49	5.45	5.18	4.62	3.87	2.78	2.17	1488.2	4.1
LATITUDE :		2.67	3.28	3.98	4.50	4.76	5.01	5.04	4.98	4.67	4.13	3.10	2.44	1478.5	4.1
LATITUDE +15:		2.78	3.28	3.82	4.11	4.20	4.35	4.43	4.54	4.49	4.17	3.24	2.59	1400.4	3.8
PITTSBURGH	PA	LATITUDE:	40 DEGREES	30 MINUTES											
LATITUDE -15:		1.75	2.36	3.32	4.30	4.96	5.34	5.23	4.97	4.36	3.63	2.20	1.68	1343.8	3.7
LATITUDE :		1.88	2.45	3.33	4.12	4.59	4.87	4.83	4.77	4.41	3.87	2.42	1.91	1323.1	3.6
LATITUDE +15:		1.92	2.42	3.17	3.76	4.05	4.23	4.25	4.35	4.22	3.89	2.51	2.03	1243.0	3.4
WILKES-BARRE-SCRAN	PA	LATITUDE:	41 DEGREES	20 MINUTES											
LATITUDE -15:		1.98	2.69	3.54	4.39	4.92	5.33	5.41	5.00	4.37	3.70	2.18	1.67	1376.0	3.8
LATITUDE :		2.16	2.82	3.55	4.20	4.55	4.86	4.99	4.79	4.41	3.94	2.40	1.84	1355.7	3.7
LATITUDE +15:		2.22	2.80	3.39	3.83	4.01	4.22	4.39	4.37	4.22	3.97	2.49	1.93	1274.1	3.5
KOROR ISLAND	PN	LATITUDE:	7 DEGREES	20 MINUTES											
LATITUDE -15:		4.14	4.72	5.04	5.36	5.05	4.72	4.65	4.70	4.73	4.56	4.29	3.93	1698.9	4.7
LATITUDE :		4.60	5.03	5.15	5.23	4.77	4.41	4.40	4.59	4.82	4.87	4.78	4.45	1736.1	4.8
LATITUDE +15:		4.84	5.10	5.00	4.85	4.27	3.91	3.95	4.27	4.68	4.95	5.04	4.75	1690.1	4.6

SITE ARRAY TILT		JAN	FEB	MAR	APR	MAY	JUN	JUL	AUG	SEP	OCT	NOV	DEC	ANNUAL TOTAL (KWH/SQ. M)	AVERAGE DAY (KWH/SQ. M)
KWAJALEIN ISLAND	PN		LATITUDE:	8 DEGREES		44 MINUTES									
LATITUDE -15:		4.69	5.32	5.57	5.50	5.23	5.20	5.16	5.34	5.01	4.64	4.35	4.28	1832.8	5.0
LATITUDE :		5.27	5.72	5.69	5.37	4.94	4.84	4.88	5.21	5.11	4.96	4.86	4.89	1876.2	5.1
LATITUDE +15:		5.58	5.82	5.53	4.97	4.41	4.27	4.36	4.82	4.96	5.04	5.13	5.26	1827.7	5.0
MAKE ISLAND	PN		LATITUDE:	19 DEGREES		17 MINUTES									
LATITUDE -15:		4.45	5.11	5.79	6.16	6.41	6.37	6.02	5.90	5.56	5.11	4.75	4.36	2007.1	5.5
LATITUDE :		5.01	5.48	5.91	5.97	5.98	5.84	5.61	5.71	5.66	5.49	5.38	5.02	2039.2	5.6
LATITUDE +15:		5.30	5.56	5.72	5.48	5.26	5.06	4.96	5.25	5.47	5.58	5.71	5.41	1969.7	5.4
SAN JUAN	PR		LATITUDE:	18 DEGREES		26 MINUTES									
LATITUDE -15:		4.33	4.96	5.69	5.95	5.68	5.67	5.86	5.80	5.33	4.89	4.47	4.05	1907.1	5.2
LATITUDE :		4.85	5.31	5.81	5.77	5.32	5.24	5.48	5.62	5.42	5.24	5.03	4.63	1938.3	5.3
LATITUDE +15:		5.12	5.38	5.63	5.31	4.72	4.58	4.85	5.17	5.24	5.32	5.33	4.97	1674.1	5.1
PROVIDENCE	RI		LATITUDE:	41 DEGREES		44 MINUTES									
LATITUDE -15:		2.30	2.95	3.72	4.53	5.14	5.38	5.25	4.95	4.43	3.77	2.48	2.01	1428.2	3.9
LATITUDE :		2.53	3.10	3.74	4.34	4.75	4.90	4.85	4.75	4.47	4.02	2.76	2.26	1415.1	3.9
LATITUDE +15:		2.64	3.10	3.58	3.96	4.18	4.26	4.27	4.33	4.29	4.06	2.88	2.39	1337.0	3.7
CHARLESTON	SC		LATITUDE:	32 DEGREES		54 MINUTES									
LATITUDE -15:		2.87	3.61	4.56	5.58	5.73	5.60	5.55	5.10	4.76	4.42	3.73	2.92	1656.9	4.5
LATITUDE :		3.17	3.82	4.62	5.38	5.33	5.14	5.15	4.91	4.82	4.73	4.19	3.31	1660.5	4.5
LATITUDE +15:		3.30	3.84	4.45	4.93	4.70	4.48	4.55	4.50	4.64	4.79	4.42	3.52	1585.6	4.3
COLUMBIA	SC		LATITUDE:	33 DEGREES		57 MINUTES									
LATITUDE -15:		3.01	3.76	4.66	5.64	5.85	5.91	5.68	5.50	4.97	4.57	3.76	3.01	1714.5	4.7
LATITUDE :		3.34	3.99	4.72	5.44	5.42	5.41	5.27	5.30	5.04	4.91	4.23	3.43	1719.3	4.7
LATITUDE +15:		3.49	4.02	4.55	4.98	4.78	4.69	4.65	4.85	4.85	4.98	4.47	3.65	1642.0	4.5
GREENVILLE-SPARTAN	SC		LATITUDE:	34 DEGREES		54 MINUTES									
LATITUDE -15:		2.93	3.66	4.60	5.50	5.67	5.82	5.65	5.50	4.88	4.50	3.66	2.84	1680.4	4.6
LATITUDE :		3.25	3.88	4.66	5.30	5.26	5.32	5.24	5.30	4.94	4.83	4.11	3.22	1683.3	4.6
LATITUDE +15:		3.39	3.90	4.48	4.85	4.64	4.62	4.62	4.85	4.75	4.90	4.34	3.43	1606.1	4.4
HURON	SD		LATITUDE:	44 DEGREES		23 MINUTES									
LATITUDE -15:		2.42	3.16	4.18	5.13	5.84	6.37	6.84	6.46	5.50	4.48	3.00	2.17	1692.0	4.6
LATITUDE :		2.70	3.34	4.22	4.91	5.38	5.78	6.29	6.20	5.60	4.83	3.37	2.46	1677.5	4.6
LATITUDE +15:		2.82	3.35	4.04	4.47	4.72	4.98	5.48	5.65	5.39	4.90	3.54	2.62	1582.2	4.3

SITE / ARRAY TILT	JAN	FEB	MAR	APR	MAY	JUN	JUL	AUG	SEP	OCT	NOV	DEC	ANNUAL TOTAL (KWH/SQ. M)	AVERAGE DAY (KWH/SQ. M)
PIERRE SD	LATITUDE: 44 DEGREES 23 MINUTES													
LATITUDE -15:	2.68	3.42	4.57	5.43	6.14	6.66	7.13	6.83	5.86	4.83	3.31	2.43	1806.6	4.9
LATITUDE :	3.00	3.63	4.63	5.21	5.65	6.03	6.56	6.57	5.98	5.22	3.74	2.78	1797.0	4.9
LATITUDE +15:	3.15	3.65	4.45	4.74	4.95	5.18	5.71	5.99	5.77	5.31	3.95	2.96	1699.3	4.7
RAPID CITY SD	LATITUDE: 44 DEGREES 3 MINUTES													
LATITUDE -15:	2.73	3.55	4.65	5.33	5.88	6.46	6.95	6.70	5.92	4.85	3.43	2.63	1799.9	4.9
LATITUDE :	3.06	3.71	4.71	5.11	5.42	5.85	6.38	6.44	6.04	5.25	3.88	3.01	1794.2	4.9
LATITUDE +15:	3.20	3.79	4.52	4.65	4.75	5.03	5.56	5.86	5.83	5.34	4.10	3.21	1700.4	4.7
SIOUX FALLS SD	LATITUDE: 43 DEGREES 34 MINUTES													
LATITUDE -15:	2.62	3.38	4.30	5.15	5.90	6.36	6.71	6.25	5.40	4.47	3.09	2.33	1704.4	4.7
LATITUDE :	2.92	3.59	4.34	4.94	5.44	5.77	6.18	6.00	5.49	4.82	3.48	2.65	1693.6	4.6
LATITUDE +15:	3.06	3.61	4.17	4.50	4.77	4.97	5.39	5.47	5.29	4.89	3.66	2.82	1601.3	4.4
CHATTANOOGA TN	LATITUDE: 35 DEGREES 2 MINUTES													
LATITUDE -15:	2.48	3.16	4.04	5.01	5.33	5.56	5.34	5.27	4.63	4.21	3.16	2.41	1540.7	4.2
LATITUDE :	2.72	3.32	4.07	4.81	4.94	5.08	4.95	5.07	4.68	4.50	3.51	2.70	1533.0	4.2
LATITUDE +15:	2.81	3.31	3.90	4.40	4.36	4.42	4.37	4.63	4.49	4.54	3.68	2.85	1454.4	4.0
KNOXVILLE TN	LATITUDE: 35 DEGREES 49 MINUTES													
LATITUDE -15:	2.48	3.21	4.12	5.19	5.56	5.77	5.56	5.41	4.84	4.31	3.15	2.40	1583.2	4.3
LATITUDE :	2.72	3.38	4.16	4.99	5.16	5.27	5.15	5.21	4.90	4.62	3.51	2.70	1576.1	4.3
LATITUDE +15:	2.82	3.38	4.00	4.57	4.55	4.58	4.54	4.76	4.71	4.67	3.68	2.86	1495.4	4.1
MEMPHIS TN	LATITUDE: 35 DEGREES 3 MINUTES													
LATITUDE -15:	2.73	3.52	4.42	5.30	5.81	6.19	6.07	5.93	5.14	4.63	3.37	2.65	1697.8	4.7
LATITUDE :	3.01	3.71	4.46	5.10	5.38	5.64	5.62	5.71	5.21	4.97	3.76	2.99	1691.6	4.6
LATITUDE +15:	3.13	3.72	4.29	4.66	4.73	4.88	4.93	5.21	5.01	5.04	3.95	3.18	1605.3	4.4
NASHVILLE TN	LATITUDE: 36 DEGREES 7 MINUTES													
LATITUDE -15:	2.30	3.06	3.91	5.01	5.62	5.94	5.83	5.66	4.91	4.32	2.95	2.18	1574.0	4.3
LATITUDE :	2.51	3.21	3.93	4.81	5.21	5.42	5.40	5.44	4.97	4.62	3.27	2.43	1559.6	4.3
LATITUDE +15:	2.59	3.20	3.77	4.40	4.59	4.69	4.74	4.97	4.77	4.67	3.42	2.55	1472.9	4.0
ABILENE TX	LATITUDE: 32 DEGREES 26 MINUTES													
LATITUDE -15:	3.63	4.34	5.42	5.94	6.28	6.70	6.59	6.32	5.49	4.90	4.03	3.57	1923.9	5.3
LATITUDE :	4.06	4.63	5.51	5.72	5.82	6.10	6.09	6.09	5.58	5.27	4.54	4.10	1933.6	5.3
LATITUDE +15:	4.27	4.68	5.32	5.24	5.12	5.26	5.34	5.57	5.38	5.36	4.80	4.40	1848.2	5.1

SITE / ARRAY TILT	JAN	FEB	MAR	APR	MAY	JUN	JUL	AUG	SEP	OCT	NOV	DEC	ANNUAL TOTAL (KWH/SQ. M)	AVERAGE DAY (KWH/SQ. M)
AMARILLO TX — LATITUDE: 35 DEGREES 14 MINUTES														
LATITUDE −15:	4.09	4.80	5.76	6.59	6.82	7.23	7.03	6.87	6.26	5.53	4.47	3.93	2112.4	5.8
LATITUDE :	4.62	5.16	5.88	6.35	6.30	6.56	6.48	6.63	6.40	5.99	5.09	4.56	2130.9	5.8
LATITUDE +15:	4.89	5.23	5.68	5.80	5.51	5.63	5.66	6.05	6.18	6.12	5.42	4.93	2041.5	5.6
AUSTIN TX — LATITUDE: 30 DEGREES 18 MINUTES														
LATITUDE −15:	3.24	4.00	4.81	5.13	5.66	6.30	6.49	6.19	5.43	4.83	3.76	3.21	1797.6	4.9
LATITUDE :	3.58	4.25	4.87	4.95	5.26	5.75	6.01	5.98	5.52	5.18	4.20	3.64	1801.9	4.9
LATITUDE +15:	3.74	4.28	4.69	4.54	4.65	4.98	5.27	5.47	5.33	5.26	4.42	3.88	1719.6	4.7
BROWNSVILLE TX — LATITUDE: 25 DEGREES 54 MINUTES														
LATITUDE −15:	3.20	3.86	4.78	5.52	5.97	6.48	6.84	6.44	5.58	4.96	3.74	3.08	1841.3	5.0
LATITUDE :	3.54	4.09	4.85	5.34	5.56	5.93	6.34	6.23	5.68	5.34	4.17	3.47	1843.4	5.1
LATITUDE +15:	3.69	4.12	4.68	4.91	4.91	5.14	5.56	5.71	5.49	5.43	4.38	3.68	1756.7	4.8
CORPUS CHRISTI TX — LATITUDE: 27 DEGREES 46 MINUTES														
LATITUDE −15:	3.23	3.97	4.73	5.23	5.77	6.39	6.75	6.35	5.62	4.99	3.81	3.11	1825.4	5.0
LATITUDE :	3.57	4.21	4.80	5.05	5.37	5.84	6.25	6.14	5.72	5.37	4.26	3.51	1829.8	5.0
LATITUDE +15:	3.73	4.24	4.62	4.64	4.75	5.06	5.49	5.62	5.53	5.46	4.48	3.73	1746.0	4.8
DALLAS TX — LATITUDE: 32 DEGREES 51 MINUTES														
LATITUDE −15:	3.21	3.91	4.86	5.23	5.82	6.47	6.54	6.30	5.47	4.77	3.73	3.19	1812.3	5.0
LATITUDE :	3.57	4.15	4.93	5.04	5.41	5.91	6.05	6.08	5.56	5.13	4.19	3.64	1816.3	5.0
LATITUDE +15:	3.74	4.18	4.75	4.62	4.77	5.11	5.31	5.56	5.37	5.21	4.42	3.89	1732.7	4.7
DEL RIO TX — LATITUDE: 29 DEGREES 22 MINUTES														
LATITUDE −15:	3.57	4.27	5.30	5.43	5.64	6.16	6.34	6.20	5.32	4.87	3.99	3.47	1844.1	5.1
LATITUDE :	3.97	4.54	5.39	5.24	5.25	5.63	5.88	5.98	5.41	5.24	4.48	3.96	1855.9	5.1
LATITUDE +15:	4.17	4.58	5.21	4.80	4.64	4.88	5.17	5.48	5.22	5.32	4.73	4.23	1777.7	4.9
EL PASO TX — LATITUDE: 31 DEGREES 48 MINUTES														
LATITUDE −15:	4.49	5.51	6.62	7.65	8.02	8.12	7.56	7.39	6.89	6.20	5.06	4.30	2367.7	6.5
LATITUDE :	5.10	5.96	6.79	7.39	7.39	7.34	6.97	7.14	7.07	6.77	5.80	5.01	2395.1	6.6
LATITUDE +15:	5.42	6.08	6.59	6.75	6.43	6.26	6.08	6.52	6.86	6.94	6.20	5.43	2297.8	6.3
FORT WORTH TX — LATITUDE: 32 DEGREES 50 MINUTES														
LATITUDE −15:	3.13	3.90	4.82	5.20	5.83	6.53	6.65	6.41	5.59	4.84	3.74	3.13	1820.3	5.0
LATITUDE :	3.48	4.14	4.89	5.01	5.42	5.95	6.15	6.19	5.70	5.20	4.20	3.56	1823.3	5.0
LATITUDE +15:	3.64	4.17	4.71	4.59	4.78	5.15	5.39	5.66	5.50	5.29	4.43	3.80	1738.1	4.8

SITE ARRAY TILT		JAN	FEB	MAR	APR	MAY	JUN	JUL	AUG	SEP	OCT	NOV	DEC	ANNUAL TOTAL (KWH/SQ. M)	AVERAGE DAY (KWH/SQ. M)
HOUSTON	TX	LATITUDE:		29 DEGREES		59 MINUTES									
LATITUDE -15:		2.84	3.64	4.33	4.86	5.48	5.79	5.65	5.39	4.94	4.57	3.46	2.76	1635.1	4.5
LATITUDE :		3.11	3.85	4.38	4.69	5.11	5.31	5.25	5.21	5.01	4.90	3.85	3.10	1636.8	4.5
LATITUDE +15:		3.24	3.86	4.22	4.31	4.53	4.62	4.64	4.78	4.83	4.97	4.04	3.29	1561.7	4.3
KINGSVILLE	TX	LATITUDE:		27 DEGREES		31 MINUTES									
LATITUDE -15:		3.29	4.02	4.75	5.28	5.77	6.22	6.51	6.12	5.39	4.88	3.76	3.12	1800.0	4.9
LATITUDE :		3.63	4.26	4.81	5.10	5.37	5.69	6.04	5.92	5.48	5.24	4.20	3.53	1804.6	4.9
LATITUDE +15:		3.80	4.29	4.64	4.69	4.75	4.94	5.31	5.42	5.29	5.32	4.42	3.75	1722.6	4.7
LAREDO	TX	LATITUDE:		27 DEGREES		32 MINUTES									
LATITUDE -15:		3.46	4.15	5.02	5.49	6.03	6.33	6.58	6.41	5.67	4.94	3.79	3.28	1862.0	5.1
LATITUDE :		3.84	4.40	5.10	5.30	5.61	5.79	6.10	6.19	5.78	5.31	4.23	3.72	1868.4	5.1
LATITUDE +15:		4.02	4.44	4.92	4.87	4.95	5.02	5.36	5.67	5.58	5.40	4.45	3.96	1784.3	4.9
LUBBOCK	TX	LATITUDE:		33 DEGREES		39 MINUTES									
LATITUDE -15:		4.25	5.04	6.18	7.04	7.40	7.69	7.44	7.19	6.39	5.65	4.68	4.05	2222.0	6.1
LATITUDE :		4.82	5.43	6.32	6.80	6.83	6.97	6.86	6.94	6.54	6.13	5.34	4.71	2242.0	6.1
LATITUDE +15:		5.11	5.52	6.11	6.21	5.96	5.97	5.98	6.34	6.33	6.27	5.70	5.09	2147.3	5.9
LUFKIN	TX	LATITUDE:		31 DEGREES		14 MINUTES									
LATITUDE -15:		2.99	3.84	4.65	5.20	5.76	6.24	6.18	5.98	5.20	4.95	3.73	3.02	1758.3	4.8
LATITUDE :		3.30	4.06	4.71	5.01	5.35	5.70	5.73	5.77	5.28	5.33	4.17	3.42	1760.2	4.8
LATITUDE +15:		3.44	4.09	4.53	4.59	4.72	4.93	5.04	5.28	5.08	5.41	4.39	3.64	1678.1	4.6
MIDLAND-ODESSA	TX	LATITUDE:		31 DEGREES		56 MINUTES									
LATITUDE -15:		4.30	5.12	6.36	7.08	7.50	7.76	7.36	7.15	6.37	5.71	4.75	4.17	2241.4	6.1
LATITUDE :		4.87	5.51	6.52	6.84	6.93	7.04	6.80	6.91	6.52	6.21	5.42	4.84	2264.1	6.2
LATITUDE +15:		5.16	5.61	6.32	6.26	6.05	6.02	5.94	6.31	6.31	6.36	5.78	5.24	2170.7	5.9
PORT ARTHUR	TX	LATITUDE:		29 DEGREES		57 MINUTES									
LATITUDE -15:		2.95	3.78	4.53	5.14	5.78	6.13	5.70	5.56	5.14	4.75	3.58	2.87	1701.5	4.7
LATITUDE :		3.25	4.00	4.58	4.96	5.38	5.61	5.30	5.37	5.21	5.11	4.00	3.23	1704.2	4.7
LATITUDE +15:		3.38	4.02	4.42	4.56	4.75	4.87	4.69	4.92	5.03	5.19	4.20	3.43	1626.2	4.5
SAN ANGELO	TX	LATITUDE:		31 DEGREES		22 MINUTES									
LATITUDE -15:		3.72	4.38	5.48	5.94	6.27	6.63	6.54	6.32	5.48	4.92	4.10	3.61	1929.2	5.3
LATITUDE :		4.16	4.67	5.58	5.73	5.81	6.04	6.05	6.10	5.57	5.29	4.62	4.14	1940.3	5.3
LATITUDE +15:		4.37	4.72	5.38	5.25	5.11	5.21	5.31	5.58	5.37	5.38	4.88	4.45	1856.2	5.1

SITE ARRAY TILT		JAN	FEB	MAR	APR	MAY	JUN	JUL	AUG	SEP	OCT	NOV	DEC	ANNUAL TOTAL (KWH/SQ. M)	AVERAGE DAY (KWH/SQ. M)
SAN ANTONIO	TX	LATITUDE:	29 DEGREES	32 MINUTES											
LATITUDE -15:		3.32	4.08	4.85	5.15	5.85	6.30	6.54	6.23	5.52	4.85	3.79	3.24	1818.1	5.0
LATITUDE :		3.68	4.33	4.92	4.96	5.44	5.75	6.06	6.01	5.61	5.21	4.24	3.68	1823.7	5.0
LATITUDE +15:		3.85	4.37	4.74	4.56	4.80	4.98	5.32	5.51	5.42	5.29	4.47	3.92	1741.4	4.8
SHERMAN	TX	LATITUDE:	33 DEGREES	43 MINUTES											
LATITUDE -15:		3.15	3.82	4.69	5.19	5.71	6.42	6.41	6.26	5.48	4.80	3.73	3.09	1789.0	4.9
LATITUDE :		3.50	4.05	4.75	5.00	5.30	5.85	5.93	6.03	5.58	5.16	4.19	3.53	1792.4	4.9
LATITUDE +15:		3.66	4.08	4.57	4.58	4.68	5.06	5.20	5.52	5.38	5.25	4.43	3.76	1709.5	4.7
WACO	TX	LATITUDE:	31 DEGREES	37 MINUTES											
LATITUDE -15:		3.18	3.95	4.84	5.17	5.48	6.41	6.56	6.31	5.47	4.78	3.72	3.20	1799.0	4.9
LATITUDE :		3.53	4.19	4.91	4.98	5.09	5.85	6.08	6.09	5.56	5.14	4.17	3.64	1803.3	4.9
LATITUDE +15:		3.69	4.22	4.73	4.57	4.50	5.06	5.33	5.57	5.37	5.22	4.39	3.89	1720.8	4.7
WICHITA FALLS	TX	LATITUDE:	33 DEGREES	58 MINUTES											
LATITUDE -15:		3.47	4.18	5.10	5.70	6.22	6.73	6.68	6.39	5.57	4.91	3.93	3.39	1895.9	5.2
LATITUDE :		3.87	4.46	5.18	5.49	5.77	6.13	6.17	6.16	5.67	5.29	4.44	3.90	1903.8	5.2
LATITUDE +15:		4.07	4.50	5.00	5.03	5.07	5.29	5.41	5.63	5.48	5.39	4.70	4.18	1818.4	5.0
BRYCE CANYON	UT	LATITUDE:	37 DEGREES	42 MINUTES											
LATITUDE -15:		4.14	4.97	6.13	7.05	7.60	8.03	7.51	7.14	7.09	6.10	4.74	3.98	2267.0	6.2
LATITUDE :		4.70	5.36	6.27	6.80	7.01	7.26	6.92	6.89	7.28	6.67	5.43	4.63	2288.9	6.3
LATITUDE +15:		4.99	5.45	6.07	6.21	6.11	6.20	6.03	6.29	7.06	6.84	5.80	5.02	2192.4	6.0
CEDAR CITY	UT	LATITUDE:	37 DEGREES	42 MINUTES											
LATITUDE -15:		3.97	4.71	5.94	6.92	7.65	8.19	7.76	7.43	7.29	6.07	4.61	3.78	2262.1	6.2
LATITUDE :		4.50	5.06	6.07	6.67	7.05	7.39	7.14	7.17	7.50	6.63	5.27	4.38	2278.0	6.2
LATITUDE +15:		4.77	5.14	5.87	6.09	6.14	6.30	6.22	6.54	7.28	6.80	5.63	4.74	2176.3	6.0
SALT LAKE CITY	UT	LATITUDE:	40 DEGREES	46 MINUTES											
LATITUDE -15:		2.96	4.07	5.40	6.32	7.36	7.76	8.08	7.61	7.05	5.64	3.86	2.84	2100.2	5.8
LATITUDE :		3.31	4.35	5.50	6.08	6.78	7.02	7.42	7.34	7.25	6.14	4.38	3.26	2096.1	5.7
LATITUDE +15:		3.48	4.39	5.31	5.55	5.92	6.00	6.45	6.69	7.03	6.29	4.65	3.48	1986.1	5.4
NORFOLK	VA	LATITUDE:	36 DEGREES	54 MINUTES											
LATITUDE -15:		2.83	3.55	4.50	5.46	5.83	6.07	5.72	5.48	4.93	4.22	3.50	2.77	1669.8	4.6
LATITUDE :		3.14	3.76	4.55	5.26	5.40	5.53	5.30	5.27	5.00	4.52	3.94	3.15	1668.1	4.6
LATITUDE +15:		3.28	3.78	4.38	4.81	4.76	4.79	4.66	4.82	4.81	4.58	4.15	3.36	1587.1	4.3

SITE ARRAY TILT		JAN	FEB	MAR	APR	MAY	JUN	JUL	AUG	SEP	OCT	NOV	DEC	ANNUAL TOTAL (KWH/SQ. M)	AVERAGE DAY (KWH/SQ. M)
RICHMOND VA	LATITUDE:	37 DEGREES	30 MINUTES												
LATITUDE -15:		2.65	3.35	4.25	5.10	5.44	5.68	5.47	5.22	4.77	4.05	3.16	2.52	1572.6	4.3
LATITUDE :		2.93	3.54	4.29	4.90	5.04	5.18	5.07	5.01	4.83	4.32	3.53	2.85	1567.2	4.3
LATITUDE +15:		3.05	3.54	4.12	4.48	4.44	4.50	4.46	4.58	4.64	4.36	3.71	3.03	1488.6	4.1
ROANOKE VA	LATITUDE:	37 DEGREES	19 MINUTES												
LATITUDE -15:		2.78	3.44	4.35	5.15	5.45	5.71	5.54	5.28	4.81	4.25	3.31	2.64	1604.6	4.4
LATITUDE :		3.08	3.64	4.39	4.95	5.05	5.21	5.13	5.08	4.86	4.55	3.71	2.99	1601.9	4.4
LATITUDE +15:		3.22	3.64	4.22	4.52	4.45	4.52	4.51	4.64	4.67	4.60	3.90	3.18	1523.8	4.2
BURLINGTON VT	LATITUDE:	44 DEGREES	28 MINUTES												
LATITUDE -15:		1.78	2.46	3.45	4.30	4.89	5.24	5.36	4.94	4.21	3.17	1.71	1.54	1311.3	3.6
LATITUDE :		1.94	2.57	3.46	4.11	4.51	4.77	4.94	4.73	4.24	3.36	1.86	1.76	1286.8	3.5
LATITUDE +15:		2.00	2.55	3.30	3.74	3.97	4.13	4.33	4.30	4.06	3.37	1.92	1.87	1204.5	3.3
OLYMPIA WA	LATITUDE:	46 DEGREES	58 MINUTES												
LATITUDE -15:		1.39	2.04	3.14	4.21	5.10	5.14	6.00	5.28	4.51	2.80	1.67	1.27	1297.0	3.6
LATITUDE :		1.54	2.11	3.14	4.02	4.70	4.67	5.52	5.06	4.56	2.96	1.83	1.44	1266.6	3.5
LATITUDE +15:		1.60	2.08	2.99	3.65	4.12	4.05	4.83	4.60	4.37	2.97	1.89	1.53	1179.3	3.2
SEATTLE-TACOMA WA	LATITUDE:	47 DEGREES	27 MINUTES												
LATITUDE -15:		1.38	2.02	3.19	4.35	5.36	5.46	7.10	5.56	4.50	2.95	1.69	1.23	1366.3	3.7
LATITUDE :		1.52	2.10	3.19	4.16	4.93	4.96	6.51	5.32	4.55	3.13	1.86	1.39	1330.4	3.6
LATITUDE +15:		1.58	2.07	3.04	3.78	4.32	4.28	5.67	4.84	4.36	3.14	1.92	1.48	1234.3	3.4
SPOKANE WA	LATITUDE:	47 DEGREES	38 MINUTES												
LATITUDE -15:		1.54	2.65	4.05	5.10	6.03	6.34	7.47	6.81	5.88	4.03	2.15	1.38	1628.7	4.5
LATITUDE :		1.68	2.80	4.09	4.88	5.55	5.74	6.86	6.55	6.01	4.33	2.39	1.54	1596.9	4.4
LATITUDE +15:		1.73	2.79	3.92	4.44	4.85	4.94	5.96	5.96	5.80	4.39	2.50	1.61	1489.2	4.1
WHIDBEY ISLAND WA	LATITUDE:	48 DEGREES	21 MINUTES												
LATITUDE -15:		1.60	2.29	3.53	4.57	5.52	5.51	6.24	5.50	4.69	3.02	1.92	1.27	1392.0	3.8
LATITUDE :		1.79	2.40	3.55	4.36	5.07	5.00	5.73	5.27	4.75	3.21	2.13	1.41	1361.1	3.7
LATITUDE +15:		1.88	2.38	3.39	3.96	4.44	4.32	5.00	4.79	4.56	3.22	2.22	1.48	1267.9	3.5
YAKIMA WA	LATITUDE:	46 DEGREES	34 MINUTES												
LATITUDE -15:		1.81	2.91	4.34	5.45	6.31	6.59	7.44	6.88	6.01	4.19	2.36	1.58	1702.5	4.7
LATITUDE :		1.98	3.07	4.39	5.22	5.80	5.96	6.83	6.61	6.14	4.51	2.63	1.78	1674.1	4.6
LATITUDE +15:		2.05	3.07	4.21	4.75	5.07	5.12	5.94	6.02	5.93	4.58	2.76	1.87	1565.2	4.3

THE NEW SOLAR ELECTRIC HOME

SITE ARRAY TILT		JAN	FEB	MAR	APR	MAY	JUN	JUL	AUG	SEP	OCT	NOV	DEC	ANNUAL TOTAL (KWH/SQ. M)	AVERAGE DAY (KWH/SQ. M)
EAU CLAIRE	WI	LATITUDE:		44 DEGREES	52 MINUTES										
LATITUDE -15:		2.23	3.20	4.10	4.77	5.24	5.67	5.89	5.48	4.54	3.63	2.23	1.76	1484.4	4.1
LATITUDE :		2.48	3.39	4.14	4.57	4.84	5.16	5.43	5.26	4.59	3.88	2.48	1.98	1466.5	4.0
LATITUDE +15:		2.58	3.40	3.97	4.16	4.25	4.46	4.75	4.79	4.41	3.91	2.58	2.09	1380.3	3.8
GREEN BAY	WI	LATITUDE:		44 DEGREES	29 MINUTES										
LATITUDE -15:		2.20	3.06	4.14	4.81	5.35	5.78	5.89	5.47	4.62	3.57	2.30	1.79	1491.8	4.1
LATITUDE :		2.43	3.24	4.18	4.60	4.94	5.25	5.42	5.25	4.67	3.81	2.55	2.02	1472.1	4.0
LATITUDE +15:		2.53	3.24	4.01	4.19	4.33	4.54	4.75	4.78	4.48	3.84	2.66	2.13	1384.0	3.8
LA CROSSE	WI	LATITUDE:		43 DEGREES	52 MINUTES										
LATITUDE -15:		2.32	3.22	4.10	4.75	5.33	5.78	5.93	5.62	4.68	3.73	2.42	1.87	1514.7	4.1
LATITUDE :		2.58	3.41	4.14	4.55	4.92	5.25	5.46	5.39	4.74	3.99	2.69	2.10	1498.5	4.1
LATITUDE +15:		2.69	3.42	3.96	4.15	4.33	4.54	4.79	4.91	4.55	4.02	2.81	2.22	1412.3	3.9
MADISON	WI	LATITUDE:		43 DEGREES	8 MINUTES										
LATITUDE -15:		2.48	3.37	4.21	4.64	5.41	5.90	6.01	5.73	4.89	3.92	2.42	1.94	1550.6	4.2
LATITUDE :		2.75	3.57	4.25	4.44	4.99	5.36	5.54	5.50	4.95	4.20	2.68	2.18	1534.5	4.2
LATITUDE +15:		2.87	3.58	4.07	4.04	4.38	4.63	4.85	5.01	4.75	4.24	2.80	2.30	1446.4	4.0
MILWAUKEE	WI	LATITUDE:		42 DEGREES	57 MINUTES										
LATITUDE -15:		2.24	3.01	4.01	4.79	5.50	5.99	6.11	5.77	4.92	3.88	2.52	1.85	1540.7	4.2
LATITUDE :		2.47	3.18	4.04	4.59	5.08	5.45	5.63	5.55	4.99	4.16	2.80	2.08	1522.9	4.2
LATITUDE +15:		2.57	3.18	3.87	4.18	4.46	4.71	4.94	5.06	4.80	4.20	2.93	2.19	1433.8	3.9
CHARLESTON	WV	LATITUDE:		38 DEGREES	22 MINUTES										
LATITUDE -15:		2.02	2.64	3.52	4.41	5.06	5.39	5.20	4.95	4.52	3.84	2.63	1.89	1402.6	3.8
LATITUDE :		2.20	2.75	3.53	4.23	4.69	4.92	4.81	4.75	4.56	4.09	2.91	2.10	1386.2	3.8
LATITUDE +15:		2.26	2.73	3.37	3.86	4.14	4.28	4.24	4.34	4.38	4.12	3.04	2.19	1307.2	3.6
HUNTINGTON	WV	LATITUDE:		38 DEGREES	22 MINUTES										
LATITUDE -15:		2.16	2.87	3.74	4.72	5.28	5.59	5.46	5.17	4.65	3.98	2.75	2.04	1474.5	4.0
LATITUDE :		2.36	3.00	3.76	4.53	4.89	5.10	5.05	4.96	4.70	4.24	3.05	2.28	1459.7	4.0
LATITUDE +15:		2.44	2.99	3.60	4.14	4.31	4.43	4.45	4.53	4.51	4.28	3.19	2.40	1378.1	3.8
CASPER	WY	LATITUDE:		42 DEGREES	55 MINUTES										
LATITUDE -15:		3.49	4.41	5.49	6.23	6.88	7.59	7.95	7.62	6.85	5.55	4.05	3.31	2114.3	5.8
LATITUDE :		3.95	4.74	5.60	5.99	6.34	6.86	7.30	7.36	7.04	6.04	4.63	3.84	2121.9	5.8
LATITUDE +15:		4.18	4.80	5.40	5.46	5.54	5.88	6.35	6.71	6.83	6.19	4.93	4.15	2021.2	5.5

SITE ARRAY TILT		JAN	FEB	MAR	APR	MAY	JUN	JUL	AUG	SEP	OCT	NOV	DEC	ANNUAL TOTAL (KWH/SQ. M)	AVERAGE DAY (KWH/SQ. M)
CHEYENNE	WY	LATITUDE:		41 DEGREES		9 MINUTES									
LATITUDE -15:		3.77	4.50	5.34	5.90	6.19	6.83	6.93	6.60	6.33	5.44	4.15	3.58	1995.7	5.5
LATITUDE :		4.27	4.83	5.44	5.66	5.71	6.19	6.37	6.34	6.47	5.91	4.73	4.16	2011.4	5.5
LATITUDE +15:		4.52	4.90	5.24	5.16	5.00	5.32	5.56	5.78	6.26	6.03	5.03	4.49	1925.5	5.3
ROCK SPRINGS	WY	LATITUDE:		41 DEGREES		36 MINUTES									
LATITUDE -15:		3.63	4.65	5.78	6.53	7.31	7.80	7.96	7.60	7.10	5.83	4.24	3.50	2190.3	6.0
LATITUDE :		4.11	5.01	5.90	6.28	6.73	7.05	7.31	7.33	7.30	6.36	4.84	4.07	2200.7	6.0
LATITUDE +15:		4.35	5.08	5.70	5.73	5.87	6.03	6.35	6.69	7.08	6.52	5.16	4.39	2098.0	5.7
SHERIDAN	WY	LATITUDE:		44 DEGREES		46 MINUTES									
LATITUDE -15:		2.65	3.41	4.58	5.17	5.88	6.55	7.31	6.90	5.91	4.61	3.14	2.46	1784.6	4.9
LATITUDE :		2.96	3.62	4.64	4.95	5.42	5.94	6.72	6.64	6.04	4.98	3.54	2.82	1775.2	4.9
LATITUDE +15:		3.11	3.64	4.46	4.51	4.75	5.11	5.85	6.05	5.83	5.07	3.74	3.01	1678.9	4.6

Appendix C
Conversion Factors

To Change	Into	Multiply by
BTU	cal	252
BTU	joules	1055
BTU	kcal	0.252
BTU	kWh	2.93×10^4
BTU ft^2	langleys (cal cm^2)	0.271
cal	BTU	3.97×10^5
cal	ft-lb	3.09
cal	joules	4.184
cal	kcal	0.001
cal min^1	watts	0.0698
cm	inches	0.394
cc or cm^3	in.^{3}	0.0610
ft^3	liters	28.3
in.^{3}	cc or cm^3	16.4
ft	m	0.305
ft-lb	cal	0.324
ft-lb	joules	1.36
ft-lb	kg-m	0.138
ft-lb	kWh	3.77×10^7

To Change	Into	Multiply by
gal	liters	3.79
hp	kW	0.745
inches	cm	2.54
joules	BTU	9.48×10^4
joules	cal	0.239
joules	ft-lb	0.738
kcal	BTU	3.97
kcal	cal	1000
kcal min^1	kW	0.0698
kg-m	ft-lb	7.23
kg	lb	2.20
kW	hp	1.34
kWh	BTU	3413
kWh	ft-lb	2.66×10^6
kW	kcal min^1	14.3
langleys (cal cm^2)	BTU ft^2	3.69
langleys min^1 (cal cm^2 min^1)	watts cm^2	0.0698
liters	gal	0.264
liters	qt	1.06
m	ft	3.28
lb	kg	0.454
qt	liters	0.946
cm^2	ft^2	0.00108
cm^2	in.^{2}	0.155
ft^2	m^2	0.0929
m^2	ft^2	10.8
watts cm^2	langleys min^1 (cal cm^2)	14.3

To convert °F to °C:	Example:
1. add 40 to the temperature	212°F + 40 = 252
2. multiply by 5/9 (0.555)	252 x 5/9 = 140
3. subtract 40	140 - 40 - 100
	212°F = 100°C

To convert °C to °F:	Example:
1. add 40 to the temperature	100°C + 40 = 140
2. multiply by 9/5 (1.8)	140 x 9/5 = 252
3. subtract 40	252 - 40 = 212
	100°C = 212°F

_____ Appendix D_____

Information Sources

BOOKS

Basic DC Circuits by Franklin Swan and Warren Palmer. Radio Shack. (Or check your library for any number of good basic books on DC theory and circuitry.)

The National Electrical Code, The National Fire Protection Association, Batterymarch Park, Quincy, MA 02269.

Practical Photovoltaics by Richard Komp, $18.95 postpaid from **aatec publications,** PO Box 7119, Ann Arbor, MI 48107. (Covers module and array construction and installation, theory, politics, history and future of PV. Companion to *The New Solar Electric Home.*)

RVers' Guide To Solar Battery Charging by Noel and Barbara Kirkby. $14.95 postpaid from **aatec publications,** PO Box 7119, Ann Arbor, MI 48107. (Complete information on PV for the recreational vehicle owner.)

The Solar Boat Book by Pat Rand Rose. Ten Speed Press, Berkeley, CA. (The only book on solar energy use for boaters.)

BOOKLETS

"Battery Service Manual," The Battery Council International, 111 East Wacker Drive, Chicago, IL 60601. (The best source of battery service information—a must.)

C&D Batteries, 3043 Walton Road, Plymouth Meeting, PA 19462. (Information booklet on storage batteries.)

"Facts About Storage Batteries," ESB Brands, Inc., PO Box 6949, Cleveland, OH 44101.

"The Storage Battery," Exide, 101 Gibraltar Road, Horsham, PA 19044. (Examines lead-acid batteries.)

"Stationary Battery Installation and Operating Instructions," Gould, Inc., Industrial Battery Division, 2050 Cabot Boulevard West, Langhorne, PA 19047.

"Golf Cart Battery Maintenance Manual," SGL Industries, Inc., 14650 Dequindre, Detroit, MI 48212.

PERIODICALS

Backwoods Home Magazine, 1257 Siskiyou Boulevard, #213, Ashland, OR 97520; 503/488 2053.

Home Power, PO Box 130, Hornbrook, CA 96044; 916/475 3179.

Independent Energy (formerly *Alternative Sources of Energy*) 620 Central Avenue N, Milaca, MN 56353; 612/983 6892.

The Mother Earth News, PO Box 70, Hendersonville, NC 28739.

The PV Network News, 2303 Cedros Circle, Santa Fe, NM 97505; 505/473 1067.

PV News, P.O. Box 290, Cassanova, VA 22107; 703/788 9626.

Solar Today, American Solar Energy Society, 2500 Central Avenue, G-1, Boulder, CO 80301; 303/443 3130.

The National Electrical Code

It doesn't matter if you feel that "authority" is not synonymous with competence, setting up your PV system to comply with Code requirements is wise. Not only will the system be safe and easy to service, it will also be easy to modify and expand with your changing needs and the state of the art.

As most PV systems are located off the building inspector's beaten path, some people feel that it isn't worth the time, effort, and expense to do their installation to Code. Being out of the inspector's reach is no excuse for unsafe work. To quote Herbert Richter (in *Practical Electricity and House Wiring*), "If you were about to take a trip into a jungle, and doctors recommended that you take along a supply of quinine for possible fevers, you would follow their advice and take quinine, instead of experimenting with your own ideas of remedies.... Similarly, the Code authorities are the electrical authorities, trying hard to avoid future electrical ills and to cure present ones."

The short section in the *National Electrical Code* listing provisions to insure safe photovoltaic systems is not a recognition of the possible hazards of PV systems. It is more an acknowledgment that PV is here to stay. Safe installations will help in the general acceptance of the technology.

Specifically, Article 690 provides guidelines for circuit requirements, overcurrent protection, and disconnection means, wiring methods, grounding, marking, and connection to other sources. For instance, although PV systems are like other electrical systems, the DC wires may not be contained in the same raceway or boxes of other systems. Of course, where the PV system ties into the wiring for a back-up generator or ties into the house wiring, PV wires can share the same junction boxes.

The *National Electrical Code*'s interpretation of PV use has changed since Article 690's first appearance in 1984. To update, grounding on the DC portion of a PV system is required only on systems using over 50 volts DC. The key word is "required." If you are experiencing radio or television interference or want a safely grounded system, it is a good idea to tie in the DC negative to your system chassis and AC earth ground.

The 1987 Code places greater emphasis on proper sizing of wire overcurrent devices (fuses and breakers) and disconnects. System design should be based on string or array open-circuit voltage and short-circuit current. The reason for this is to provide adequate protection in case components are removed from the system. For example, should you disconnect your battery bank to clean and paint shelves, the output of the disconnected array will no longer be at the battery load voltage. An open-circuit array nominally rated at 24 volts can place almost 44 volts on-line.

While on the subject of disconnects, be sure to allow for equipment repair and replacement. Should you need to service or recalibrate your regulator, can you safely pull it from the system and make a temporary reconnect to still keep the batteries charged? With disconnects this is simple. Switch your array off. Switch your battery bank off. Pull the regulator and use jumper wires between the two disconnect switches. Turn the switches back on and you are charging your batteries again. This may be done without covering the array, although it is not recommended.

Labeling modules is important, too. Not only should plus and minus be tagged, ratings must also be marked. The same goes for source combiners. Wire runs, connections, terminals, and boxes should be clearly labeled and tagged. Installation and servicing is simpler when everything is easily identified.

Grounding PV systems remains controversial, but Code specifically states that the negative DC and neutral or third-wire AC portions of the system must be grounded. The controversy is whether the array mount should also be hard-grounded to the DC circuit. To meet the spirit of the Code and still keep the system safe, an isolating resistor from array to earth ground can be used. If someone should touch the array positive and the mount, the isolated ground will prevent shock. Even a slight shock is surprising and can cause loss of footing and a fall from the roof.

It is recommended to locate earth ground (ground rod or metal water pipe) as close to the array as possible to better protect against lightning-induced voltage surges. In fact, a straight line from the highest point on the array mount or the highest DC negative wire is a good idea.

Grounding consists of (1) ground on the DC negative, (2) standard ground on the AC, (3) array frame ground, and (4) equipment chassis ground. All these grounds are connected to the same grounding electrode (earth ground rod) or grounded metal water pipe. Should you require more than one ground rod because the PV system is spread out over the site, or because the earth potential is greater than 25 ohms (under 5 ohms in dry weather is better), all grounds must be tied together.

A new section on batteries has been added to the Code. This section requires provisions for adequate ventilation of gases, but does not mandate mechanical ventilation. To quote the code, "Hydrogen disperses rapidly and requires very little air movement to prevent accumulation. Unrestricted natural air movement in the vicinity of the battery, together with normal air changes for occupied spaces or heat removal, will normally be sufficient. If the space is confined, mechanical ventilation may be required in the vicinity of the battery. Ventilation can be a fan, roof ridge vent, or louvered areas."

Accessible connected battery cells should not exceed 50 volts DC. "Live parts of battery systems for dwellings shall be insulated against accidental contact, regardless of voltage." The Code acknowledges that batteries require maintenance and the checking of electrolyte and that insulated wires, cables, and connectors will help prevent hazardous accidents.

The updated Code requires control of the state of charge of the battery bank. The exception to this ruling is self-regulated systems. More importantly, charge controller or regulator adjustments shall only be accessible to qualified persons. While the Code does not define who is qualified, it can be assumed that a knowledgeable user may be just as capable of adjustments as an experienced and licensed electrician.

Public libraries will have a copy of the *National Electrical Code*, as will electricians and the building inspection office. The responsibility is yours. The reward is a safe PV system.

NATIONAL ELECTRICAL CODE
Article 690
Solar Photovoltaic Systems

A. General

690-1. Scope. The provisions of this article apply to solar photovoltaic electrical energy systems including the array circuit(s), power conditioning unit(s) and controller(s) for such systems. Solar photovoltaic systems covered by this article may be interactive with other electric power production sources or stand alone, with or without electrical energy storage such as batteries. These systems may have alternating- or direct-current output for utilization.

690-3. Other Articles. Wherever the requirements of other articles of this Code and Article 690 differ, the requirements of Article 690 shall apply.

690-4. Installation.

(a) Photovoltaic System. A solar photovoltaic system shall be permitted to supply a building or other structure in addition to any service(s) of another electricity supply system(s).

(b) Conductors of Different Systems. Photovoltaic source circuits and photovoltaic output circuits shall not be contained in the same raceway, cable tray, cable, outlet box, junction box or similar fitting as feeders or branch circuits of other systems.
Exception: Where the conductors of the different systems are separated by a partition or are connected together.

(c) Module Connection Arrangement. The connections to a module or panel shall be so arranged that removal of a module or panel from a photovoltaic source circuit does not interrupt a grounded conductor to another photovoltaic source circuit.

B. Circuit Requirements

690-7. Maximum Voltage.

(a) Voltage Rating. In a photovoltaic power source and its direct-current circuits, the voltage considered shall be the rated open-circuit voltage.

(b) Direct-Current Utilization Circuits. The voltage of direct-current utilization circuits shall conform with Section 210-6.

(c) Photovoltaic Source and Output Circuits. Photovoltaic source circuits and photovoltaic output circuits which do not include lampholders, fixtures or standard receptacles shall be permitted up to 600 volts.

(d) Circuits Over 150 Volts to Ground. In one- and two-family dwellings, live parts in photovoltaic source circuits and photovoltaic output circuits over 150 volts to ground shall not be accessible while energized to other than qualified persons.
(FPN): See Section 110-17 for guarding of live parts, and Section 210-6 for voltage to ground and between conductors.

690-8. Circuit Sizing and Current.

(a) Ampacity and Overcurrent Devices. The ampacity of the conductors and the rating or setting of overcurrent devices in a circuit of a solar photovoltaic system shall not be less than 125 percent of the current computed in accordance with (b) below. The rating or setting of overcurrent devices shall be permitted in accordance with Section 240-3, Exception No. 1.
Exception: Circuits containing an assembly together with its overcurrent device(s) that is listed for continuous operation at 100 percent of its rating.

    (1) Photovoltaic Source Circuits. The sum of parallel model operating current ratings.
    (2) Photovoltaic Output Circuit. The photovoltaic power source current rating.
    (3) Power Conditioning Unit Output Circuit. The power conditioning unit output current rating.
Exception: The current rating of a circuit without an overcurrent device, as permitted by the Exception to Section 690-9(a), shall be the short-circuit current, and it shall not exceed the ampacity of the circuit conductors.

(b) Computation of Circuit Current. The current for the individual type of circuit shall be computed as follows:

690-9. Overcurrent Protection.

(a) Circuits and Equipment. Photovoltaic source circuit, photovoltaic output circuit, power conditioning unit output circuit, and storage battery circuit conductors and equipment shall be protected in accordance with the requirements of Article 240. Circuits connected to more than one electrical source shall have overcurrent devices so located as to provide overcurrent protection from all sources.

Exception: A conductor in a photovoltaic source circuit, photovoltaic output circuit, or power conditioning unit output circuit having an ampacity not less than the maximum available current under short-circuit or ground-fault conditions with the condition of a shorted blocking diode shall be permitted without an overcurrent device.

(FPN): Possible backfeed of current from any source of supply including a supply through a power conditioning unit into the photovoltaic output circuit and photovoltaic source circuits, must be considered in determining whether adequate overcurrent protection from all sources is provided for conductors and modules.

(b) Power Transformers. Overcurrent protection for a transformer with a source(s) on each side shall be provided in accordance with Section 450-3 by considering first one side of the transformer, then the other side of the transformer as the primary.

Exception: A power transformer with a current rating on the side connected toward the photovoltaic power source not less than the short-circuit output current rating of the power conditioning unit shall be permitted without overcurrent protection from that source.

(c) Photovoltaic Source Circuits. Branch-circuit or supplementary type overcurrent devices shall be permitted to provide overcurrent protection in photovoltaic source circuits. The overcurrent devices shall be accessible, but shall not be required to be readily accessible.

C. Disconnecting Means

690-13. All Conductors. Means shall be provided to disconnect all current-carrying conductors of a photovoltaic power source from all other conductors in a building or other structure.

690-14. Additional Provisions. The provisions of Article 230, Part F shall apply to the photovoltaic power source disconnecting means.

Exception No. 1: The disconnecting means shall not be required to be suitable as service equipment and shall be rated in accordance with Section 690-17.

Exception No. 2: Equipment such as photovoltaic source circuit isolating switches, overcurrent devices, and blocking diodes shall be permitted on the photovoltaic power source side of the photovoltaic power source disconnecting means.

690-15. Disconnection of Photovoltaic Equipment. Means shall be provided to disconnect equipment, such as a power conditioning unit, filter assembly and the like from all ungrounded conductors of all sources. If the equipment is energized (live) from more than one source, the disconnecting means shall be grouped and identified.

690-16. Fuses. Disconnecting means shall be provided to disconnect a fuse from all sources of supply if the fuse is energized from both directions and is accessible to other than qualified persons. Such a fuse in a photovoltaic source circuit shall be capable of being disconnected independently of fuses in other photovoltaic source circuits.

690-17. Switch or Circuit Breaker. The disconnecting means for ungrounded conductors shall consist of a manually operable switch(es) or circuit breaker(s): (1) located where readily accessible, (2) externally operable without exposing the operator to contact with live parts, (3) plainly indicating whether in the open or closed position, and (4) having ratings not less than the load to be carried. Where disconnect equipment may be energized from both sides, the disconnect equipment shall be provided with a marking to indicate that all contacts of the disconnect equipment may be live.

Exception: A disconnecting means located on the direct-current side shall be permitted to have an interrupting rating less than the current-carrying rating when the system is designed so that the direct current switch cannot be opened under load.

690-18. Disablement of an Array. Means shall be provided to disable an array or portions of an array.

(FPN): Photovoltaic modules are energized while exposed to light. Installation, replacement, or servicing of array components while a module(s) is irradiated may expose persons to electric shock.

D. Wiring Methods

690-31. Methods Permitted.

(a) Wiring Systems. All raceway and cable wiring methods included in this Code and other wiring systems and fittings specifically intended and identified for use on photovoltaic arrays shall be permitted. Where wiring devices with integral enclosures are used, sufficient length of cable shall be provided to facilitate replacement.

(b) Single Conductor Cable. Type UF single conductor cable shall be permitted in photovoltaic source circuits where installed in the same manner as a Type UF multiconductor cable in accordance with Article 339. Where exposed to direct rays of the sun, cable identified as sunlight-resistant shall be used.

690-32. Component Interconnections. Fittings and connectors which are intended to be concealed at the time of on-site assembly, when listed for such use, shall be permitted for on-site interconnection of modules or other array components. Such fittings and connectors shall be equal to the wiring method employed in insulation, temperature rise and fault-current withstand, and shall be capable of resisting the effects of the environment in which they are used.

690-33. Connectors. The connectors permitted by Section 690-32 shall comply with (a) through (e) below.

(a) Configuration. The connectors shall be polarized and shall have a configuration that is noninterchangeable with receptacles in other electrical systems on the premises.

(b) Guarding. The connectors shall be constructed and installed so as to guard against inadvertent contact with live parts by persons.

(c) Type. The connectors shall be of the latching or locking type.

(d) Grounding Member. The grounding member shall be the first to make and the last to break contact with the mating connector.

(e) Interruption of Circuit. The connectors shall be capable of interrupting the circuit current without hazard to the operator.

690-34. Access to Boxes. Junction, pull and outlet boxes located behind modules or panels shall be installed so that the wiring contained in them can be rendered accessible directly or by displacement of a module(s) or panel(s) secured by removable fasteners and connected by a flexible wiring system.

E. Grounding

690–41. System Grounding. For a photovoltaic power source, one conductor of a 2-watt system rated over 50 volts and a neutral conductor of a 3-wire system shall be solidly grounded.

Exception: Other methods which accomplish equivalent system protection and which utilize equipment listed and identified for the use shall be permitted.

(FPN): See the first Fine Print Note under Section 250–1.

690–42. Point of System Grounding Connection. The direct-current circuit grounding connection shall be made at any single point on the photovoltaic output circuit.

(FPN): Locating the grounding connection point as close as practicable to the photovoltaic source will better protect the system from voltage surges due to lightning.

If other-than-solid grounding is utilized as permitted by Section 690–41, Exception, the connections should be made in accordance with the markings on the equipment or its installation instructions.

690–43. Size of Equipment Grounding Conductor. The equipment grounding conductor shall be no smaller than the required size of the circuit conductors in systems: (1) where the available photovoltaic power source short-circuit current is less than twice the current rating of the overcurrent device, or (2) where overcurrent devices are not employed as permitted in the Exception to Section 690–9(a). In other systems, the equipment grounding conductor shall be sized in accordance with Section 250–95.

690–44. Common Grounding Electrode. Exposed noncurrent-carrying metal parts of equipment and conductor enclosures of a photovoltaic system shall be grounded to the grounding electrode that is used to ground the direct-current system. Two or more electrodes that are effectively bonded together shall be considered as a single electrode in this sense.

F. Marking

690–51. Modules. Modules shall be marked with identification of terminals or leads as to polarity, maximum overcurrent device rating for module protection and with rated: (1) open-circuit voltage, (2) operating voltage, (3) maximum permissible system voltage, (4) operating current, (5) short-circuit current, and (6) maximum power.

690–52. Photovoltaic Power Source. A marking, specifying the photovoltaic power source rated: (1) operating current, (2) operating voltage, (3) open-circuit voltage, and (4) short-circuit current, shall be provided at an accessible location at the disconnecting means for the photovoltaic power source.
(FPN): Reflecting systems used for irradiance enhancement may result in increased levels of output current and power.

G. Connection to Other Sources

690–61. Loss of System Voltage. The power output from a power conditioning unit in a solar photovoltaic system that is interactive with another electric system(s) shall be automatically disconnected from all ungrounded conductors in such other electric system(s) upon loss of voltage in that electric system(s) and shall not reconnect to that electric system(s) until its voltage is restored.

(FPN): For other interconnected electric power production sources, see Article 705.

An interaction system shall be permitted to operate as a stand-alone system to supply premises wiring.

690-62. Ampacity of Neutral Conductor. If a single-phase, 2-wire power conditioning unit output is connected to the neutral and one ungrounded conductor (only) of a 3-wire system or of a 3-phase, 4-wire wye-connected system, the maximum load connected between the neutral and any one ungrounded conductor plus the power conditioning unit output rating shall not exceed the ampacity of the neutral conductor.

690-63. Unbalanced Interconnections.

(a) Single-Phase. The output of a single-phase power conditioning unit shall not be connected to a 3-phase, 3- or 4-wire electrical service derived directly from a delta-connected transformer.

(b) Three-Phase. A 3-phase power conditioning unit shall be automatically disconnected from all ungrounded conductors of the interconnected system when one of the phases opens in either source.
Exception for (a) and (b): Where the interconnected system is designed so that significant unbalanced voltages will not result.

690-64. Point of Connection. The output of a power production source shall be connected as specified in (a) or (b) below.

(FPN): For the purposes of this section a power production source is considered to be the output of a power conditioning unit when connected to an alternating current electric source, the photovoltaic output circuit when interactive with a direct current electric source.

(a) Supply Side. To the supply side of the service disconnecting means as permitted in Section 230-82, Exception No. 6.

(b) Load Side. To the load side of the service disconnecting means of the other source(s), if all of the following conditions are met:

(1) Each source interconnection shall be made at a dedicated circuit breaker or fusible disconnecting means.

(2) The sum of the ampere ratings of overcurrent devices in circuits supplying power to a busbar or conductor shall not exceed the rating of the busbar or conductor.
Exception: For a dwelling unit the sum of the ampere ratings of the overcurrent devices shall not exceed 120 percent of the rating of the busbar or conductor.

(3) The interconnection point shall be on the line side of all ground-fault protection equipment.

Exception: Connection shall be permitted to be made to the load side of ground-fault protection provided that there is ground-fault protection for equipment from all ground-fault current sources.

(4) Equipment containing overcurrent devices in circuits supplying power to a busbar or conductor shall be marked to indicate the presence of all sources.

Exception: Equipment with power supplied from a single point of connection.

(5) Equipment such as circuit breakers, if back-fed, shall be suitable for such operation.

H. Storage Batteries

690–71. Installation.

(a) General. Storage batteries in a solar photovoltaic system shall be installed in accordance with the provisions of Article 480.

Exception: As provided in Section 690–73.

(b) Dwellings.

(1) Storage batteries for dwellings shall have the cells connected so as to operate at less than 50 volts.

Exception: Where live parts are not accessible during routine battery maintenance, a battery system voltage in accordance with Section 690–7 shall be permitted.

(2) Live parts of battery systems for dwellings shall be insulated to guard against accidental contact, regardless of voltage.

(FPN): Batteries in solar photovoltaic systems are subject to extensive charge-discharge cycles and typically require frequent maintenance, such as checking electrolyte and cleaning connections.

690–72. State of Charge. Equipment shall be provided to control the state of charge of the battery. All adjusting means for control of the state of charge shall be accessible only to qualified persons.

Exception: Where the design of the photovoltaic power source is matched to the voltage rating and charge current requirements for the interconnected battery cells.

690–73. Grounding. The interconnected battery cells shall be considered grounded where the photovoltaic power source is installed in accordance with Section 690–41, Exception.

GLOSSARY

AC—Alternating current; the electric current which reverses its direction of flow. The standard current used by utilities in the U.S. is 60 cycles per second.

acceptance angle—The total range of sun positions from which sunlight can be collected by a system.

AH—*see* **ampere hours**

Air Mass 1 (AM1)—The amount of sunlight falling on the earth at sea level when the sun is shining straight down through a dry clean atmosphere. (A close approximation is the Sahara Desert at high noon.) The sunlight intensity is very close to 1 kilowatt per square meter ($1kW/m^2$).

alternating current—*see* **AC**

ampere—A unit of electrical current or the rate of flow of electrons. One volt across one ohm of resistance causes a current flow of one ampere. One ampere equals 6.25×10^{18} electrons per second passing a given point in a circuit; amp.

ampere hour—A current of one ampere running for one hour.

array—A set of modules or panels assembled for a specific application; may consist of modules in series for increased voltage, or in parallel for increased current, or a combination of both.

battery, marine—A deep-discharge battery used on boats; capable of discharging small amounts of electricity over long time periods.

battery, stationary—For use in emergency standby power systems, a battery with long life but sometimes poor deep-discharge capabilities.

battery, storage—A secondary battery, rechargeable electric storage unit that operates on the principle of changing electrical energy into chemical energy by means of a reversible chemical reaction. The lead-acid automobile battery is the most familiar type.

battery autonomy—The amount of battery storage required to provide power during cloudy days and at night for a specific load at a specific location.

battery capacity—Expressed in ampere hours, the total amount of electricity that can be drawn from a fully charged battery until it is discharged to a specific voltage.

battery capacity, available—The total ampere hours that can be drawn from a battery under specific operating conditions of discharge rate, temperature, initial state of charge, age, and cut-off voltage.

battery capacity, energy—The total watt hours (kilowatt hours) that can be drawn from a fully charged battery. This varies with temperature, rate, age, and cutoff voltage.

battery capacity, installed—The total ampere hours that can be drawn from a new battery when discharged to the specified maximum depth of discharge.

battery capacity, rated—The manufacturer's conservative estimate of ampere hours that can be drawn from a new battery under specific conditions

battery cell—The simplest operating unit in a storage battery; one or more positive electrodes, an electrolyte that permits ionic conduction, one or more negative electrodes, and separators enclosed in a single container.

battery cycle life—The number of cycles to a specified depth of discharge a battery can undergo before efficiency is affected.

battery life—The period when a battery is operating above specific efficiency levels. Measured in either cycles or years, depending on intended use.

blocking diode—A device that prevents current from running backward through an array, thereby draining the storage battery.

BTU—British Thermal Unit; the unit of heat energy sufficient to raise the temperature of one pound of water 1°F.

charge rate—The current applied to a cell or battery to restore its available capacity. This rate is commonly normalized with respect to the rated capacity of the cell or battery.

charging—Conversion of electrical energy into chemical potential energy within a cell by the passage of a direct current in the direction opposite to that of discharge.

concentration ratio—The ratio between the area of clear aperture (opening through which sunlight enters) and the area of the illuminated cell.

converter—A device that changes AC to DC or reduces a DC voltage to a lower DC voltage.

cycle life—The number of cycles, to a specified depth of discharge, that a cell or battery can undergo before failing to meet its specified capacity or efficiency performance criteria.

DC—Direct current; electric current that always flows in the same direction. Photovoltaic cells and batteries are all DC devices.

deep-discharge cycles—Cycles in which a battery is nearly completely discharged.

depth of discharge (DOD)—The ampere hours removed from a fully charged cell or battery expressed as a percentage of rated capacity. For example, the removal of 25 ampere hours from a fully charged 100 ampere-hour rated cell results in a 25% depth of discharge.

direct current—*see* **DC**

discharge rate—The current removed from a cell or battery. This rate can be expressed in amperes, but more commonly is normalized to rated capacity (C), and expressed as C/X. For example, drawing 20 amperes from a cell with a rated capacity of 100 ampere hours is referred to as the $C/5$ discharge rate (100 AH/20 amps). Similarly, discharge currents of 5, 10, and 33.3 amperes would be designated as the $C/20$, $C/10$, and $C/3$ rates, respectively.

efficiency—The ratio of the useful output to the input.

ampere hour (coulombic) efficiency—The ratio of the ampere hours removed from a cell or battery during a discharge to the ampere hours required to restore the initial capacity.

voltage efficiency—The ratio of the average discharge voltage of a cell or battery to the average charge voltage during the subsequent restoration of an equivalent capacity.

energy (watt hour) efficiency—The ratio of the energy delivered by a cell or battery during a discharge to the total energy required to restore the initial state of charge. The watt hour efficiency is approximately equal to the product of the voltage and ampere hour efficiencies. This is sometimes referred to as the roundtrip efficiency. Roundtrip energy efficiencies usually do not include energy losses resulting from self-discharge, auxiliary equipment (parasitic losses), or battery equalization.

electric current—The rate at which electricity flows through an electrical conductor; expressed in amps.

electrolyte—The medium which provides the ion transport mechanism between the positive and negative electrodes of a cell. In some cells, such as the lead-acid type, the electrolyte may also participate directly in electrochemical charge/discharge reactions.

equalization—The process of restoring all cells in a battery to an equal state of charge. For lead-acid batteries this is a charging process designed to bring all cells to 100% state of charge. Some battery types may require a complete discharge as a part of the equalization process.

equalizing charge—A continuation of normal battery charging, at a voltage level slightly higher than the normal end-of-charge voltage, in order to provide cell equalization within a battery.

full sun—*see* **Air Mass 1**

gassing—The evolution of gas from one or more of the electrodes in a cell. Gassing commonly results from local (self-discharge) or from the electrolysis of water in the electrolyte during charging.

grid—The utility network of transmission lines used to distribute electricity.

hybrid system—A system that produces both usable heat (e.g., to heat water) and electricity.

insolation—The amount of sunlight striking a given area.

inverter—A device that converts DC to AC.

inverter, synchronous—A device that converts DC to AC in synchronization with the power line. Excess power is fed back into the utility grid.

kilowatt hour—Unit of energy used to perform work; kWh.

load—A device (or combination of devices) which consumes electrical power.

motive power cell or **battery**—A cell or battery that is intended to power electrically operated mobile equipment (e.g., forklift trucks) and is designed to be operated in a daily deep-cycle regime at moderate discharge rates.

open-circuit voltage—The voltage produced by a solar cell when exposed to standard sunlight conditions, and with no load.

overcharge—Charging of a cell continued after 100% state of charge has been reached. Overcharging does not increase the energy stored in a cell and usually results in gassing and/or excessive heat generation, both of which reduce battery life.

peak sun hours—The daily kilowatt hour direct current production of a one square meter PV array for a specific location.

primary cell or battery—A cell or battery whose initial capacity cannot be significantly restored by charging and is therefore limited to a single discharge.

regulator—A device that prevents overcharging of batteries.

secondary cell or battery—A cell or battery that is capable of being charged repeatedly.

self-discharge rate—The rate at which a battery will discharge on standing; affected by temperature and battery design.

silicon—The second most abundant element in the earth's crust; intermediate-grade silicon (less costly than electronic-grade) is used in the manufacture of silicon solar cells.

SLI battery—Starting, lighting, and ignition battery (standard automotive battery) designed primarily for high rate, low depth of discharge operation typical of automobile engine starting.

solar cell—The photovoltaic device that converts sunlight directly into electricity.

state of charge—A battery's available capacity, stated as a percentage of rated capacity.

sulfation—A condition which afflicts unused and discharged batteries; large crystals of lead sulfate grow on the plate, instead of the usual tiny crystals, making the battery extremely difficult to recharge.

volt—Unit of electrical potential difference across which a current flows.

watt—Unit of power, power being the rate at which energy is used to do work.

Index